MUIRHEAD LIBRARY OF PHILOSOPHY

An admirable statement of the aims of the Library of Philosophy was provided by the first editor, the late Professor J. H. Muirhead, in his description of the original programme printed in Erdmann's *History of Philosophy* under the date 1890. This was slightly modified in subsequent volumes to take the form of the following statement:

'The Muirhead Library of Philosophy was designed as a contribution to the History of Modern Philosophy under the heads: first of Different Schools of Thought—Sensationalist, Realist, Idealist, Intuitivist; secondly of different Subjects—Psychology, Ethics, Aesthetics, Political Philosophy, Theology. While much had been done in England in tracing the course of evolution in nature, history, economics, morals and religion little had been done in tracing the development of thought on these subjects. Yet "the evolution of opinion is part of the whole evolution".

'By the co-operation of different writers in carrying out this plan it was hoped that a thoroughness and completeness of treatment, otherwise unattainable, might be secured. It was believed also that from writers mainly British and American fuller consideration of English Philosophy than it had hitherto received might be looked for. In the earlier series of books containing, among others, Bosanquet's *History of Aesthetic*, Pfleiderer's *Rational Theology since Kant*, Albee's *History of English Utilitarianism*, Bonar's *Philosophy and Political Economy*, Brett's *History of Psychology*, Ritchie's *Natural Rights*, these objects were to a large extent effected.

'In the meantime original work of a high order was being produced both in England and America by such writers as Bradley, Stout, Bertrand Russell, Baldwin, Urban, Montague, and others, and a new interest in foreign works, German, French and Italian, which had either become classical or were attracting public attention, had developed. The scope of the Library thus became extended into something more international, and it is entering on the fifth decade of its existence in the hope that it may contribute to that mutal understanding between countries which is so pressing a need of the present time.'

The need which Professor Muirhead stressed is no less pressing today, and few will deny that philosophy has much to do with enabling us to meet it, although no one, least of all Muirhead himself, would regard that as the sole, or even the main, object of philosophy. As Professor Muirhead continues to lend the distinction of his name to the Library of Philosophy it seemed not inappropriate to allow him to recall us to these aims in his own words. The emphasis on the history of thought also seemed to me very timely; and the number of important works promised for the Library in the very near future augur well for the continued fulfilment, in this and other ways, of the expectations of the original editor.

H. D. LEWIS

MUIRHEAD LIBRARY OF PHILOSOPHY

General Editor: H. D. Lewis
Professor of History and Philosophy of Religion in the University of London

Action by SIR MALCOLM KNOX
The Analysis of Mind by BERTRAND RUSSELL
Belief by H. H. PRICE
Brett's History of Psychology edited by R. S. PETERS
Clarity is Not Enough by H. D. LEWIS
Coleridge as a Philosopher by J. H. MUIRHEAD
The Commonplace Book of G. E. Moore edited by C. LEWY
Contemporary American Philosophy edited by C. P. ADAMS and W. P. MONTAGUE
Contemporary British Philosophy first and second Series edited by J. H. MUIRHEAD
Contemporary British Philosophy third Series edited by H. D. LEWIS
Contemporary Indian Philosophy edited by RADHAKRISHNAN and J. H. MUIRHEAD 2nd edition
Contemporary Philosophy in Australia edited by ROBERT BROWN and C. D. ROLLINS
The Discipline of the Cave by J. N. FINDLAY
Doctrine and Argument in Indian Philosophy by NINIAN SMART
Essays in Analysis by ALICE AMBROSE
Ethics by NICOLAI HARTMANN translated by STANTON COIT 3 vols
The Foundations of Metaphysics in Science by ERROL E. HARRIS
Freedom and History by H. D. LEWIS
The Good Will: A Study in the Coherence Theory of Goodness by H. J. PATON
Hegel: A Re-examination by J. N. FINLAY
Hegel's Science of Logic translated by W. H. JOHNSTON and L. G. STRUTHERS 2 vols
History of Aesthetic by B. BOSANQUET 2nd edition
History of English Utilitarianism by E. ALBEE
History of Psychology by G. S. BRETT edited by R. S. PETERS abridged one volume edition 2nd edition
Human knowledge by BERTRAND RUSSELL
A Hundred Years of British Philosophy by RUDOLF METZ translated by J. N. HARVEY, T. E. JESSOP, HENRY STURT
Ideas: A General Introduction to Pure Phenomenology by EDMUND HUSSERL translated by W. R. BOYCE GIBSON
Identity and Reality by EMILE MEYERSON

Imagination by E. J. FURLONG
Indian Philosophy by RADHAKRISHNAN 2 vols revised 2nd edition
Introduction to Mathematical Philosophy by BERTRAND RUSSELL 2nd edition
Kant's First Critique by H. W. CASSIRER
Kant's Metaphysic of Experience by J. H. PATON
Know Thyself by BERNADINO VARISCO translated by GUGLIELMO SALVADORI
Language and Reality by WILBUR MARSHALL URBAN
Lectures on Philosophy by G. E. MOORE
Lecturers on Philosophy by G. E. MOORE edited by C. LEWY
Matter and Memory by HENRI BERGSON translated by N. M. PAUL and W. S. PALMER
Memory by BRIAN SMITH
The Modern Predicament by H. J. PATON
Natural Rights by D. G. RITCHIE 3rd edition
Nature, Mind and Modern Science by E. HARRIS
The Nature of Thought by BRAND BLANSHARD
Non-Linguistic Philosophy by A. C. EWING
On Selfhood and Goodhood by C. A. CAMPBELL
Our Experience of God by H. D. LEWIS
Perception by DON LOCKE
The Phenomenology of Mind by G. W. F. HEGEL translated by SIR JAMES BAILLIE revised 2nd edition
Philosophy in America by MAX BLACK
Philosophical Papers by G. E. MOORE
Philosophy and Illusion by MORRIS LAZEROWITZ
Philosophy and Political Economy by JAMES BONAR
Philosophy and Religion by AXEL HÄGERSTROM
Philosophy of Space and Time by MICHAEL WHITEMAN
Philosophy of Whitehead by W. MAYS
The Platonic Tradition in Anglo-Saxon Philosophy by J. H. MUIRHEAD
The Principal Upanisads by RADHAKRISHNAN
The Problems of Perception by R. J. HIRST
Reason and Goodness by BLAND BLANSHARD
The Relevance of Whitehead by IVOR LECLERC
The Science of Logic by G. W. F. HEGEL
Some Main Problems of Philosophy by G. E. MOORE
Studies in the Metaphysics of Bradley by SUSHIL KUMAR SAXENA
The Theological Frontier of Ethics by W. G. MACLAGAN
Time and Free Will by HENRI BERGSON translated by F. G. POGSON
The Transcendence of the Cave by J. N. FINDLAY
Values and Intentions by J. N. FINDLAY
The Ways of Knowing: or the Methods of Philosophy by W. P. MONTAGUE

Muirhead Library of Philosophy

EDITED BY H. D. LEWIS

HYPOTHESIS AND PERCEPTION

By the same author

Fundamentals of Philosophy, 1969

The Foundations of Metaphysics in Science, 1965

Annihilation & Utopia, 1966

Revelation through Reason, 1958

Nature, Mind and Modern Science, 1954

HYPOTHESIS AND PERCEPTION

THE ROOTS OF SCIENTIFIC METHOD

ERROL E. HARRIS

LONDON · GEORGE ALLEN & UNWIN LTD
NEW YORK · HUMANITIES PRESS INC.

FIRST PUBLISHED IN 1970

BRITISH ISBN 0 04 121013 1

PRINTED IN GREAT BRITAIN
in 11 on 12 pt Imprint
BY UNWIN BROTHERS LIMITED
WOKING AND LONDON

PREFACE

In 1956–57 I gave two seminars to graduate students at Yale University on the Philosophy of Nature and the Theory of Knowledge. These marked the beginning of the research which eventually came to fruition in *The Foundations of Metaphysics in Science* and in the essay that follows. The work might have been completed several years sooner if the routine of teaching and administering a Department of Philosophy had not slowed it up between 1957 and 1962, when appointment to the Roy Roberts Professorship in the University of Kansas gave me additional time and opportunity for writing. Assistance from the Bollingen Foundation of New York helped considerably with the earlier book and enabled me to make a start on this one, and a rather curtailed grant from the National Science Foundation in Washington aided its progress towards completion. My sincere thanks are due to these organizations for their support.

Meanwhile, the completely independent work of Russell Norwood Hanson and of Thomas Kuhn, in rather different fields of the history and philosophy of science, had exercised an influence upon the writers whose theories I had long been criticizing, so that what was until very recently regarded as 'the accepted view' of scientific method is no longer as popular, nor as current, as when my own work began. It is this 'accepted view' that is criticized in Part I of this volume. The extent to which Sir Karl Popper's theory was a revolt against that same view of science I have indicated in the text, as well as the degree to which his revisions of it remain subject to its limitations.

My own theory of scientific systems as successive dialectical phases of a continuous development has been in some degree anticipated by Thomas Kuhn in his *Structure of Scientific Revolutions*, and this anticipation has not simply been acknowledged in Chapter VII (below), but has made it unnecessary for me to develop *ab initio* a theory of successive conceptual schemes. I have been able simply to adopt what I found satisfactory in Kuhn's theory and to build upon that. But what I find deficient in Kuhn's account is his failure to set out the logic of transition from one so-called 'paradigm' to the next, or to explain the source of new hypotheses—which he seems simply to find proliferating at the appropriate periods in the history of science. Why they should have proliferated he makes tolerably understandable in his account of the progressive break-down of obsolescent theories; but the logical links between such collapses of the old and the emergence of the new remain, in his description, obscure. If he does not explicitly agree with those who hold that discovery is a purely psychological matter, which it is beyond the competence of the logician to explain, he does leave us to assume that a new hypothesis occurs to a scientist with little or no prompting from the facts and merely as a fortunate or mysteriously inspired hunch.

I have attempted to remedy this defect, with the help of suggestions derived from C. S. Peirce and R. G. Collingwood, by elaborating a view of scientific progress as a dialectical development, and of the logic of scientific discovery as a logic of construction.

Meanwhile, I find the erstwhile 'accepted view' being somewhat hesitantly abandoned by its pundits and becoming modified in the Kuhnian direction, while an entirely new theory of scientific method is developing from the side of Phenomenology, inspired by Husserl and Heidegger which converges in large measure towards the theory I have tried to advocate in this book.

ERROL E. HARRIS

High Wray
December, 1969

CONTENTS

PLATES

PART I

CRITICAL

CHAPTER I

PREVALENT VIEWS OF SCIENCE

i. THE POPULAR VIEW

Of all fields of human endeavour science is the only one in which past progress can confidently be claimed and in which advances continue in the present. Some doubt may legitimately be entertained about the occurrence and even about the possibility of progress in religion, morality, art or philosophy, and such as has been alleged has usually been the consequence of scientific discovery. But why this should be so, what special character of science makes progress possible in its case which is doubtful in the case of other human activities, is far from clear. Moreover, even though most people would readily agree that scientific knowledge has advanced and continues to do so, the way in which science progresses and the special character of its methods are subject of some dispute and the most prevalent opinions on these matters are, on close investigation, the least defensible.

The popular conception of science, its methods, the reasons for and the nature of its advance are somewhat as follows. Whereas other types of human speculation are based upon mere opinion, science pursues and sticks to the facts. These facts are ascertained by direct observation—by sense-perception—and they supply the scientist with his data. He collects as large a mass of data as he can, classifies them and proposes hypotheses to explain their nature and occurrence, which can then be tested by further observation and by experiments devised to render specific observations more precise, more selective and more easily obtainable. The outcome of this method is a body of scientific laws, systematically related to one another, by reference to which the phenomena investigated can be explained.

The steady advance of science is assured by this method because it adheres to the facts and abjures prejudices, preconceptions, and wishful thinking and so avoids self-deception; and the nature of the advance is the steady accumulation of fresh knowledge by the methods described. But because science does not begin *in vacuo*, and in the past men have been prone to believe false and groundless theories which have resulted at worst from the entertainment of unscientific notions, wild imaginings, animistic fears, myths and the rationalization of repressed desires, or at best from ignorance and the faulty or partial application of scientific methods, the advance of knowledge has also been marked by the disproof and abandonment of false hypotheses and their replacement by others supported more adequately by empirical evidence.

This account of science is so familiar and so generally approved, not only by laymen but also by philosophers and often by scientists themselves, that to call it in question seems perverse and almost sacrilegious, yet the investigation to be undertaken in subsequent chapters will show that it is misleading, if not altogether wrong, in almost every detail.

ii. PHILOSOPHICAL VIEWS

The prevalence of the view and its common appeal is due mainly to the work of an eminent group of philosophers, from Francis Bacon to the present day, and it is, in fact, a popular version of a philosophical doctrine originally propounded in the seventeenth and eighteenth centuries but developed in more recent times into a highly sophisticated and technical body of theory. Even the opinion of practising scientists about their own methods and procedures is sometimes no more than a reflection, or watered-down version, of philosophers' theories; for the immediate task of the scientist is not to reflect upon his own scientific activities, but to perform them. His primary interest is in the investigation of his subject-matter, and theories of the methods of discovery and principles of argument are the work of the philosopher and the logician. In these matters the scientist is a layman rather than an expert; but he is a privileged layman with direct experience of the processes upon which the philosopher pronounces, while the philosophical and logical expert often lacks the experience which would be a valuable guide to his theories. Consequently the pronouncements of scientists on scientific method must be regarded with caution and critical discrimination, and those of the philosophers require careful examination in the light of actual practice of scientists. It follows that the popular view of science which I have outlined above can only be effectively examined and criticized if attention is directed upon its more elaborate and professional forms, and that we must address ourselves forthwith to the philosophical theory.

Francis Bacon was responsible for the idea that the first task of the scientist is to seek the facts in direct observation freed from preconception and prejudice, to describe and classify these directly observed facts, and to propound hypotheses cautiously only after a prodigious number of such facts have been accurately recorded and correlated. Newton was so impressed by this doctrine that he professed to abstain altogether from the invention of hypotheses. But Newton's disclaimer is generally recognized as unwarranted and misconceived, and Bacon's theory has long been seen to be one-sided and to give an exaggerated role to the collection and classification of data without the directing influence of a working hypothesis. Nevertheless, the spirit of his doctrine survives and has been more systematically expressed in the work of Locke and of Hume.

The acknowledged importance of observation and experiment, freshly recognized in the startling results of the scientific revolution in the sixteenth and seventeenth centuries, led these philosophers (Locke in particular) to assert that human knowledge of matters of fact—of the nature and events of the world at large—was exclusively derived from sense-given particulars and that all reliable knowledge was the result either of the direct inspection and comparison of these or of relating, classifying, and generalizing from, their perceived properties and occurrence. According to Hume all our ideas are derived from (are in fact fainter copies of) impressions of sensation and reflection and all our reasoning is either demonstrative, by comparison of our ideas and impressions, or inductive, by the use of the principle of causation.

The sciences are thus to be divided into the exact or demonstrative sciences, concerned with the relation and comparison of 'ideas' immediately before the mind, and the empirical sciences based on causality and using inductive argument. The procedure of the former is purely deductive and its results are certain. That of the latter is by generalization from particular instances and its results are never more than probable in a degree determined by the frequency and volume of the favourable evidence.

For our present purpose the importance of Hume's teaching is his statement of the central problem of method in the empirical sciences, the problem of induction, a statement which, for penetration and cogency, has never been superseded (though Bertrand Russell has repeated it with unerring insight).[1] This

[1] *The Problems of Philosophy*, Ch. VI.

problem was left by Hume as virtually insoluble, for the solution he offers is not logical but psychological and leaves the validity of natural science open to the most serious doubt. Modern philosophers have offered solutions which will presently be examined. What I wish to emphasize at the outset is that with success or failure in the solution of this problem the theory of method and advance in the natural sciences must stand or fall, because if scientific procedure is that of generalization from sense-given particulars the validity of inductive reasoning is indispensable to it.

Hume's statement of the problem is so well known that only the briefest summary of it need be given. His argument is that no connexion between perceived events is ever apparent in the actual perception of them. They occur consecutively and contiguously and in conjunction and that is all we ever experience. If in the case of any two events of specifiable kinds such conjunction occurs constantly we are apt to conclude that it is 'necessary' and to affirm that it will occur always. But for such a conclusion there are and can be no rational grounds. To deny it would not be self-contradictory, so it is not demonstrable; nor do we learn it from experience, for the frequency of the occurrence in the past of a conjunction of particulars gives no evidence that it will recur in the future. However frequent and copious the past 'evidence' may be this cannot, without circularity, be used as a ground even for the probability of future repetition, because our estimate of probability is itself dependent on the belief that future experience will resemble past, for which we have absolutely no warrant.

The main force of Hume's argument lies in this last point which is frequently overlooked in modern elaborations of his position. The past occurrence of B in conjunction with A, however frequent, not only provides no evidence for their future conjunction but does not even make it probable. To assert that, though B has followed A *x* times in our past experience, there is no probability of its doing so again, involves no contradiction, and there is thus no logical requirement that we conclude from past conjunctions to the probability of their repetition. The fact that we commonly do so conclude, as Hume recognizes, is due to our inveterate habit of expecting the future to resemble the past for which there can be neither rational ground nor, in the nature of the case, empirical evidence. But without either of these there is no warrant at all for concluding from the frequent (or constant) past conjunction of events to any degree of probability of their future conjunction.

Inductive argument, therefore, by which we infer from particular cases to a general conclusion is no satisfactory epistemological basis for the belief even in the probability of the general statement to which we infer, let alone to its truth.

This is the impasse inherited from Hume by all philosophers of science who adopt the empiricist foundation on which Hume built. If Hume is right, natural science is throughout a texture of belief without a shred of logical cogency, founded entirely upon the peculiar psychological character of the human mind and is in no way distinguishable from superstition and prejudice.[1] This Hume tells us in so many words: 'Thus all probable reasoning is nothing but a species of sensation. 'Tis not solely in poetry and music, we must follow our taste and sentiment, but likewise in philosophy.'[2] It is for this reason that I have said that the theory of scientific method stands or falls with its success in solving the problem of induction. For if no clear distinction is possible between science and prejudice, no coherent account can be given of how scientific advance overcomes and leads to the rejection of errors based upon superstition or wishful thinking. I shall try to show later that if the traditional account of inductive argument is accepted we should be in precisely that predicament.

It is to be noticed however that the problem arises directly from the position laid down *ab initio* that all our knowledge of matters of fact is, and can only be, derived from sense-given particulars, and is inseparably connected in Hume's theory with his insistence that every idea originates in some impression (or group of impressions) of sensation or reflection. If we are not indissolubly wedded to that philosophical foundation, or to some variant of it, we are not inevitably involved in the problem of induction—or, at least, not in this form. And if we are wedded to the empiricist foundation we cannot escape the problem and, as will shortly appear, we shall be hard put to resolve it satisfactorily.

In recent years the best known and most widely publicized writings on the philosophy of science have been produced by prominent logical positivists or logical empiricists or men closely associated with the type of philosophy nowadays usually styled 'analytic'. Of course, all philosophers claim to use a method in some sense analytic and the special sense in which empiricists,

[1] Cf. *Treatise of Human Nature*, Bk I, Pt III, Sects ix and x.

[2] *Ibid.* (Selby-Bigge's edition), p. 103. In Hume's day the term 'philosophy' covered what today we should call 'science' as well.

from Bertrand Russell and Wittgenstein to Ryle and Nelson Goodman, are called 'analysts' has never been made clear or precise. It would be more accurate simply to classify them as empiricists, for they would all accept the title in some sense, and whether they accept it or not they all propound views which conform to empiricist principles. In particular, those who have written on scientific method openly accept and insist upon the Humean doctrine of constant conjunction as the basis of scientific laws. This doctrine is explicitly repeated by R. B. Braithwaite,[1] Ernest Nagel,[2] Arthur Pap,[3] Hans Reichenbach,[4] Philipp Frank,[5] to mention only these, and is assumed by many others who seem to consider it so universally acknowledged as to need no further reiteration.

The Humean elements that such writers adopt are clear and usually admitted. If they are full-blooded positivists they adhere to the verifiability theory of meaning which, as they acknowledge, is akin to, if not in all respects the same as, Hume's insistence that the significance of any idea is traceable to the impression from which it is derived. Most of them, nowadays, recognize the numerous difficulties in which the verifiability theory is involved and do what they can to meet, or by modifying the doctrine, to avoid them. Their success in this regard is dubious, but I shall not enter here into a discussion of these manœuvres, though it may be necessary at a later stage to refer again to this doctrine when we come to consider the role of observation in scientific method. For the present we may simply take as typical of all writers of this group Braithwaite's statement that: 'A scientific hypothesis is a general proposition about all the things of a certain sort. It is an empirical proposition in the sense that it is testable by experience; experience is relevant to the question as to whether or not the hypothesis is true, i.e., as to whether or not it is a scientific law.'[6] Experience is (presumably) understood as sense-given, and acceptability of the hypothesis is taken to be wholly dependent upon the empirical evidence. Its truth can be established in no

[1] See *Scientific Explanation* (Cambridge, 1953), p. 10.

[2] See *The Structure of Science* (New York, 1961), pp. 55–6, *et seq.*

[3] See *An Introduction to the Philosophy of Science* (New York, 1962), pp. 252 ff.

[4] See *The Rise of Scientific Philosophy* (Berkeley and Los Angeles, 1957), Ch. 5. Cf. *A Modern Philosophy of Science* (London, 1959); *Experience and Prediction* (Chicago, 1938).

[5] *Philosophy of Science* (New York, 1957).

[6] *Op. cit.*, p. 2.

other way than by reference, in the last resort, to the experience of sensible particulars. It is this dependence on experience of empirical generalizations for their establishment (understanding 'experience' as sense-given particulars) that makes the problem of inductive reasoning inescapable, for to decide upon their degree of probability we are compelled to conclude from the particular to the general, which formal logic forbids.

What follows naturally from this empiricist starting-point is the division of propositions into two main classes, (i) empirical propositions, about synthetic matters of fact, which are (or should be, if they are to have literal meaning) testable by experience, and (ii) those which are purely analytic, the function of which is to elucidate the use and meaning of terms, but which give no information about the world. The truth or falsity of the latter depends solely on their self-consistency and the law of non-contradiction, whereas of the former self-consistency, though necessary, is not a sufficient condition of truth.[1] Accordingly there are two main types of science, exact science on the one hand comprising logic and mathematics, concerned with analytic truths and using purely deductive reasoning; and empirical science on the other seeking laws which are generalizations from particular experiences and are verifiable (or, more strictly, 'probabilifiable') only by observation and experiment.

The conception of logical (and mathematical) truth to which empiricists are then committed is that it is purely analytic and tautological, and all deductive reasoning is similar. It can proceed validly only to conclusions which are analytically entailed by its premises according to the rules of deductive inference which it prescribes. It can never deduce from any premises, whether empirically or logically true, any conclusion which is synthetic with respect to them; in other words, it cannot empower us to deduce distinct matters of fact one from another or from any mere truths of logic. The only kind of inference available for concluding to matters of fact which are unknown, in the sense that they neither

[1] The dichotomy is not confined to Empiricism. A similar distinction was made also by Leibniz, but his was based on a somewhat different criterion. Eternal truths depend, on his view, solely on the law of non-contradiction, but contingent truths upon sufficient reason and compossibility. As he also taught that an adequate idea of any subject (substance or monad) must contain within it, or necessarily imply, all the predicates truly assertable of it, the distinction between analytic and synthetic propositions is undercut and seems at best to be merely provisional.

are being or have previously been perceived, is inductive inference. This is not to say that no use is ever made of deductive inference in the empirical sciences, but only that such use of it as is made does not lead us to fresh factual truths. It enables us to subsume particular cases under general laws and to develop the analytically entailed consequences of empirical propositions, but it can lead to no material advance or enlargement of empirical knowledge whatsoever. That requires, if it can be accomplished at all, the use of inductive reasoning, which, as Hume taught, is the sole means of inferring from what is empirically known to what is empirically unknown, from what is present to observation to what is not present, from what has been experienced in the past to what is to be expected in the future.

The logical justification of induction is therefore vital and indispensable to the validation of empirical science, for it is the sole method offered for the establishment and advancement of knowledge. Unless it is logically sound the grounds for believing in the truth of scientific theory could only be either the authority of the scientist or the psychological set of the individual believer, and science would not be distinguishable from dogmatism or superstition, and so would have no title to the epithet 'scientific'. It is not surprising, therefore, that contemporary philosophers of empiricist persuasion devote a large portion of their time and ingenuity to attempts at the justification of the sort of inductive generalization and argument that they assert to be the sole sound method in empirical science. The way in which they seek to justify it is not in all cases the same, and some are even prepared to declare that it is in no need of justification. We shall begin by asking (i) whether the disclaimers of any need for justification can possibly be sustained, if not (ii) whether the attempts at justification are successful, and, finally, (iii) whether the doctrine as a whole is coherent.

CHAPTER II

INDUCTION

i. DISCLAIMER OF NECESSITY FOR JUSTIFICATION

Any reader of empiricist persuasion is likely at this point to cast the book aside. 'The problem of induction', he will say, 'is a pseudo-problem. Properly speaking, there is no problem at all, and the volume of writing attempting to justify inductive argument or to show that this or that attempt has failed has long been consigned to the rubbish heap. Any writer who still imagines that there is an issue here must be hopelessly behind the times.'

'Attempts to justify induction', he will continue, 'are attempts to do either the impossible or the undesirable. You have already revealed that this is the case by your own statement. What makes the problem inescapable, you allege, is that inductive argument concludes from the particular to the general, which formal logic forbids. By "formal logic" you mean presumably "deductive logic", and if inductive argument could be made to conform to the rules of deductive logic it would not be inductive, so of course the alleged "problem" is insoluble. The attempt to "solve" it is an attempt to do what is impossible by definition. And, as deductive reasoning can give no new information about the world not already entailed in its premises, to reduce induction (which is the sole means of extrapolating beyond the given fact) to deduction, if it were possible, would be undesirable.

'Or again, if you ask what *justifies* our making the predictions that we do, you are seeking a way of knowing in advance what predictions will be fulfilled, and there is no way of knowing this without prevision. Or, if you ask what justifies our thinking a prediction *probable*, you are seeking a way of knowing beforehand what the future frequency will be of favourable cases, and there is no way of doing this.'[1]

Others will say that to predict future events on the ground of past regularities is precisely what we ordinarily mean by being 'reasonable'; it is then futile to ask what reasonable (or rational) grounds we have for doing so. What kind of 'justification', they

[1] Cf. Nelson Goodman, *Fact, Fiction and Forecast* (Harvard, 1955), pp. 65 f.

will ask, would satisfy? What is being demanded? And they would allege that these questions are unanswerable.

Such arguments are very telling and plausible, but they are no more than a triumphal disguise for an admission of failure. If you make the fundamental epistemological presupposition, that knowledge of facts is supplied only by particular atomic sense-experiences, and if that forces you to make a sharp separation between deductive reasoning (confined to tautologies) and empirical knowledge (restricted to the accumulation of perceptual data, from none of which can any other be deduced), it goes without saying that 'inductive' argument neither can nor should be reduced to 'deductive'. Likewise, if the sole source of knowledge of facts is perception, then obviously there is no other way of knowing facts. Nor can we perceive them before we do. If we assay to predict what we shall perceive, it can (*ex hypothesi*) only be on the grounds of what we do or have already experienced. But, as Hume insisted, our principles make it impossible that past experience could provide any grounds. It gives no knowledge of what it does not include and nothing that is not already included in it can be deduced from it. Yet we do predict on the basis of past experience. On empiricist principles this should not be possible and is explicable, if at all, only psychologically. To say that we do it and that to predict on the basis of past experience is precisely what we mean by being 'reasonable', is only to repeat (by implication) this psychological explanation. That we habitually accept past occurrence as evidence is no reason why it should count as evidence. To show what logical claim it can have to count as such is the problem of induction. It is not disposed of simply by restating it—simply by saying that we do predict on the basis of past experience and that this needs no explanation or 'justification'—for on empiricist principles there are no logical grounds why any prediction should ever be credible.

What is still worse, the facts we claim to know are not the bare perceived data, and can be derived from what is directly perceived only by a process of construction which relies on past experience, not simply for anticipation of the future, but even for interpretation of the present.[1] The recognition of a material object present to perception involves at least the awareness that certain simple data (as Locke has it) 'go constantly together', which is a generaliza-

[1] Cf. E. E. Harris, *The Foundations of Metaphysics in Science* (London, 1965), pp. 407–16.

tion from past regularities. Thus the apprehension of any present fact, involving as it does the use of funded past experience, is a kind of immediate inductive inference. And past experience is not 'presented', unless, in some sense, by memory. Its relation to the present cannot be perceived in any direct way. Nor (*ex hypothesi*) can it be deduced from anything that is presented. Consequently our knowledge of it must depend (as Hume well knew) upon some form of causal or associative argument which is inductive in general character.

It follows that all factual knowledge worth the name depends upon some degree of 'projection' (as Goodman calls it)—upon some way of going beyond the barely given to what either no longer is presented or has not yet been. I shall have occasion later to go into more detail about this requirement and the ways in which it is met. The point I am making here is that without 'induction' or something akin to it no factual knowledge is possible. The question of the validity of induction, therefore, is the question (especially for the empiricist) of the validity of all factual knowledge. It is, in brief, the question of the validity of Empiricism.

Nor is it correct to say that generalizing from experienced regularities is precisely what we mean by being reasonable. As I shall show later it is often precisely what we should regard as being unreasonable. What we consider reasonable is only *justifiable* generalization from past experience, so that the question of justification is inescapable, not simply because the generalization must follow from adequate evidence and evidence of the right kind, but because some sorts of experienced regularities do not justify generalizations of the scientific or 'nomic' sort at all. Thus it is no escape to say, with Strawson,[1] that the only possible kind of rational justification of inductive procedures is conformity to inductive principles; for nobody has yet made clear what inductive (or deductive) principle enables us to distinguish those cases of constant conjunction which do, from those which do not, entitle us to generalize nomically. The semantic appeal to a definition of 'rational', which equates it with the observance of inductive methods, is, in effect, a form of the pragmatic justification of induction that is later to be examined; for to say that it is reasonable to base an empirical generalization on experienced regularities is as much as to say that such procedure is usually successful. If it were not it would not be reasonable to pursue it.

[1] See *Introduction to Logical Theory* (London, 1952), pp. 256–63.

Max Black maintains that the so-called problem of induction is illegitimate because it has been so framed as to be not merely difficult but impossible of solution.[1] This may well be the case, but we must remember that it is the empiricist, by his insistence that sense-given particulars are the sole source of factual knowledge, who forces the philosopher seeking to think consistently with that premise to frame the problem in this insoluble way. If we are allowed to abandon Empiricism, the problem may well take on a new aspect; it may become soluble or even cease to be a problem at all. My present thesis is that those philosophers who refuse to abandon Empiricism either find themselves in an impasse (which Black has done much to clarify); or, if they claim to have 'solved' the problem of induction in some way, have begged the question, or have surreptitiously abandoned their professed principles while asserting that they have not been violated, or have abandoned them openly (even if unconfessedly) while they still cling tenaciously to some features of the position they are deserting, inconsistent though those features are with their new positions.

The empiricist predicament is plainly revealed in Goodman's explanation of how what he calls the 'old problem' of induction (for he recognizes new ones) has been 'solved or dissolved'.[2] He asserts with some justice that induction is validated in the same sort of way as deduction. If the inference follows the rules it is valid, if not it is invalid. But whereas we have elaborate and precise formulations of deductive rules, we have not, up to the present, formulated the rules of induction at all precisely. He then confesses that (for the empiricist, at least) this kind of justification is flagrantly (though for some undisclosed reason not viciously) circular, because our choice of rules depends upon their 'validity' and that again upon our readiness to accept the conclusions of inferences made in accordance with them. If we are reasonable, however, we should accept conclusions only if they are validly drawn; that is, only if they are drawn in accordance with the rules. 'A rule is amended', declares Professor Goodman, 'if it yields an inference we are unwilling to accept; an inference is rejected if it violates a rule we are unwi lling to amend.'[3] On what grounds we should be willing to do either he does not say: presumably there are no *logical* grounds, only the inclinations of 'taste and sentiment'.

[1] *Problems of Analysis* (New York, 1954), Ch. X, p. 189.
[2] *Op. cit.*, Ch. III.
[3] *Op. cit.*, p. 67.

If deduction is in so parlous a case, should we be surprised to find induction similarly placed? Here the rule, broadly speaking, is that if a conjunction of facts or events has occurred constantly or frequently in the past we may expect it to occur constantly or frequently in the future. We accept this rule because we find that predictions based upon it are commonly successful, and we believe that future predictions will be similarly successful because we accept the rule. In fine, there is no more 'justification' available for induction than there is for deduction, and at best we can give only a psychological explanation for what we habitually do—for our acceptance of conclusions and of the rules that enable us 'validly' to infer to them. And if all our logic is, in the last resort, but psychology, on what logic can the psychologist depend for the validation of his conclusions? This is altogether to abandon logic and rationality either as a source of, or even as a means to, reliable knowledge. It cannot be a source of knowledge, for, whether deductive or inductive, it originates nothing. It cannot be a means to knowledge for the rules which validate procedures are subject to predilections determined by e-logical motives in the persons claiming to know, and not by universal principles intrinsic to the objects known. Once again the barrier between science and prejudice is broken down, and psychology itself is deprived of those canons of validity that can alone give title to a systematic discipline.

If there can be no rational justification of the methods we use in science (inductive methods) and no rational justification of the conclusions we reach by their means, there can be no science, and any epistemological theory that forces us to this conclusion must at least be suspect—must at least call for re-examination. To argue that the empiricist conclusion is unavoidable would be to argue that the premises from which it is derived are true. To allege that it could be avoided simply by amending the logical rules according to which the inference was made, would be to license cheating so that any theory would be as sound as any other and scepticism would be rampant. No appeal to 'experience' is admissible in the absence of inductive justification. So the empiricist must either make good his case or capitulate. Goodman's statement is valuable in that it underlines the mutual involvement of induction and deduction in the same problem of justification. To what he calls the 'new riddle of induction' we shall attend in due course; but to claim as some do that the old one has been solved because in fact it is insoluble is to confess bankruptcy, and to disclaim the need for justification because none can be found is to conceal failure in bravado. Hume was more modest when he recommended 'carelessness and inattention', and more honest in his admission of scepticism.

ii. INDUCTION AND PROBABILITY

Attempts at justifying inductive argument are sometimes criticized for implicitly making the demand that it should lead to conclusions which are certain. The thesis to which the objection is made runs somewhat as follows: If an argument from true premises is valid its conclusion must be true—if true, then certain; so if induction is valid, since its observed premises are true its generalized conclusion should be. But to argue thus, the objector will say, is altogether to misunderstand the nature of induction, which essentially differs from deduction, and never can, and never claims to, reach conclusions other than probable. To justify inductive argument, therefore, all we need is a sound theory of probability upon which we can ground our inductive conclusions.

But this criticism, with the consequently proffered solution, is itself a misunderstanding of the problem—an *ignoratio elenchi.* What has to be justified (as Hume was well aware) is no claim to certainty, or even reliability, but is any claim to any degree of probability. For any belief in the probability of an event based upon how (or how often) similar events have occurred hitherto depends on the implicit assumption that the unobserved future will to some extent resemble the experienced past. It is therefore probability which depends on induction and not the converse. If this is not recognized any appeal to probability theory is bound to beg the question. I shall examine two notable attempts made in recent times to justify inductive reasoning by means of a mathematical (i.e. deductive) theory of probability, and shall argue that the both fail because they mistakenly reverse the logical relation between probability and induction.

Of the various meanings and interpretations that have been given to probability there are two main types—subjective and objective. The first takes probability to be a state of mind or a condition of knowledge. It is regarded as a degree of confidence or conviction with which we hold a belief, or a measure of the amount of evidence at our disposal. For the purpose of the present discussion, the intensity of belief is not relevant, as it is a psychological condition dependent upon causes which may have nothing whatever to do with evidence. In fact, there are some who can maintain the strength of a conviction in inverse proportion to the amount of evidence available and in the teeth of evidence to the contrary. This propensity is sometimes revered as Faith. *Credo quia impossible* is its maxim; but it is not typical of the scientific temper of our age. The notion of probability as a measure of the evidence we have so far ascertained is certainly relevant to our discussion, for this is a matter of logic and mathematics and is subjective only accidentally, in as much as the amount of evidence at our disposal is accidental. In other respects this conception belongs to the objective type of interpretation which makes probability a property of the event or the manner of its occurrence, or of the relationship between evidence and conclusion in an argument.

Mathematical probability belongs to the latter type and is a calculation of chances based on the assumption that the alternatives are all equally possible; or that there is no more reason to expect one than another. The essential presupposition of a calculus of chances is therefore indeterminacy and it is only in the realm of indeterminacy that it applies.[1] The basis of the calculation is then the known or determinable number of equal possibilities, and the probability of any one or any combination of these is expressed as the ratio of the number of possibilities in the combination to the total number of possibilities. The so-called frequency interpretation is no exception, for here the total number of possibilities is the total number of events in the observed series, and the number of appearances of the property under examination relative to the total number of events gives the ratio mentioned above. Here too we have a known number of equal possibilities and the ratio to this number of a determinate selection from them.

The utility of such calculation for wagering upon the future occurrence of indeterminate events is not and cannot be established by the calculus itself. That can be determined only in practice. That it will lead to greater and more consistent success than, say, wild guess-work or the indulgence of subjective prejudices can be discovered only by experience—in short only by inductive argument. It is, therefore, impossible for any mathemati-

[1] Cf. C. S. Peirce, 'The Probability of Induction', *Collected Papers* (Harvard, 1966), Vol. II, Ch. 7, p. 426.

cal theory of probability to justify induction, because knowledge of its efficacy as an instrument of prediction depends, and must depend, upon induction and not upon mathematics.

If human experience is, as Hume alleged, a succession of particular impressions and ideas between any two of which no connexion can be found, then the order of their occurrence is indeterminate and, whatever past experience may have been, what the future will bring is always a matter of chance. A mathematical calculation of chances, based upon past experiences would thus be possible; but the efficacy of its use in predicting the future would remain a matter of inductive argument, the logical basis of which is independent of the calculus. It cannot be established by the calculus, because success in predicting with the help of the figured probability can be discovered *only* by experience.

This is the view of the most noted writers on the subject; for instance, Charles Sanders Peirce says:

'In any problem in probabilities, we have given the relative frequency of certain events, and we perceive that in these facts the relative frequency of another event is given in a hidden way. This being stated makes the solution. This is therefore mere explicative reasoning, and is evidently entirely inadequate to the representation of synthetic reasoning, which goes out beyond the facts given in the premisses. There is, therefore, a manifest impossibility in so tracing out any probability for a synthetic conclusion.'[1]

And Bertrand Russell, writing in *Human Knowledge* of the mathematical basis of Bernoulli's law:

'Why this purely logical fact should be regarded as giving us good ground for expecting that, when we toss a penny a great many times, we shall in fact attain an approximately equal number of heads and tails, is a different question, involving laws of nature in addition to logical laws. . . .

'I want to lay stress on the fact that, in the above interpretation there is nothing about possibility, and nothing which essentially involves ignorance. There is merely a counting of the members of a class B and determining what proportion of them also belong to a class A.'[2]

[1] C. S. Peirce, *Collected Papers*, Vol. II, p. 423 f.
[2] *Human Knowledge* (London, 1948), p. 374.

And again:

'If, then, the . . . extension to an infinite series is to be used in empirical series, we shall have to invoke some kind of inductive axiom. Without this there is no reason for expecting the later parts of such a series to continue to exemplify some law which the earlier parts obey.'[1]

Two important points are made here. First, that as every probability estimate is the statement of a ratio between the numbers of two classes, no such estimate can be stated if one of the classes is infinite in number. Secondly, if the future is involved, with the consequent possibility of unknown change with the passage of time, prediction must depend on induction and is not determinable by any calculus of probabilities. Nevertheless, the argument of some eminent contemporaries, who have opposed this opinion and have striven to establish the inductive principle by mathematical reasoning, are sufficiently alluring to justify some examination.

One of the best attempts to ground induction upon probability has been made by Donald Williams.[2] His argument, in summary, is as follows. Probable reasoning is as cogent, in its own way, and as 'necessary' as demonstrative reasoning; it differs only in stating definitely the proportion of cases in which M is P, when it is known that neither All M is P nor No M is P. The two last-mentioned cases give us demonstrative reasoning, but in the intermediate case, in which a proportion of things which are M are also P, and we can discover the precise proportion (e.g. m/n), we can assert with complete aplomb that the probability that any M will be P is m/n.[3] As Russell maintains, probability is a matter of class relationship.

Williams then demonstrates mathematically that in any population of given composition (such as a bag of marbles, some of which are black and some white) the number of possible samples of a sufficient size which match the population in composition is always in the majority, and those which almost match it are in the vast majority. This is the case whatever the com-

[1] *Human Knowledge*, p.376. Cf. Karl Popper, *The Logic of Scientific Discovery*, Ch. X, 81; R. B. Braithwaite, *Scientific Explanation*, pp. 258–9; Kneale, *Probability and Induction* (Oxford, 1949), pp. 211 ff., 223; J. O. Wisdom, *Foundations of Inference in Natural Science* (London, 1952), pp. 210–11; Arthur Pap, *op. cit.*, p. 238.

[2] See *The Ground of Induction* (Harvard, 1947).

[3] This is called the 'proportional syllogism'.

position of the population may be, and it is known *a priori*—it is a mathematical truth. It follows that if we have in our possession (know from observation) the precise composition of any large sample of any population (of whatever size, short of infinity; or of whatever composition, between 100 per cent and zero) we can assert a precise calculable probability that the composition of the total population will be approximate to that of the sample. The figures are impressive. If the number of the sample is 2,500, the standard deviation *cannot* be more than 0·01. There must be a probability of at least 0·6826 that the sample composition does not deviate from that of the population by more than 1 per cent; and there must be a probability of at least 0·9545 that it does not deviate more than 2 per cent. (There is no need to repeat the unimpeachable calculations by which Williams reaches this indisputable conclusion.)

When we know that the probability that M is P is *m/n*, because, of *n* observed instances, we have found *m* to be P—i.e., when we know that *m/n* of MQ (observed cases of M) are P—then we know positively that there is a 95 per cent chance of not more than 2 per cent deviation from a probability of *m/n* that any *M* is P. Where we have this kind of assurance that there is a high probability of *M*s being P, it would certainly be unreasonable to harbour unfounded doubts about predictions based upon it—*on two conditions*: that we can be sure (or at least have the same degree of confidence as we have in the calculated probability) that the population we are testing is not infinite in number, or that its composition does not change with time.

The fundamental assumption of William's argument is that the tested population, or class, is fixed and static; that, as a whole, it has a definite composition which is more or less simulated in samples drawn from it. The bag of marbles is a paradigm case. The proportion of white rabbits among an existent population of rabbits would be another. But if the number of marbles or rabbits was infinite, the basis of all probability calculations would disappear. Or if surreptitiously somebody poured in more marbles, or if the rabbits were very prolific, and the proportion of colours was varied between the drawing of one sample and the next, our experiments would be ruined. If we are to rely upon the ground for prediction offered by Williams, then, we must know that the population does not change with time. But we cannot discover that by the same reasoning. True, we may examine a large number of successive samples to see whether their composition varies; but if it does not, and we argue from this fact to the constancy of the composition of the population, we are using the traditional form of enumerative induction and not comparing a sample with a population from which it is drawn. It would be extravagant to suggest that a series of successive samples showing a series of compositions (either relatively constant, or varing in some specific manner) was itself a sample of a population of samples, and to base our argument on this. For example, we might say that as the variation in composition from one sample to the next was such-and-such, a continued selection of samples, or all possible samples which could subsequently be drawn would display a similar variation, because the super-population of samples would have approximately the same properties as the sample of samples drawn. Not only is the super-population potentially infinite in number but this argument would in any case be very implausible, for now we are dealing not with a proportion of Ms which are P, but with a succession of sub-classes of M, each of which contains a different proportion of P, and we are seeking to conclude that the proportion of all Ms which are P, not simply has varied similarly, but will continue to vary similarly; for which we have no warrant. And even if we did argue in this way, we should be involved in a vicious infinite regress, because we should then have to be assured afresh that the super-population of samples was not itself changing its composition with time.

In short, we should still be in the predicament described by Hume of basing our estimate of probability on our belief that the population does not change its composition, or in other words, that what has been observed in the past will continue to be the case in the future. This is the inductive principle, which requires validation, but is not validated by the theory of probability offered, because the application of that theory must, in the last resort, rest upon the assumption of the validity of induction, not *vice versa*; and this is the case with any theory of probability.

The second main attempt to reverse this relationship has been made by Reichenbach. His exposition of the theory is less lucid than Williams', but the theory is more subtle; yet it fails for similar reasons. Reichenbach actually has two different arguments claiming to justify inductive reasoning. One is slipped in surreptitiously and is not claimed as the main support of his case. It is the one most commonly used by apologists for induction and we shall return to it anon. The other, on which he lays most stress and thinks most cogent, is, in effect, the claim to justify induction by a deductive proof based upon the theory of probability. 'Although the inductive inference is not a tautology,' he says, 'the proof that it leads to the best posit is based on tautologies only'.[1] The tautologies on which it is based are presumably contained in the mathematical theory of probability, hence the claim is to have justified the use of induction deductively by means of probability-theory. We must examine this proof and the import of the above quotation.

Reichenbach reduces all forms of probability to relative frequency and asserts that only by so doing can 'the meaning of probability statements . . . be determined in such a way that our behaviour in utilizing them for action can be justified'.[2] When we do not know the outcome of action, but are nevertheless forced to act, the best we can do is to count on the most probable occurrence as determined by the frequency interpretation of probability. This procedure, he says, will prove to be the most favourable on the whole, and will, in the long run, lead to the greatest number of successes.[3]

If this were all, we should be entertaining what has been called the pragmatic justification of induction, attributed to Peirce and offered by Braithwaite, Kneale and some others. This is the argument slipped in, to which reference was made above, but on which Reichenbach does not (at least in appearance) claim to rely. Yet, as we shall see, it is really the only one that will serve his purpose, and it is fallacious. Before we attended to it in detail, however, we must examine his main contention.

The relative frequency theory of probability, in brief, is that in any given series of events, some with the property A and the rest without it (non-A, or $\bar{A}$) the ratio of A to the total number of events in the series is the relative frequency of the event whose probability is sought. This gives us a value between 0 and 1 which Reichenbach calls the 'weight' of the proposition asserting the occurrence of the event. What this ratio will be will (or may) depend on the length of the series. As it is prolonged the ratio may change in any way, but if it tends to vary within progressively decreasing limits, the series is said to be convergent, and the ratio,

[1] *Experience and Prediction* (Chicago, 1938), p. 359.
[2] *Ibid.*, p. 309.
[3] Cf. *ibid.*, p. 310.

or relative frequency, to converge to a limit. Thus suppose that in a series of throws of a die the relative frequency with which the 2 appears uppermost is $\frac{6}{34}$; as the series increases in length, the ratio of 2's to the total number of throws will converge to $\frac{1}{6}$. In technical terms (as Reichenbach states it), if we symbolize the relative frequency of a particular type of event in a series of n events as $h^n = m/n$ then 'for any further prolongation of the series as far as s events ($s>n$) the relative frequency will remain within a small interval around h^n; i.e. we assume the relation $h^n - \epsilon \leq h^s \leq h^n + \epsilon$ where ϵ is a small number.'[1]

Reichenbach calls this assumption (that the series is convergent) 'the principle of induction'. But strictly it is not; it is simply the result of a calculation performed on examined cases. If I examine n cases and of them I find m to be A, $h^n = m/n$. If I then examine further cases (say r, such that $n + r = s$) and find that $h^s = h^n \pm \epsilon$, I have discovered that up to s the series appears to be convergent. No inductive argument has so far been used. If, however, I were then to project the empirical series into the future and infer that, because the series up to s appears to be convergent, it will, if prolonged, converge to the limit hitherto calculated, the inference will be inductive. The principle of induction more accurately stated would, therefore, be the assumption that a series of examined cases which appears to be convergent, will, if prolonged into the future, prove to be convergent. The limit to which the relative frequency tends in the examined series is known, but it is further assumed that future prolongation of the series will result in its tendency to the same limit, or one near it.

The questions now to be raised are whether, and if so how, this assumption may be justified. Reichenbach states Hume's problem with admirable accuracy and clarity,[2] but as soon as he addresses himself to its solution he misinterprets it. He alleges that Hume thought the justification of induction impossible because he sought proof that it must be successful, or, in other words, because he sought a demonstration that the conclusion of the inductive inference is true. But, we have already seen that this is false (and, oddly enough, its falsity is clear from Reichenbach's statement of Hume's argument a few pages earlier). Hume asserts that not only is it impossible to demonstrate the truth of the inductive assumption, but also, and more relevantly, it is impossible to justify it by

[1] *Experience and Prediction*, p. 340.
[2] *Ibid.*, pp. 341–2.

experience—because that would be to use it as the principle of justification—or, in other words, *it is impossible to show that the inductive conclusion is even probable.*

Proof, says Reichenbach, is a sufficient condition of justification, but not a necessary condition. The sufficient condition is unattainable, but a necessary condition he thinks he can give. All we need show is that inductive inference furnishes us with the best assumption concerning the future, and that, unless it is applicable, prediction is impossible. In short, we must prove that induction is the necessary (but not sufficient) condition of prediction.

This he proceeds to do by alleging that the assumption of a probability value (or weight) for a future event is not an assertion of a proposition as true, but is simply a 'posit' in accordance with which we are prepared to act. The probability calculus (presumably) ensures that it is the best posit, because it is the limit of relative frequency to which the series of examined cases converges; and unless we adopt it, prediction would be impossible. Not even clairvoyance can help us out, because to discover whether the clairvoyant is reliable we should have to calculate the relative frequency of his accuracy and use inductive inference to establish its constancy.

Reichenbach concedes that 'the character of induction as a necessary condition of success must be demonstrated in a way which does not presuppose induction'. He believes himself to do this, but he fails. His demonstration runs as follows: If we wish to predict, we must define what we mean by prediction, and all we can sensibly mean is that we assume that there is a limit to which the future prolongation of our observed series converges. If there is such a limit, there is an element in the series from which by induction we can arrive at the true value of the limit. If our series is too short for this purpose, we can extend it, and by continuing this process, if there is a limit at all, we shall continually approximate to it. Our first posit (m/n) (before prolongation) is 'blind', because the assertion that it is the limit can be given no weight. Nevertheless, it is the best we can get, and by prolongation we can continuously correct it, and we know mathematically that by this method we shall eventually reach a true value. Therefore, *if there is a limit*, the necessary condition of finding it is to use induction. The applicability of induction is the necessary condition of predictability. He confesses that we do not know if there is a limit, but if there were none prediction would be impossible.

This proof has two defects. One mathematical and the other logical. First let us be clear that to say 'if there is a limit' can mean only 'if the series is convergent'. If it is not, there is no limit to approach, and *vice versa*.

(i) Now it is common ground that our series of examined cases is always incomplete. It is always (for the purposes of any argument to justify induction) an 'incomplete enumeration'.[1] But any partial series of this kind is a selection from a larger series which, mathematically, may converge to a different limit from that to which the selection converges, or may not converge at all. Moreover, in any long or infinite series, it is possible to select elements which form a convergent series though the larger series does not converge, or converges to a different limit. The method of selection is therefore important, and it is not always true that prolongation of the series must result in progressive correction of the estimate of the limit. Some rule must be added governing selection, to the effect that it should be 'random', whatever that may mean in the context, and should afford a 'fair' sample. To discover a satisfactory rule is the statistician's problem.

(ii) Suppose, however, that we can formulate such a rule satisfactorily, we are still far from having justified induction non-inductively. Reichenbach defines the aim of induction as 'to find a series of events whose frequency of occurrence converges toward a limit'.[2] But this is easy enough in all the relevant cases, for the limit is calculated from observed instances, and so our *observed* series of events is such a series. The aim of induction goes further. It is to determine whether, as the series continues empirically, the frequency of occurrence of events of the relevant sort will continue to converge, and, moreover, to converge to the same limit. If we assume that it does we can predict, and the assumption would justify the prediction. But what is to justify the assumption? Not the calculation establishing the convergence of the observed frequency, for that gives no grounds for assuming similar convergence in the *continuation* of the empirical series. To assume that there is a limit is simply to assume that the mathematical series and the empirical series are the same and that the latter as it continues does converge. To adopt the relative frequency as a 'posit', is no more than to expect the series to continue as it has begun. If we say that it is the 'best' posit, we

[1] Induction by complete enumeration requires no justification.

[2] *Experience and Prediction*, p. 305.

mean, again, that we expect the limit of convergence to remain near to it. These are all ways of assuming but one thing: that the future will resemble the past. But that is the inductive principle which has to be justified, not the justification. To say that we cannot predict unless we make this assumption is only to say that we cannot argue inductively except on the principle of induction, which would, indeed, be a tautology, but no justification.

Consequently, there are but two alternatives open to Reichenbach. He must either claim that the posit is justified by the mathematical calculus of probability—which is true as far as it goes, but is irrelevant, because the calculus which determines the limit to which an observed series tends cannot justify any assumption about its continuance—or he must claim (as he did earlier, by implication) that inductive procedure is justified by its success in the long run. But this is the pragmatic justification, which is circular, for the degree of success in the long run (which is future) can only be assessed inductively. We shall examine this horn of the dilemma more closely in the next section.

iii. THE PRAGMATIC JUSTIFICATION OF INDUCTION

Braithwaite attributes to Peirce the argument that induction is justified by its 'truth-producing virtue', a phrase which he translates as 'productive reliability'. There is, however, much more to Peirce's theory than Braithwaite and empiricist writers like him seem to recognize. It includes elements of Kantianism and even of Hegelianism, to comment upon which will not serve our present purpose. Peirce (even as quoted by Braithwaite) says that 'the processes by which our knowledge has been derived are such as must generally have led to true conclusions'.[1] If this is no more than a report of past conditions, nothing follows from it as to future events, but it is not clear that Peirce intends it so. The word 'must' may have a deeper significance, which is apparently overlooked by empiricist commentators, who, almost to a man, insist that the inductive procedure is precisely that of which we can never say that it *must* (even 'generally') lead to true conclusions, but only that it probably will. Peirce holds that the foundation of inductive inference is that it is somehow involved in the condition of experience; which is as much as to say that it is somehow *a priori*.[2]

[1] R. B. Braithwaite, *Scientific Explanation*, p. 265.

[2] *Collected Papers*, Vol. II, Ch. 7. If this is not Peirce's intention, his theory falls under the criticism presently to be offered of the pragmatic justification.

It will be remembered that Bertrand Russell holds a similar view.[1] The pragmatic justification is different, and, as Braithwaite states it most fully and in its most sophisticated form, it will be best to examine his version of it.

In common with other writers,[2] Braithwaite reduces all inductive procedures to simple enumeration, for though he distinguishes that from what he calls eliminative procedures, and asserts that all other forms are either modifications of one or a mixture of both of these, he admits that eliminative arguments depend on a suppressed major premise that one and only one of a finite number of possible hypotheses is true and that all possibilities have been examined. This major premise can be established, if at all, only by an inductive procedure of simple enumeration.[3] We may therefore confine ourselves to consideration of the latter alone.

Induction by simple enumeration is that form according to which an empirical hypothesis is taken to be established 'if it has not been refuted by experience and has been confirmed by not fewer than *n* positive instances' (*loc. cit.*). The justification offered for this procedure is that in the great majority of cases in which it is used it leads to success in prediction. The criterion of this 'predictive reliability', as Braithwaite gives it, is 'the proportion among the inferences from true premises covered by the policy of those inferences which lead to true conclusions'. The claim that the policy is successful in most cases is moderated by Braithwaite to no more than that it is successful in many cases, and that hypotheses have been confirmed by its use within a specified period (e.g. since Archimedes, or since Galileo). This provision, he admits, is arbitrary and is inserted in order to restrict the criterion to methods used by 'reputable scientists'. Such a policy is called 'effective-in-the-past', and to accept it as valid is said to be 'reasonable'; but, should it fail in the future, while it would not be 'reasonable' in the same sense to accept it, it might still be 'reasonable' in some other sense. If hypotheses established in the past were disproved, we might reject the policy altogether; but if only new hypotheses failed to be confirmed, we might say that the old policy had been effective-in-the-past but could no longer serve the needs of science, and that a new one was required.[4]

[1] *The Problems of Philosophy*, *loc. cit.*, and *Human Knowledge*.

[2] Cf. W. Kneale, *Probability and Induction* (Oxford, 1949), and Reichenbach, *op. cit.*

[3] Braithwaite, *op. cit.*, p. 260.

[4] *Op. cit.*, p. 269.

Dubiousness infects this theory at many points. First, what can be meant by 'reputable' as applied to scientists who use a particular inductive method, other than that they have been successful? If the method is good, the science that uses it will be reputable, but not otherwise. So the introduction of 'reputable scientists' is irrelevant, apart from their prestige value.

Secondly, if a method by which many hypotheses have been confirmed in the past fails to confirm subsequent hypotheses, should we conclude that the method is bad, or simply that the unconfirmed hypotheses are false? It is hardly 'reputable' procedure to change one's methods to suit one's theories, and if the theories prove unreliable, it would surely be unscientific always to impugn the methods of testing. There would seem to be some confusion here between methods of arriving at hypotheses (what Peirce called 'abduction') and those of testing, or confirming, them. We may suspect the former if the latter repeatedly give negative results. But to indict the latter, disappointment with results is not enough; we must find some logical flaw in their procedure. Again, if previously confirmed hypotheses subsequently prove false, is that because the inductive policy by which they were formerly confirmed was unreliable, or that by which they were subsequently disconfirmed? Or, alternatively, is it simply that what, on the basis of some evidence, seemed true, on other evidence proved untrue? Further, is it suggested that there might be some inductive procedure, other than simple enumeration, which could be adopted, and which, if predictions made on the basis of simple enumeration failed, might succeed better?

Braithwaite's conclusion is that no other method has, in fact, proved more successful than enumerative induction. He finds Kneale's contention (and incidentally Reichenbach's), that there is no other systematic way of making predictions, too sweeping, but asserts that experience has taught us that other methods (e.g. free association or soothsaying) are less successful.

This is clearly and confessedly an inductive argument. If experience has taught us that policy π (enumerative induction) has been more successful than policies ρ or σ, and we thence conclude that π is the most reliable, this can only mean that we argue from the superior success in past instances of its use to the general rule that it is the most dependable policy. In short, we use enumerative induction to justify enumerative induction—or, more correctly, we assume that enumerative induction is predictively reliable in order to establish that it is. This is obviously a viciously circular argument, a blatant *petitio principii*.

But Braithwaite denies that it is so, because, he says, all that is needed to make the argument valid is either (i) the belief that enumerative induction is valid—a belief that need not be rationally held (i.e. need not be supported by argument), or (ii) that the proposition stating the predictive reliability of induction should be true; nor is it necessary that this proposition be used as a premise in inductive argument. An argument works like a machine, which may produce a valid conclusion from premises fed into it, working according to certain rules of inference without circularity; and if the conclusion should happen to be a proposition about the character of its own rules of inference, it would still be validly drawn. Finally, he says, reasonable belief is like right action. The belief in the premises and the validity of the principles of inference need not be objective; so long as they are held the inference is reasonable—just as action may be morally good so long as the agent believes it to be right. Let us take each of these arguments in turn.

First, if I believe that induction is a valid method of argument, I shall argue inductively, but my argument will be invalid if the belief is false. If it is true, but I am not able to show this by any rational method, then I cannot justify induction and cannot know that it is valid. If I attempt to do so by using as a premise to my justificatory argument my belief in its validity, I shall beg the question. This is the essential point. Inductive argument, in any particular case, proceeding from past evidence to future occurrence (or to a generalization—like the conclusion that all ravens are black), need not be circular; but any argument to *justify* enumerative induction which uses simple enumeration to establish its conclusion must be circular, for if the method were invalid in the first place the justification would be fallacious. If, on the other hand, the principle of induction is sound, its truth must be established by some other means. It cannot be established inductively. Braithwaite's defence, therefore, like Reichenbach's, commits the fallacy of *ignoratio elenchi*, for what is alleged is not that inductive argument is circular and question-begging, but only that any attempt to demonstrate the validity of inductive procedure on the strength of empirical evidence must be.

Next, if a machine is programmed to proceed according to a certain principle of inference, all its data will be processed in accordance with that principle, so that it will be presupposed in all its conclusions. If any conclusion is a statement validating the principle, it could only be reached by the equivalent of a viciously circular argument. But, in fact, no machine could perform such a fallacious operation. A machine is a mechanical calculus, and no calculus can be used to prove its own axioms or justify its own transformation rules.

Thirdly, it is not true that 'reasonable' belief is like moral action. Action may be held morally right if the agent believes it to be right; but argument is not valid if the theorist only believes it to be valid, any more than the belief is true because it is believed. If the inductive principle is valid, induction will be sound procedure, but my belief that it is so will neither make it so nor logically justify my procedure. Reasonable belief must be grounded on valid reasons, and a viciously circular argument can provide no valid reason for any belief.

Max Black attempts to accommodate himself to this situation by arguing that the circularity involved in the empirical justification of induction is not vicious.[1] He calls it a self-supporting argument, considering self-support a desideratum. It is, that is to say, an argument which follows the rule for the success of which it argues. This sounds well enough, but is none the less fallacious. An argument which is sound may be used to support a claim that, because it is sound, it will derive true conclusions from true premises. But we cannot validly argue that an argument is sound because, if it were valid, it would be successful. Though the hypothetical is true, it does not prove the categorical form of the protasis. p does not follow from $p \supset q$. In the case of enumerative induction, what we argue, in effect, is that so long as the principle is presumed valid it demonstrates its own validity; but the presumption is not thereby established, and to refute the justificatory argument all that is needed is the rejection of the principle.

Black, however, supports his case by alleging that all valid argument is circular (pre-eminently deductive argument), because if it is valid the conclusion must be entailed by the premises. A hoary criticism of the syllogism proceeds in the same fashion. It relies on the empiricist dogma that universal propositions are no more than summaries of particular facts, and takes them in extension. It then alleges that the major premise of a syllogism in

[1] *Op. cit.*, Ch. XI.

Barbara must contain its conclusion (as well as its minor). But this contention serves only to underline the implicit scepticism in empiricist epistemology and the necessity, if we are to render our experience intelligible, of adopting a conception of logic which does not reduce all deduction to tautology and all induction to an e-logical mental propensity.

Hume, therefore, is finally vindicated. The inductive procedure, which he and other empiricists consider to be the only one available to us for the acquisition of factual knowledge, can be validated neither by demonstration nor by experience. The position remains as sceptical as he left it and has not been saved by modern arguments. It does not follow from this that there is no valid empirical procedure characteristic of scientific investigation, or that no satisfactory account can be given of scientific method; only that the account which is given by traditional and modern Empiricism is incoherent and would if true render science as a form of respectable knowledge impossible.

iv. INSTRUMENTALISM

Little need be said of other expedients used by empiricists to escape the impasse of inductive argument. Herbert Feigl[1] admits all the strictures we have so far offered. He concedes the insolubility of the problem of validation and that no demonstration can be given that the principle of induction has the least probability. He also rejects as irrelevant any attempt to justify inductive procedures psychologically. The principle, he maintains, is not a proposition but a precept or regulative maxim. But this expedient is of little avail. Why, we may ask, should the precept be followed or the maxim adopted? Feigl's answer is that we have no option if we wish to produce an orderly account of the world, from which we can make successful predictions. We return then to the position that induction is justified because it is the sole (or the best) method of successful prediction, but this is the pragmatic justification over again. Moreover, that it *is* the sole method is simply asserted—not only by Feigl, but also by Kneale, Reichenbach and the rest—it is not demonstrated. Whether or not there are other methods and whether the traditional inductive procedure actually is the one used in science has still to be

[1] See 'The Logical Character of Induction' in *Readings in Philosophical Analysis* (ed. Feigl and Sellars).

investigated. Feigl and Reichenbach tell us that induction gives no guarantee of success, though they recommend it as the best method available. We have seen (and Feigl confesses) that not even probable success can be demonstrated. Why, then, it should be 'reasonable' to accept the maxim becomes altogether obscure.

V. NEW PUZZLES FOR OLD

All empiricist discussions of such problems are committed to the sharp distinction between *a priori* (analytic) truth, dependent upon deduction, as a string of tautologies, and *a posteriori* (synthetic, or factual) truth dependent upon induction.[1] The former gives us 'nomic' universal statements (which are however only conventional) and the latter only factual (or accidental) universal statements which cannot function as scientific laws. No empiricist has yet succeeded in producing any satisfactory theory of the way in which we arrive at law-like statements from empirical evidence, as opposed to merely contingent general statements. This essentially is the problem of induction for Empiricism; and Goodman, in asserting that 'the old puzzle of induction' has been disposed of,[2] is misled by the clamour of those who try to evade the problem by denying the necessity to justify inductive argument. This 'old puzzle of induction' has never been solved and the so-called 'new puzzle of induction' is only the old one in more crucial form. The reason why inductive argument gives trouble to the empiricist logician is that he can allow no logical connexion between particular (observable) facts. The discovery of any such connexion would remove the problem, for it would enable us to enunciate 'nomic' universals. The 'new puzzle', as explained by Goodman, is to discover a criterion among constant conjunctions for distinguishing those which do from those which do not justify the enunciation of nomic universals. In short, he seeks the criterion which, if it were discernible, would justify inductive inference—the procedure of establishing a general law stating necessary (or universal) connexion between particulars frequently observed in conjunction. On the Humean basis there can be no way of doing this, yet in actual scientific practice there is and must be. It follows that no satisfactory

[1] Quine's critique of this dichotomy seems, as yet, to have had little effect on current theories of scientific method. See *From a Logical Point of View*, Ch. II.

[2] *Op. cit.*, pp. 63–8.

account of scientific method can be given if the Humean position is stubbornly retained.

vi. THE UNREASONABLENESS OF INDUCTION

Curiously enough, the method commonly approved as inductive is not much used by scientists, as critics of the doctrine whose views we are shortly to consider have asserted, and is used, if at all, only tentatively and only at those stages of scientific research in which the scientist is groping for guide-lines to a satisfactory solution to a problem. It is much more typical of low grade thinking and of superstition. Such theories as that bad luck is the result of walking under ladders or of breaking mirrors are usually supported by allegations that on numerous occasions this has been experienced in the past, as when Auntie walked under a ladder and was hit on the head by a brick, or when Grandma broke the hall mirror and suffered a long series of disasters. Superstition, in short, is prone to rely on this sort of argument and would qualify as scientific if induction were also typical of science. What must be noted once again is that Hume reaches the only consistent conclusion to which a strict empiricist epistemology can bring us: a sceptical conclusion, incompatible with respect and admiration for science or any deference to its authority.

There are, moreover, good grounds for holding that induction, so far from being the only reasonable procedure, is the very opposite of reasonable, even on the basis laid down by classical Empiricism. It was argued above that the essential presupposition of the empiricist position was indeterminacy in the sequence of empirical facts as we experience them. Where no connexions can be detected predictions must be based on the calculation of chances. But on that assumption the more reasonable procedure should be what has been called counter-induction. We should argue most reasonably that what has happened frequently up to the present is less probable in future proportionately to the frequency of its past occurrence. The probability of throwing a six at dice is $\frac{1}{6}$; the probability of throwing two sixes in succession is $\frac{1}{6^2}$; of throwing three in succession is $\frac{1}{6^3}$; and so on. So it might be (and often is) thought that if three successive sixes have turned up the chances against a fourth would be very high. This is in fact a fallacy, for the chance of the fourth six is still only $\frac{1}{6}$ (if the first three have already shown up). Nevertheless, it is quite unwarranted to conclude from a long sequence of similar chance events that the probability of the next event's being similar is greater (that the chance of the fourth six is more than $\frac{1}{6}$). The gambler's fallacy is an understandable and an excusable lapse, but the gambler who imagines that, because he has had a run of luck, the chance of its continuing is enhanced, is unreasonably optimistic. Yet this must be the underlying assumption of inductive argument on the empiricist theory. So little can we justify induction by saying that it is the only reasonable procedure in dealing with matters of fact, if matters of fact were related as empiricists assume, it would be altogether unreasonable.

Worse than this, confirmation of probable hypotheses is not possible in the manner alleged. For if the probability of AB is m/n the occurrence of AB does not confirm the hypothesis that this is the probability, nor does its non-occurrence. Not even approximately m future occurrences out of n trials amounts to confirmation, because the probability might in fact be different and, nevertheless the actual frequency of the event *in any limited series* could still be m/n. Or the probability might be m/n and yet the actual frequency *in a limited series* could differ, without contradicting the hypothesis. It does not follow from the statement that $P = m/n$ that the event of which P is the probability must always occur m times out of n trials. If we assume that $P = m/n$ because in the past the proportion of occurrences in an observed series of trials has been m/n, this, as Reichenbach has said, is a mere posit. It has no logical or other connexion with the actual frequency of events in any other series, and therefore no actual occurrence or frequency of occurrence can rank as evidence for or against it—except on one condition.

If there could be some kind of connexion between particular matters of fact, then the experience of a frequent conjunction would be presumptive evidence of such connexion, and constant conjunction would be almost conclusive. One or two observations of conjunction we might discount as due to chance coincidence, but a large number we should attribute to something more binding. Accordingly, it would then be reasonable to assess the probability of a connexion more highly in proportion to the number of past conjunctions observed. The inductive form of argument would therefore be reasonable and justified only on the assumed feasibility of a necessary connexion of some sort, and the procedure of both the statistician and the empirical scientist bear this out. The former eliminates as mere chance a conjunction which occurs with a frequency below a certain minimum and allows as significant one which is more constant. The latter goes on from this point to seek for a connexion which can be expressed in a formulable law.

CHAPTER III

THE EMPIRICIST TREATMENT OF DEDUCTION AND NECESSITY

The conception of induction we have been examining is the indispensable foundation of an empiricist philosophy of science, and the insolubility of the problem it raises is bound up with the denial of necessary connexion between matters of fact. Yet the purport of scientific laws is the assertion of necessity in the interrelation of facts and this has somehow to be explained away. Modern philosophers are not much enamoured of Hume's recourse to psychological causes to account for our belief in necessity, and other methods are used to explain its persistence in scientific thinking. For the empiricist, necessity belongs only to logic and mathematics the deductions of which are all purely analytic, and nowadays the appearance of necessity in science is attributed wholly to the mathematical and deductive element freely admitted as part of its method. There are two ways in which this is done. First it is alleged that universal propositions function only as stipulative definitions (which are always analytic), or as conventional principles of inference (again analytic) for the purpose of prediction. Secondly, empirical generalizations derived by inductive inference are held to be introduced into deductive systems, either by the substitution of empirical terms for mathematical or logical variables, or as postulates (or axioms), so that other empirical hypotheses may be deduced from them. Or alternatively, hypotheses involving purely theoretical concepts may be formulated and introduced into a deductive system in such a way that empirically testable lower level hypotheses can be deduced from them. Such a deductive system is said to be explanatory, and to explain a scientific law is to show how it may be deduced from higher level hypotheses in such a system. The first of these two expedients is called 'conventionalism' and it is not altogether unconnected with the second; but neither of them is satisfactory.

i. CONVENTIONALISM

The futility of conventionalism as a device for explaining the necessary character of scientific laws should be obvious at once, and it is surprising that anybody has taken it seriously. Scientific laws are empirical and essentially synthetic. If they were merely conventional rules of inference or stipulative definitions they could state no facts. Moreover, empirical evidence would have no bearing upon them and experience could never prove them false. Truth and falsity would be inapplicable to them, for whatever was found not to conform to a law would simply not come under the law; and whatever was not in accordance with a definition would not be what the definition defined. But scientific laws are laws of nature, for they state conditions which hold of the objects and events of the actual world—at least this is their purport. If they were merely conventional principles, they would concern nature as little as do the rules of chess and would be as irrelevant to natural science.

Karl Popper objects to conventionalism, not because he finds in it any logical flaw, but because it is not compatible with the scientific ideal of devotion to truth. Implicit definition of key terms and axiomatic use of fundamental laws are (as we have said) analytic and neither they nor their logical consequences can therefore be refuted by experience. They may serve scientific investigation, he thinks, when theories are accepted and no awkward examples occur, but in times of stress the temptation to ignore awkward facts, or exclude them by disqualification, is too great, and obstructs scientific progress. Conventionalism enables the theorist to evade falsification, which, for Popper, is the unforgivable sin in science.

But there is yet another objection, which should be still more serious for the strict empiricist. The practising scientist does not regard scientific laws as mere conventions but as statements of fact and he treats them, moreover, as necessary. The philosopher who asserts that they are conventional does so, in order to make them conform to his own epistemological theory. For they can be necessary principles, according to that theory, only if they are analytic. But a truly empirical philosophy should be one which conforms its theory to the facts and not the facts to the theory. The facts of scientific practice (as we shall later observe) do not conform to the requirements of traditional Empiricism, but when the empiricist encounters this conflict he adopts, presumably as a convention, the postulate that scientific laws are only conventions, in order to preserve the theory at the expense of the facts.[1]

[1] I owe this point to my colleague, Dr Richard Cole.

Popper's objection to conventionalism, that it runs counter to the scientific spirit, applies therefore to Empiricism as such, for it tempts the philosopher as well as the scientist to dogmatize in the face of contrary evidence.

ii. DEDUCTION AND EXPLANATION

It is not therefore surprising that even among empiricist writers conventionalism is rarely adopted and the second course listed above is preferred. This adheres to the view that natural laws are empirical and state only constant conjunctions, and explains the appearance of necessity by the incorporation of these laws into deductive systems which, being purely formal in themselves, are logically necessary.

But the prevailing doctrine of logical necessity as tautological is incompatible with the alleged explanatory function of scientific deductive systems; for what is deducible by strict logic (or mathematics) from higher level hypotheses cannot be new facts and the lower level empirical hypotheses must, therefore, on this view, be synonymous with all, or with some part of, the higher level hypotheses. It is difficult to see what explanatory function the latter could serve in that case. The theory claims, however, that the higher level laws are of wider scope and from them can be deduced other lower level hypotheses as yet otherwise undiscovered. These can be empirically tested and, if confirmed, will add strength to the entire system, if they are disconfirmed the system will require modification or must be rejected altogether.

It must be remembered that, for strict Empiricism, no matter of fact is logically deducible from any other. This is because, on the one hand, there is no logical connexion between facts, and, on the other, logical deduction is analytic and tautological. Now, either the higher level laws in a deductive system are factual or they are mere logical antecedents of the lower level hypotheses. If they are factual no further factual hypotheses could be deduced from them by pure logical processes, and equally if they are only the logical major premises from which the known and hitherto confirmed lower level hypotheses follow, no new factual material could be deduced from them. It may be said that together with other empirical premises derived from observation, fresh hypotheses could be deduced from the higher level formulae. But then the confirmation or disconfirmation of these new hypotheses would have no logical bearing on the system such as is envisaged, because there is supposed to be no logical connexion between the empirical propositions concerned—i.e. the earlier confirmed hypotheses and the empirical premises added for the new deduction. The fact that these deductions proceed from major premises, from which in conjunction with suitable minors the earlier confirmed hypotheses also follow, only means that the earlier and the subsequently deduced hypotheses all have some synonymous terms in common. Whether or not they are true or probable depends only on observed facts which are all logically independent. Their having some terms in common, therefore, cannot forge any logical link between their several probabilities.

The deductive system, which is supposed to represent a scientific theory, could, therefore, serve in no way to predict hitherto unobserved events (for anything deduced by it must be synonymous either with the theoretical major premises or with the already observed empirical minors, neither of which can entail any new facts); nor could it serve any explanatory purpose because it provides no information or insight into factual matters already investigated and established. It could not help to discover or elucidate any factual connexions between such matters, not only because no such factual connexions are supposed to exist, but also because, if there were any, they too would be matters of fact and could not be deduced prior to observation or from other facts previously observed.

This is well illustrated by Braithwaite's explanation of the role of deduction in scientific procedure.[1] He defines a deductive system as a series of propositions which follow, according to logical principles, and in precise order, from a set of 'initial propositions'. This may be represented as a calculus, in which certain stated rules of procedure, called 'transformation rules', correspond to the principles of deductive inference; and the terms of the propositions are symbolized by elements from which are constructed formulae corresponding to the propositions themselves. The calculus may be operated without reference to any interpretation of the symbols simply by manipulation of the symbols according to the rules; but as the rules correspond to principles of inference, the result is the logical consequent of a corresponding train of deductive reasoning. A deductive system may begin from propositions which are logically necessary and if

[1] *Op. cit.*, Chs II and III.

it contains no contingent propositions it is said to be 'pure'. If the initial propositions are contingent it is an 'impure' deductive system, and if some are necessary and some contingent it is said to be 'mixed'. A mixed or an impure system is called an 'applied deductive system'. In any such mixed system the purely logical or mathematical portion can be segregated from the empirical portion by the use of variables, for which may be substituted any elements of the impure portion. The accompanying table, taken from Braithwaite, is an example of a mixed deductive system of this sort.

[4]	$(\xi\xi) \leftrightarrow \xi$		[1]	$\alpha \leftrightarrow (\lambda\mu)$	
[5]	$(\xi\eta) \leftrightarrow (\eta\xi)$		[2]	$\beta \leftrightarrow (\mu\nu)$	
[6]	$(\xi(\eta\zeta)) \leftrightarrow ((\xi\eta)\zeta)$		[3]	$\gamma \leftrightarrow (\nu\lambda)$	
.	.	.	.	.	.
[7]	$\xi \leftrightarrow \xi$	from [4] and [4]			
[8]	$(\alpha\beta) \leftrightarrow (\alpha\beta)$	from [7]	[9]	$(\alpha\beta) \leftrightarrow ((\lambda\mu)\beta)$	from [8] and [1]
			[10]	$(\alpha\beta) \leftrightarrow ((\lambda\mu)(\mu\nu))$	from [9] and [2]
[11]	$((\lambda\mu)(\mu\nu)) \leftrightarrow (((\lambda\mu)\mu)\nu)$	from [6]	[12]	$(\alpha\beta) \leftrightarrow (((\lambda\mu)\mu)\nu)$	from [10] and [11]
[13]	$(\lambda(\mu\mu)) \leftrightarrow ((\lambda\mu)\mu)$	from [6]	[14]	$(\alpha\beta) \leftrightarrow ((\lambda(\mu\mu))\nu)$	from [12] and [13]
[15]	$(\mu\mu) \leftrightarrow \mu$	from [4]	[16]	$(\alpha\beta) \leftrightarrow ((\lambda\mu)\nu)$	from [14] and [15]
[17]	$((\lambda\mu)\nu) \leftrightarrow (\nu(\lambda\mu))$	from [5]	[18]	$(\alpha\beta) \leftrightarrow (\nu(\lambda\mu))$	from [16] and [17]
[19]	$(\lambda\lambda) \leftrightarrow \lambda$	from [4]	[20]	$(\alpha\beta) \leftrightarrow (\nu(\lambda\lambda)\mu)$	from [18] and [19]
[21]	$(\lambda(\lambda\mu)) \leftrightarrow ((\lambda\lambda)\mu)$	from [6]	[22]	$(\alpha\beta) \leftrightarrow (\nu(\lambda(\lambda\mu)))$	from [20] and [21]
[23]	$(\nu(\lambda(\lambda\mu))) \leftrightarrow ((\nu\lambda)(\lambda\mu))$	from [6]	[24]	$(\alpha\beta) \leftrightarrow ((\nu\lambda)(\lambda\mu))$	from [22] and [23]
			[25]	$(\alpha\beta) \leftrightarrow (\gamma(\lambda\mu))$	from [24] and [3]
			[26]	$(\alpha\beta) \leftrightarrow (\gamma\alpha)$	from [25] and [1]
[27]	$(\gamma\gamma) \leftrightarrow \gamma$	from [4]	[28]	$(\alpha\beta) \leftrightarrow ((\gamma\gamma)\alpha)$	from [26] and [27]
[29]	$(\gamma(\gamma\alpha)) \leftrightarrow ((\gamma\gamma)\alpha)$	from [6]	[30]	$(\alpha\beta) \leftrightarrow (\gamma(\gamma\alpha))$	from [28] and [29]
			[31]	$(\alpha\beta) \leftrightarrow (\gamma(\alpha\beta))$	from [30] and [26]

(from R. B. Braithwaite, *Scientific Explanation*, p. 39)

The rules of procedure are:

I. We may write a new formula which exactly repeats an

existing formula, except for the substitution for one of the elements, wherever it occurs, of the element opposite to it in some other existing formula. E.g. if $(\alpha\beta) \leftrightarrow ((\lambda\mu)\beta)$ and $\beta \leftrightarrow (\mu\nu)$ are existing formulae, $(\alpha\beta) \leftrightarrow ((\lambda\mu)(\mu\nu))$ may be obtained by substituting for β occurring on the right-hand side of the first formula the right element of the second formula, in which $(\mu\nu)$ is opposite to β.

II. Any element whatever may be substituted in a new formula for a variable in an existing formula so long as the same element is substituted for the same variable wherever it occurs. This is the ordinary substitution rule in algebra.

The formulae in the left-hand column of the table are all logically necessary and this portion constitutes the algebraical part of the calculus. The formulae in the right-hand column are contingent and the elements of which they are constructed are to be interpreted as representing empirical terms. The interpretation suggested by Braithwaite is one which makes each element of the contingent formulae represent a class: thus α represents the class of things with property A (or the class A), β the class of things with property B (the class B), γ the class of things with property C (the class C), and so on. Conjunction of elements represents the logical product of the corresponding classes.

The initial contingent formulae thus may be interpreted as the propositions:

(i) Everything which is A is both L and M and *vice versa*;
(ii) Everything which is B is both M and N and *vice versa*;
(iii) Everything which is C is both N and L and *vice versa*.

The conclusion is:

(iv) Everything which is both A and B is also C.

It may be deduced in similar fashion that

(v) Everything which is both A and C is also B and that
(vi) Everything which is both B and C is also A.

It is not easy to find empirical classes which will satisfy these formulae and which make the deduction seem very significant, but by carefully stacking one's cards one may produce an example which fits. Let us take the class A to be the class of human beings, the class L to be that of things with the property of having no feathers (featherless things), M that of things with the property of having two legs (bipeds); then proposition (i) becomes

(vii) Everything that is a human being is a featherless biped (and *vice versa*).

Let B stand for the class of fliers, N for the class with the capacity of being airborne (volatile things) and C for the class of airmen. Then propositions (ii) and (iii) become

(viii) Everything that is a flier both has two legs and is volatile (and *vice versa*);
(ix) Everything that is an airman is both featherless and volatile, etc.

The conclusions deduced are:

(x) Everything that is both human and a flier is an airman;
(xi) Everything that is both human and an airman is a flier;
(xii) Everything that is both a flier and an airman is a human being.

It is readily apparent that if we take the initial propositions as mere conventional definitions no empirical confirmation of the conclusions need be sought, but if we regard them as empirically supported hypotheses, no further empirical confirmation of the conclusions will strengthen them, because the conclusions must already have been known in the confirmation of the premises. If all human beings are featherless bipeds and *vice versa*, and this has been empirically confirmed, then we already know that all airmen are featherless bipeds, and if it has also been empirically discovered that all fliers have the capacity of becoming airborne, and that all airmen are volatile as well as featherless, then it is already known that all airmen are fliers. So we know from the start that all human airmen are fliers, and that all human fliers are airmen, and all flying airmen are human. As the deductive process is tautological throughout it should not surprise us that it issues in a conclusion that we already know, and it is not apparent how in any way the initial propositions or the conclusions have been explained. In fact, the elaborate symbolic array that has produced these results appears to be epistemologically altogether worthless and otiose.

Defenders of Braithwaite and the theory of science he propounds might complain that this criticism is unfair, for the example he has given is deliberately highly simplified and is intended only to illustrate the relationship in scientific theory of the formal mathematical part to the empirical part. True though this may be, it is for this reason either a bad example to serve Braithwaite's purpose, or else clear proof that the theory as a whole is unsound. The relationship of deductive reasoning to empirical confirmation which it is alleged to illustrate is one by virtue of which the deductive system provides an explanation of the lower level empirical hypotheses, inductive confirmation of which strengthens our belief in the higher level laws. But the example shows only that the deductive system provides no explanation and either renders empirical evidence irrelevant or its own contribution nugatory. According to the theory it should be possible to deduce new hypotheses from the initial propositions which could then be empirically tested, but in the example it is not apparent how this might usefully be done.[1]

In a subsequent chapter (Ch. III) Braithwaite asserts that theoretical terms have no meaning other than that given to them by their place in the calculus. We give meaning directly only to the propositions about observable entities, and to the theoretical terms only indirectly through their relation in the system to the empirical terms. The function of the theoretical terms is said to be to explain the empirical relationships. In the above example of a deductive system the propositions that

Everything which is A and B is also C and
Everything which is B and C is also A

may have been discovered by observation or experiment. We then explain the facts by relating them deductively to the theoretical terms L, M and N, as defined in the interpreted calculus.

[1] P. K. Feyerabend has argued that, as a consequence of the condition in this theory of explanation, that descriptive terms at each level must be either the same or mutually related via an empirical hypothesis, only such explanatory theories are admissible as *contain*, or are at least consistent with, those already in use in the particular domain. Scientific progress will consequently be inhibited. See 'Problems of Empiricism' in *Beyond the Edge of Certainty*, ed. R. G. Colodny (Englewood Cliffs, 1965), pp. 163–4.

Suppose, then that we have had no experience of featherlessness, or two-leggedness, or the property of becoming airborne, but that we have discovered empirically that whatever is a human being and also a flier (which must now be understood as some creature or entity that periodically disappears from the surface of the earth and again reappears) is also an airman (presumably so-called because he has this propensity) and we have also discovered that whatever is a flier and also an airman is a human being. It is improbable that we should seek an explanation of the humanity of airmen though we might well wonder what enabled them in common with other fliers to disappear from time to time. Suppose that we invent the theoretical term 'volatile' to refer to this capacity. It would be credible that we might then postulate that

(xiii) Everything that is an airman and also a flier is volatile.

But we should need no elaborate deductive system or abstract calculus to arrive at this conclusion. There would be no need to invoke two-leggedness or featherless nudity to explain the observed phenomena. Nor would the proffered explanation be more enlightening than that given by Molière's scholar of the soporific effects of opium.

The example, therefore, does not seem to answer to the actual character of scientific procedure, which is not wholly occupied with the obvious or with the provision of fatuous explanations. Could Braithwaite have given us more direct illustrations from scientific theory itself he might have been more convincing, but this, with one exception, he fails to do. Let us, then, consider the single example he provides. This is Galileo's theory of the acceleration of falling bodies, which Braithwaite represents as a deductive system as follows:

I. Initial hypothesis: Every body near the earth freely falling towards the earth falls with an acceleration of 32 feet per second, per second ($d^2s/dt^2 = 32$).

II. (Deducible by the integral calculus): Every body starting from rest and freely falling towards the earth falls $16t^2$ feet in t seconds, whatever number t may be.

From II there follow an infinite set of empirically testable hypotheses:

III(a) Every body starting from rest and freely falling for one second towards the earth falls a distance of 16 feet.

III(b) Every body starting from rest and freely falling for two seconds towards the earth falls a distance of 64 feet.

And so on.

What this example seems most evidently to illustrate is the process of empirical generalization; for, having observed that a body falling from rest for one second falls 16 feet, that one falling from rest for two seconds falls 64 feet, and for three seconds 144 feet, one may generalize and so arrive at hypothesis I. It is then obviously possible to deduce II from I and III *a*, *b*, etc. from II. It is true also that we could then deduce further cases III *x*, *y*, *z* which would be empirically testable, and if the results were positive they would help to establish the initial hypothesis. But this would be the familiar process of induction, the theoretical difficulties of which were explored in the last chapter. We shall later examine Galileo's actual procedure and argument to see whether it conforms to this pattern, but what is relevant at this stage is the question whether hypothesis I and the subsequent steps of the deductive series contribute anything by way of explanation or any method of discovery not already included in the process of inductive generalization; and that they do is not apparent.

Despite the ingenuity and elegance of Braithwaite's examples, therefore, they fail to account for any element of necessity in scientific law-like statements, which is compatible, as Braithwaite claims, with the empirical doctrine of constant conjunction,[1] and is at the same time scientifically significant.

iii. COUNTERFACTUAL CONDITIONALS

Yet necessity cannot lightly be abandoned, and the persistent adherence of contemporary philosophers to the Humean epistemology ensnares them in yet another tangled web. How may they consistently explain the function performed in science, as well as in ordinary perception, of unfulfilled subjunctive, or counterfactual conditional propositions? Scientific argument bristles with instances in which the investigator reasons: if this were (or had

[1] *Op. cit.*, p. 10.

been) the case (which it is, or was, not), so-and-so would have occurred (though, in fact, it did not), and no such argument is valid or useful unless it proceeds from a universal proposition with the character of necessary law. Also, even empiricist theories of perception, which are directly germane to the role of observation in science, require arguments of the form: If such-and-such conditions had obtained, such-and-such sense-data (or, if preferred, percepts) would have been experienced. Accordingly, empiricists are constrained to find some way of reconciling counterfactual conditionals with the doctrine that general statements which are not purely analytic can only be empirical hypotheses inductively reached.

The characteristic form of a general statement is All As are Bs and, according to the doctrine, this can only be asserted of empirical matters if at least a number of As have been observed and have been found to be Bs. Probability statements which say this with qualifications are only variations on the same theme. To confirm any such hypothesis the only effective method is that of seeking new instances and the degree of confirmation will depend on the absence (relative or complete) of negative cases. It follows that the truth of factual conditional statements, like 'If x is A then x is B', will depend upon the truth values of their antecedents and consequents; so that only if 'x is B' is found to be true whenever 'x is A' is found to be true (whether or not it be true in other circumstances) can the conditional statement be held true. Whether this conjunction holds, moreover, could be ascertained only by observation of the relevant instances.

Such factual conditionals, therefore, would all conform to the rules of material and formal implication and we may write:

All As are Bs is equivalent to $(x)\,(x \text{ is A} \supset x \text{ is B})$
or $(x)\,(\phi x \supset \psi x)$
or $(x) \sim (\phi x \sim \psi x)$

The confirmation of any propositions resulting from substitution for the variables would have to depend on observation of particular instances.

The traditional problem of inductive reasoning arises from the familiar fact that empirical generalizations of the sort described are only factually universal so that they do not warrant prediction. For example, the fact that all the books now in my library were published in Britain does not warrant the prediction that the next, or any future, additions to my library will be British publications. For prediction we need at least to presume what has been termed nomic universality: that is, the generalization must have the force of a universal law. As we know, the Humean position, that experience reveals only constant conjunction and never connexion between attributes or events, provides no warrant for the enunciation of nomic universals, apart from a psychological propensity to treat as such what are by origin, and can on the strength of any available evidence, only be, factual generalizations. But nomic universality is also required as the basis of counterfactual conditional statements which, as has been said, are not only indispensable to scientific reasoning but are also implied by all dispositional statements, and so are deeply involved in theories of perception, and the nature of observation itself.

The trouble reveals itself in the fact that a counterfactual conditional is not, and is not intended to be, truth functional and the rules of material and formal implication cannot be applied to it. With ordinary conditionals, whenever the antecedent is false, by the rules of material implication the conditional as a whole is true; but in counterfactuals the antecedent is always false, yet we would never accept the statement as true merely on that account. So also it should be true if, as is usually the case, both antecedent and consequent are false, but we should not normally accept this as a valid criterion. Again as the antecedent is false, a true conditional would result with the contradictory of the consequent, which for counterfactuals is altogether inadmissable. A counterfactual conditional statement is intended to express a connexion which holds independently of the truth values of its components, and we believe that it holds on the basis of a general statement having the status of a law of nature. Thus if it is a law of nature that all As are Bs we are prepared to say that if something had been an A it would have been a B. But as good empiricists we should claim to know only that all the observed As are Bs and can point to no evidence relevant to what would be the case if something observed not to be A had been otherwise.

The problem is, of course obviously connected with that of induction, for it is what Nelson Goodman calls the problem of the 'projection' (or 'projectibility') of an hypothesis from observed instances to those which have not been, but might have been, observed, and to others which, though they could not have been observed, yet if they had been would have revealed what the consequent of the conditional states. It is in principle the same as that of projection from the already observed to the not yet observed—i.e. the problem of prediction. I wish to contend that on a strictly Humean basis (such as has been assumed above) the problem is insoluble and that some who have claimed to solve it without damage to their Humean foundation, have surreptitiously abandoned that position despite their explicit avowals. I disagree with Goodman only in prognostication, for though he admits that the problem is so far unsolved, he continues to hope that persistent efforts will produce a solution, whereas my conviction is to the contrary. Current solutions differ only in detail and terminology and seem all to agree in principle that what the counterfactual conditional expresses is the logical relation holding within a given system between certain propositions in that system. It is said to be or to imply a metasystematic statement to the effect that the implication is tautological within the system.[1]

The argument is that if we adopt a set of axioms, postulates, and rules of inference, we can develop a deductive system in which either we can substitute for some of the variables empirical terms, so that our law of nature is derivable as a theorem within the system, and from that again, in conjunction with further premises, we can derive the consequent of our counterfactual; or else we can add the law to the axiom set so that the implication between antecedent and consequent becomes tautological within the system.

I take this to be substantially what Nagel intends when he says that

'a counterfactual can be interpreted as a *metalinguistic* statement (i.e. a statement about *other* statements, and in particular about the logical relations of these other statements) asserting that the indicative form of its consequent clause follows logically from the indicative form of its antecedent clause, when the latter is conjoined with some law and requisite initial conditions for the law.'[2]

He claims that the criticism stemming from the problem of counterfactuals does not undermine the Humean analysis of nomic

[1] Cf. Henry Hiz, 'On the Inferential Sense of Contrary-to-Fact Conditionals', *Journal of Philosophy*, XLVIII, 19, 1951; R. B. Braithwaite, *op. cit.*, Chs II, III and IX; Ernest Nagel, *The Structure of Science*, pp. 68–73; B. K. Milmed, 'Counterfactual Statements and Logical Modality', *Mind*, LXVI, pp. 453–70.

[2] *Op. cit.*, p. 72.

universality, though it does bring out 'the important point that a statement is usually classified as a law of nature because the statement occupies a distinctive position in the system of explanations in some area of knowledge'.[1] This is virtually identical with the position maintained by Braithwaite who makes a similar claim.[2]

This doctrine seems to me to fail because the status of the law of nature and what it is entitled to imply is overlooked. *If Hume is to be taken seriously* then every universal (or general) statement can be taken only in extension, and 'All As are Bs' can mean strictly only that all the examined As have been found to be Bs. To this (according to Hume) we add the expectation that the unexamined As will also be Bs, but the conjunction of A-ness with B-ness can never be more than factual and is in no case logical (or 'necessary'). If it were so, 'All As are Bs' would have to be a tautology (like 'All bachelors are unmarried'), and no factual universal can be that. Consequently, whatever part the law plays in any 'system of explanation', in the statement 'If x is A it is B' there can be *no* logical relation between 'x is A' and 'x is B' but only a factual one. But we have seen that factual universals cannot support counterfactual conditionals because the fact that all examined As are Bs gives no ground for conclusions to what things would be which are not As even if they were (*ex hypothesi*) unexamined As.

It is thought that the difficulty can be circumvented by conjoining the general statement with the antecedent thus:

$$(\text{All As are Bs and } x \text{ is an A}) \supset (x \text{ is a B})$$

and so long as x *is* an A this is tautological, especially if all As are Bs is taken as axiomatic in the system. But in the counterfactual, 'If x (which is not A) were A . . .' the hypothesis is being projected and the tautological statement (All As are Bs and x is an A) $\supset$ (x is a B), is not a projection and is of no use for prediction, for it can hold only if x is subsumable under the law, and the law (if Humean) tells us only what is true of examined As. Thus x must be one of the examined As or the implication fails.

To state this somewhat more technically, we may point out that what can be deduced formally in any system must be tautologically

[1] *Op. cit.*, p. 70.

[2] *Scientific Explanation*, p. 11 and pp. 315–16.

derivable from the axioms and postulates. Thus if a so-called law is added to an axiom set, whatever is derived from it deductively must be tautologically equivalent to it alone, or to some part of it, or to a conjunction of it with other statements. If in such a system $p \,.\, q \supset r$ is tautological, r can tell us nothing that is not already stated in $p \,.\, q$. It is, therefore, impossible for r to be a prediction—i.e. for $p \,.\, q \supset r$ to be a synthetic statement.

Suppose in a given system, including among its axioms those of logic, we adopt as an additional axiom 'all copper conducts electricity'. Let us express this as

(x) (If x is copper x conducts)

(Note that this statement is not itself tautological). We may then write:

(x) (If x is copper x conducts) $\supset$ (if a is copper a conducts).

If this statement is to be tautological, then 'a' must be taken simply as a substitution for x. It cannot be interpreted as referring to a particular observed case, because what will be empirically observed in any particular instance cannot be determined purely formally. 'If a is copper a conducts' can therefore be regarded legitimately only as a restatement in the particular form of the generalized statement '(x) (If x is copper x conducts)'. This is all that pure logic permits by way of deduction from axioms. On the other hand, if the generalized statement is an empirical (Humean) 'law', a can be subsumed legitimately only if it has already been observed. It must be *known* to fall under the law. If this is not known, or if it is found not to be the case, the implication (if a is copper a conducts) cannot be stated.

The essential point is that if the generalization is only a factual universal (as all these writers admit it to be if the Humean position is to be preserved), even if it is added to a deductive system either by substitution of empirical terms for variables, or as an axiom, the logical relation revealed between antecedent and consequent of a conditional which arises out of it can never be other than material implication. If 'All things which are copper conduct electricity' is equivalent only to 'All the things hitherto observed to be copper have been observed to conduct', then the truth of '(a is copper) $\supset$ (a conducts)' will depend on the conjoint truth or falsity of antecedent and consequent according to the rules of material implication. No other logical relation has been or could be revealed by incorporating the general statement into a deductive system which did not alter its Humean character, for in accordance with that, there is no logical relation between '*a* is copper' and '*a* conducts'.

Moreover, it does not help to postulate a 'physical' or a 'causal' relation; for unless one is invoking the sort of 'power' or occult influence which Hume and most modern philosophers emphatically reject, and which, as Hume demonstrated, would still fail to remedy the situation, a causal relation would be, once more, analysed into a constant conjunction plus a psychological propensity to predict. The latter we may ignore, for as Goodman has shown, we have no criterion for deciding which constant conjunctions justify the expectation alleged by Hume and which (for there are admittedly many) do not. No constant conjunction reveals any necessary or logical or other connexion between the conjuncts. Therefore, if Nagel is right in saying that the counterfactual is a metalinguistic statement asserting that the indicative form of the consequent clause follows logically from the indicative form of the antecedent clause, either the metalinguistic assertion is always false or Hume must have been wrong. We must relinquish either the Humean analysis or Nagel's solution of the problem and others similar to it.

iv. CONCLUSION

The upshot of our criticism so far is that the empiricist position can neither dispense with a logical justification of inductive argument nor provide one that will hold water, and that its attempts to relate observation to theory, to explain the cogency and success of scientific reasoning and to account for its explanatory power is hollow and incoherent. Thus it altogether fails to provide a theory of scientific method which permits to science the position that it holds in the popular imagination. That the superiority of science over prejudice and superstition is at least partly due to its adherence to observed fact is not to be denied, but a doctrine which makes observation reducible to a succession of sense-given particulars falsifies that opinion, because it renders the principle of generalization powerless to produce anything distinguishable from prejudice or superstition.

CHAPTER IV

EMPIRICIST REFORMERS

i. DILUTION OF EMPIRICISM

In recent years the internal incoherence of empiricist theories of science, as well as their inadequacy (to be displayed more fully later) to account for the actual procedure and practice of scientists, has become apparent to several writers on the philosophy of science—in particular, Karl Popper, J. O. Wisdom and William Kneale. Each in his own way has abandoned some salient tenet of the doctrine and has attempted to develop a new theory; but equally each has retained important features of Empiricism which undermine and frustrate the improvements made. The primary tenet, that scientific knowledge originates in sense-given particulars is either assumed tacitly or embraced explicitly by all of them, though not always consistently with other views that they advocate. Consequently, the prevalent conception of generalization is not materially changed (though Kneale vacillates on this issue), and the unmanageable problem of inductive confirmation still hampers the attempts at reconstruction. Likewise, the notion of deductive logic, as a linear procession of analytic steps from proposition to proposition (or statement to statement) impotent to reveal any new fact, persists and ruins the best efforts to evolve a theory of scientific procedure with any degree of verisimilitude. Another writer, N. R. Hanson,[1] virtually abandons the central dogma of Empiricism in a devastating critique of the current assumptions about observation, yet fails to develop, by any conclusive argument, a satisfactory conception of theory construction.

All these philosophers make significant advances upon the more usual empiricist theories of science, and all display penetrating insights into the nature of scientific knowledge, to which later reference will be made. This chapter will confine itself to adverse criticism with the object only of showing that the residuary elements of Empricism are fatal to the theories proposed, and so to clear away the rubbish in the path of further progress.

[1] Cf. *Patterns of Discovery* (Cambridge, 1958).

ii. KNEALE ON NECESSITY, PERCEPTION AND CONSILIENCE

That universal law proves on final analysis legitimately to state no more than constant conjunction is rejected by Kneale. He maintains that scientific laws are 'principles of necessitation' and associates the necessity which they prescribe with the nature of perception (i.e. he connects it with observation). These are undoubtedly moves in the right direction, but he gives no clear explanation of necessity, nor any adequate theory of perception and returns, for the solution of the most crucial problems of induction, rather disappointingly, to the traditional doctrines.

There is much in Kneale's contentions to encourage acceptance, his rejection especially of the phenomenalist presumption that perceptual object statements can be reduced to a multiplicity of mutually independent statements recording the occurrence of sensa. This, in effect, if he adhered consistently to all its implications, would be the rejection of the empiricist dogma that perceptual knowledge is a conglomeration of sensuously given particular data. He also displays clearly the dependence of probability theory as well as phenomenalist theories of perception on the tacit assumption of principles of necessitation. This enables him not only to avoid, but explicitly to expose, the error of any attempt to found a justification of induction upon probability theory.

He fails, however, to enlighten us as to the kind of necessitation involved in natural law, except to say that it neither is nor is not like the necessity involved in logical entailment, his conception of which is apparently no different from the usual empiricist view described above. His final appeal seems to be to linguistic usage and the exhortation to 'take perceptual object terminology as we find it'.[1] This tendency to succumb to the contemporary Oxford addiction to 'linguistic analysis' and the solution (or dissolution) of all problems by deference to ordinary language is unfortunate, for it limits unnecessarily the scope of Kneale's treatment of the subject. His account of perception (though it does not claim to be exhaustive or complete) makes no notable advance upon familiar theories and seems to aim simply at showing that the various uses of the word 'see' and of perceptual object and sensum terminologies do not warrant our reduction of the former to the latter, or enable us to dispense with principles of necessitation, at any

[1] *Probability and Induction*, p. 258.

rate, as the basis of unfulfilled hypothetical sentences indispensable to every theory (phenomenalist with the rest) of perception.

The crux of the argument is that perceptual object terminology —and so, presumably, the experience of objects perceived as complex totalities—is prior to any discrimination of sensible qualities or of particular sensa, and that the elaboration of scientific laws depends on the former, whereas our description, and any account we can give, of the latter must presuppose a knowledge both of scientific laws and (*a fortiori*) of perceptual objects. As far as it goes, this argument is sound and reveals the underlying reason for the paralogisms involved in so many versions of the representative theory of perception inspired by physiological discoveries. But to point this out, important as it is, is to solve no problems either of perception or of scientific explanation. Kneale's failure to do more seems consequent upon his conviction that 'the question [of the relation of sensa to material objects] cannot be settled by arguments from physics or physiology, but only by reflection on the way in which we use words'.[1] Arguments from physics and physiology may have their limitations,[2] but to say that problems so closely related to physiology and psychology should be soluble 'only by reflection on the way in which we use words' places altogether too much strain upon the credulity of the reader.

No good purpose would be served, at this juncture, by entering into detailed discussion of Kneale's linguistic arguments (many of which are highly dubious). Suffice it to say that the theories he is combating are unsound and are rightly rejected. None the less, having rejected them, he gives no clear alternative. Though he asserts that natural laws are principles of necessitation setting boundaries to possibility, he rejects Hume's theory of causality, not because of Hume's refusal to admit necessary connexion, but because causal laws are today no longer the typical form of scientific law. He admits that laws in general all have the same character of alleging some kind of uniformity between the observed and the unobserved, acquiesces in the traditional empiricist conception of induction, demonstrating that all inductive reasoning ultimately rests on 'summative' or 'ampliative' induction (i.e. simple enumeration), and finally accepts the pragmatic justification of induction as the only means at our disposal for 'reasonable'

[1] *Op. cit.*, p. 84.
[2] See my *Foundations of Metaphysics in Science*, Ch. XIX.

prediction. Thus despite his professed refusal to accept the fundamental empiricist doctrine he retreats finally into the same empiricist impasse from which he seemed at first to be releasing himself.

Kneale's account of transcendent hypotheses and 'secondary induction' is more interesting, though, for the most part, it is similar to the theory of explanation given by Braithwaite, with which we have already dealt. The theory of 'consilience of primary inductions',[1] derived from Francis Bacon and Whewell, strives to account for the increase in probability accruing to laws supported by independent evidence when they are found to be deducible from higher level hypotheses. Thus when a generalization made in biology is subsequently found to follow from laws previously well established in physics and chemistry, it is the more readily accepted as true. This circumstance suggests a coherence theory of truth; but Kneale explicitly repudiates coherence, because (he maintains) the force of consilience depends upon the prior probability of conclusions inductively confirmed. If, however, the argument of Chapter II above is accepted, conclusions cannot be effectively confirmed by the inductive procedures contemplated, or even rendered probable; and Kneale himself admits that mere multiplicity of instances does not explain the increase in probability afforded by 'consilience'.[2] If, then, the fact that a number of applicable hypotheses can be deduced from a single principle co-ordinates them (as Kneale asserts) in such a way as to enhance their probability, this could only be because of their mutual corroboration, which is the sole virtue of consilience. One might say that if coherence were the nature of truth consilience of primary inductions would increase the probability of mutually corroborating hypotheses, if only a satisfactory account could be given of primary induction. It is significant to note besides, that Kneale commends consilience on the ground that when a new law, as yet unconfirmed empirically, can be deduced from a 'transcendent' hypothesis, and empirical evidence for it is subsequently found, this proves that the transcendent law is no mere conjunction of the primary generalizations which it is invoked to explain but does tell us 'something new and interesting about the world'.[3] If that is so, it must somehow be possible to deduce new factual information from that already known without (or prior to) empirical investigation and the deductive procedures used cannot

[1] *Op. cit.*, pp. 106–10. [2] *Op. cit.*, p. 108. [3] *Op. cit.*, p. 109.

conform to the logical theories so ubiquitously accepted in contemporary empiricist circles.

iii. POPPER ON FALSIFICATION

We have already addressed ourselves to this difficulty and shall do so again when we come to discuss the theory of so-called 'hypothetico-deductive' method, which is advocated by Karl Popper and developed by J. O. Wisdom; but before we do this it is necessary to examine the central position which Popper propounds, and on the strength of which he abandons the salient features of Empiricism.

The point of departure for Popper's theory is the positivist use of the verification principle to demarcate factually significant from metaphysical, or non-significant, statements. As this essentially is the epitome of Empiricism, Popper's repudiation of it should have far-reaching consequences, and, in fact, his readiness to pay heed to actual scientific practice brings him much nearer to a satisfactory theory of scientific procedure than other empiricists. Yet, as will appear, he cannot bring himself to abandon Empiricism completely and remains entangled in some of the more intractable problems.

Popper's first notable insight is that induction as commonly described is not a valid method of argument at all and he declares that it is not used in science.[1] He believes that empirical confirmation of universal propositions is impossible, for no conjunction of particular statements is ever equivalent to a universal. Thus induction can establish nothing of universal import. The alleged procedure of induction from particular premises to universal conclusion is never justifiable, and the attempts to justify it lead only to an infinite regress, because the inductive principle must be invoked afresh to validate the justificatory argument. Scientific hypotheses, therefore, are, and can only be, tested by deduction. Consequences deduced from them may be empirically investigated and if the evidence is negative they will be disproved or falsified; if it is consistent with the hypothesis, that will not confirm or verify it, but simply leave it standing, as yet unfalsified. This procedure is thought to be logically defensible, because though a universal proposition cannot be verified by any number

[1] *The Logic of Scientific Discovery* (London, 1959 and New York, 1961), Ch. I, esp. p. 40.

of particular affirmative propositions, one negative proposition is sufficient to contradict it. That all crows are black is conclusively disproved by the advent of a single white crow.

The difference between empirical science and metaphysics is now made to consist in the fact that statements typical of the former are in principle empirically falsifiable, while those of the latter are not. The arguments by which the familiar empiricist and positivist positions are demolished are elegant and intriguing. 'Positivists', Popper finds, 'in their anxiety to annihilate metaphysics, annihilate natural science along with it.'[1] And 'instead of eradicating metaphysics from the empirical sciences, positivism leads to the invasion of metaphysics into the scientific realm'.[2] Because his critique is so penetrating and has nevertheless been so consistently ignored by its victims, it is tempting to summarize his arguments and to reinforce his polemic. But this temptation must be resisted, for our object is to find a way to an adequate theory of scientific knowledge, not to exult over the discomfiture of the Midianites.

In spite of his admirable beginnings and the pregnant suggestiveness of his theory, Popper fails to provide any adequate criterion of scientific reliability. The reasons for his failure are his persistent adherence to two elements of empiricist doctrine. One is the final appeal to basic empirical statements as the test of truth (or falsity), and the other is the conception of deductive logic as a pure sequence of tautologies. Popper is aware of the difficulties involved in the first of these and strives vigorously to overcome them, but he comes nearest to doing so only when, on the verge of giving up the appeal to basic statements altogether, he adumbrates an entirely new theory of verification.

(*a*) *Basic statements*

The fundamental contention is that a universal proposition (e.g. a scientific law) can be refuted or falsified by a singular proposition negating the connexion of its subject to its predicate. If there is any A which is not B then 'All As are Bs' is false. On the other hand the discovery of any number of favourable instances is insufficient to verify the universal. To distinguish empirical from non-empirical (mathematical or metaphysical) universals the

[1] *Op. cit.*, p. 36. Cf. my *Nature, Mind and Modern Science* (London, 1954), pp. 330–6.

[2] *Op. cit.*, p. 37. Cf. *Nature, Mind and Modern Science*, pp. 8 and 324 ff.

falsifier must be an empirical proposition. So Popper asserts that the potential falsifiers of all empirical scientific laws are singular statements of fact recording the occurrence of a specific event. An 'occurrence' he distinguishes from 'an event' by the indication of individual space-time co-ordinates and proper names involved in the description of the former: 'It is thundering in the thirteenth district of Vienna on June 10, 1933, at 5.15 p.m.' describes an occurrence. The event is the feature common to all such statements which differ only in the individual names and co-ordinates which they contain. A potential scientific falsifier is then a statement describing an event; it has the form of a singular existential statement and the event concerned must be 'observable'. This is a basic statement.

So far we seem to have, over again, no more than the old empirical appeal to sense-perception, but Popper is far from being so naïve. First, we must remember that no amount of observation and no accumulation of favourable basic statements can establish any law. 'There is no such thing as induction.'[1] Basic statements can only falsify. Then we are reminded that, after all, a single such statement is not enough even to falsify, first because we can modify the qualifying assumptions of our theory so as to accommodate almost any exception, and, secondly, because no isolated observation can be taken seriously. 'We do not take even our own observations quite seriously, or accept them as scientific observations, until we have repeated and tested them.'[2]

Popper counters the objection that falsification can be evaded by modifying ancillary hypotheses by alleging that the scientist (or the logician of science) adopts as a methodological rule that theories shall be so framed as not to evade falsification by empirical findings. Further, theories are to be considered most acceptable when they are most easily falsifiable in this way (so long as they withstand the tests to which they are subjected). Nevertheless, we are here in the presence of a formidable obstacle to the falsifiability theory. In the first place, if, as is usually maintained, empirical hypotheses can never be more than probable in some degree they cannot be conclusively falsified by any singular statement of fact. That this A is not B does not contradict the statement that As are probably or generally B, nor the more precise statement that the probability P that A is B is m/n. Nor is the statement (to which all Humean generalizations are equivalent) that all As hitherto

[1] *Op. cit.*, p. 40. [2] *Op. cit.*, p. 45.

observed have been Bs contradicted by the fact that this (newly observed) A is not B. Secondly, it is a notorious fact about science that the most persistent observations of unfavourable cases are not considered conclusive evidence against a theory if, in other respects, it satisfies a wide variety of scientific needs.[1] Newton's theory of gravitation was very slightly (if at all) shaken by the discovery of the precession of the perihelion of Mercury, which was known for decades before the theory was superseded. Puzzling occurrences and events do not persuade scientists to abandon theories forthwith. They stimulate them to further detailed investigation and experiment, but theories are abandoned only when better explanatory substitutes can be found. What constitutes better explanation we have still to inquire, but that it is equivalent to easier falsification (as Popper implies) would be hardly plausible even without the difficulties we have already found. If the mark of an empirical theory (distinguishing it from a metaphysical one) were its falsifiability by a singular factual statement, we should have to conclude, in the light of the above considerations, that there were no such theories.

Popper's admissions concerning the insufficiency of single observations to falsify are even more serious, for if one observation is not acceptable as conclusive evidence against a scientific hypothesis, the alleged logical basis of the theory is undermined. Inductive confirmation was rejected on the ground that particular affirmatives do not add up to universal affirmatives; but empirical falsification was allowed because a particular negative does contradict a universal affirmative so that a single case should prove sufficient. But if our observations, to be taken seriously, must be 'repeated and tested', they are placed in the same case with the scientific hypothesis to be investigated. To assure ourselves that a basic statement is true (or our observation reliable) we must know that a particular event has occurred. This now becomes an empirical hypothesis in its turn, and it must be confirmed by repetition of the observations, or tested by seeking falsifying evidence. What would be a potential falsifier in this sort of situation Popper does not say. He is too wily to fall into the trap of admitting the possibility of positive confirmation by repeated favourable cases, for that would require him to readmit induction.

[1] Cf. R. G. Swinburne, 'Falsifiability of Scientific Theories', *Mind*, LXXIII, No. 291, July, 1964; and D. W. Peetz, 'Falsification in Science', *Proc. Arist. Soc.*, 1968–9. Also, below, pp. 227 ff.

He therefore takes refuge in the conclusion that 'there can be no ultimate statements in science',[1] and admits an infinite regress in his own theory such as he rejects when involved in the theory of induction.

It is here that Popper reaches the verge of an important advance beyond Empiricism and a distinct contribution to the understanding of the relation between observation and theory. But the verge is not passed, and he vacillates oddly between the assertion that a theory will be rejected as falsified 'only if we discover a *reproducible effect* which refutes it', suggesting the need for some new kind of inductive procedure to establish the falsifier, and the requirement of an 'observable event' as the means of intersubjective testing of basic statements, as well, it would seem, as of hypotheses.

An empirical basis for scientific theory is seen as a desideratum but its precise formulation remains a problem. That Popper recognizes and faces the problem is most refreshing after the long period of blinkered persistence by empiricist writers insisting on 'protocol sentences', 'ostensive propositions', and 'observation statements', accompanied by the most incoherent and self-refuting analyses of perception, or superficial accounts of linguistic usage which tacitly assume empiricist principles but decline to acknowledge them. The appeal to direct perception Popper castigates as psychologism. The conviction we experience of the factual reliability of what we immediately sense is, he contends, like any other emotional state, irrelevant as a logical ground. So we find ourselves in the 'trilemma' to which J. F. Fries drew attention. Scientific statements cannot be accepted dogmatically. If they are to be justified logically it can be only by other statements, and if these are to be similarly justified we are committed to an infinite regress. Yet if we attempt to justify some statements as 'ostensive' or basic, we are victims to psychologism.

That there can be no such ostensive propositions Popper is fully aware, and dependence on them, he holds, 'founders on the problem of induction'.

'. . . we can utter no scientific statement that does not go far beyond what can be known with certainty "on the basis of immediate experience" (this fact may be referred to as the "transcendence inherent in any description"). Every description uses

[1] *Op. cit.*, p. 47.

universal names (or symbols or ideas); every statement has the character of a theory, of a hypothesis. The statement, "Here is a glass of water" cannot be verified by any observational experience. The reason is that the *universals* which appear in it cannot be correlated with any specific sense-experience. (An "immediate experience" is *only once* "immediately given"; it is unique.) By the word "glass", for example, we denote physical bodies which exhibit a certain *law-like behaviour*, and the same holds for the word "water".'[1]

In spite of this excellent and perspicacious passage, however, Popper continues to insist on 'basic statements' as potential falsifiers of hypotheses and on the requirement that the events they describe be 'observable', to ensure that hypotheses can be inter-subjectively tested. But to avoid the charge of capitulating to that very psychologism which he has already censured, he redefines 'observable' as 'involving position and movement of macroscopic physical bodies'. 'Observable events' are to be of a 'mechanistic' or 'materialistic' kind.[2]

This is peculiar language which will not save the position. Macroscopic physical bodies are public complex objects, describable by his own admission, only with the help of universal terms which cannot be reduced without remainder to particulars. Their being observable involves the possibility of sensory perception, but the appeal to that, we have been warned, is psychologistic. Moreover, it does not help because macroscopic physical objects are not, as such, directly intuited by sensation. Present a child or any uneducated person with a complicated piece of physical apparatus and what they 'see' will be something very different from what the physicist observes. Much less will the uninitiated detect the physical effects which constitute the evidence for the physicist's theory—the spectral lines, or the Brownian movement. At a later stage we shall show reason to hold that expert perception differs from common perception only in the degree of its discrimination and interpretation and that the 'object' perceived depends more on the background of knowledge by reference to which the sensory presentation is interpreted than upon what is presented or what is immediately sensed. Popper is, indeed, aware of this for he admits in a significant footnote 'that observations, and even more so observation statements and statements of experimental

[1] *Op. cit.*, pp. 94 f.
[2] *Op. cit.*, p. 103.

results, are always *interpretations* of the facts observed; that they are *interpretations in the light of theories*'.[1] (His italics.)

If observation is in this way dependent upon interpretation, it cannot be the observable character of potential falsifiers that makes them 'basic', if in fact that term is at all appropriate. Their 'materialistic' or 'mechanistic' character would be even more irrelevant, for whether or not we consider physical events (let alone biological or psychological) 'materialistic' or 'mechanistic' depends entirely upon theoretical considerations, both scientific and metaphysical, equally whichever way we decide.

In fact, Popper does not really believe in basic statements at all. He admits quite openly that every statement is corrigible and subject to error. The most elementary, he confesses, require testing and if we actually do adopt any as 'basic', we do so by convention or arbitrary decision, which we can at any time revise. As so accepted, our basic statements are mere 'dogmas', and so far as we do not accept them we are committed to an infinite regress of tests and possible corrections.[2] Yet, if our acceptance of statements as 'basic' is only conventional and if as so accepted they are only 'dogmas', how would they differ from metaphysical statements? Might the ingenuity of the metaphysician not enable him to find statements which he could conventionally recognize as basic and regard as potential falsifiers of his theories? Clearly the criterion of demarcation has once more broken down.

Not surprisingly, in consequence, we are told that:

'Science does not rest upon rock-bottom. The bold structure of its theories rises, as it were, above a swamp. It is like a building erected on piles. The piles are driven down from above into the swamp, but not down to any natural or "given" base; and when we cease our attempts to drive our piles into a deeper layer, it is not because we have reached firm ground. We simply stop when we are satisfied that they are firm enough to carry the structure, at least for the time being.'[3]

What then can satisfy us that the foundation is firm enough (even for the time being)? What gives the structure its stability and strength? Whence comes its superior virtue? Why is not divine inspiration or 'feminine intuition' just as good? From

[1] *Op. cit.*, p. 107, n. 2. Cf. N. R. Hanson, *op. cit.*, Ch. I, and my *Foundations of Metaphysics in Science*, Ch. XX, as well as Chs VII (ii), VIII and IX, below.

[2] *Op. cit.*, pp. 104–5.

[3] *Op. cit.*, p. 111.

Popper, we do not learn the answer, yet he leads us in the right direction, and, as far as he goes, he is right; for our best science is hedged about with doubts and uncertainties and our clearest insights shade off into the mists of ignorance. No ground-rock is to be found in sense-presentation alone, nor yet in simple self-evident, intuitable truths; still less in tautologies or in linguistic conventions. But epistemological theory, the aim of which is to elucidate the foundations of such knowledge as we have and the sources of its reliability, cannot rest in the purely negative conclusion that it is limited and fragmentary. It must provide some touchstone of credibility, or abandon us to scepticism.

In defence against this criticism an advocate for Popper might point out that he insists only on falsification *in principle* and denies that it is ever conclusively possible in practice.[1] To require it, he maintains, or to insist upon strict proof is to forsake the scientific spirit. 'If you insist on strict proof (or strict disproof) in the empirical sciences, you will never benefit from experience, and never learn from it how wrong you are.'[2] These statements, however, cast doubt upon the whole position. What precisely would be falsification in principle, if there actually are no potential falsifiers (i.e. basic statements) that can be accepted without inductive reasoning, without dogmatism, or without appeal to psychologism? If we can accept statements as basic simply by convention, the whole theory of falsification must surrender to conventionalism which was previously rejected as unscientific. Again, if falsification is strictly impossible in practice and to insist upon it would be unscientific, it is hardly consistent to require falsification *in principle* (whatever that may be) as the criterion demarcating scientific from metaphysical theories which, we are told, are not susceptible to falsification. In general one may doubt both of these contentions. Experience and the evidence of history, at least, seem to suggest that falsification in science has some practical possibility and that metaphysical theories are especially vulnerable to destructive criticism. If not, one may wonder what, for the most part, philosophers have been doing during the past twenty centuries.

(*b*) *The hypothetico-deductive method*

The Baconian doctrine of the origin of scientific hypotheses, as a kind of distillation from innumerable factual observations alto-

[1] *Op. cit.*, p. 50. [2] *Ibid.*

gether devoid of theoretical bias, is seen by both Kneale and Popper to be futile in practice and untrue of science. 'The advance of science', says Popper, 'is not due to the fact that more and more perceptual experiences accumulate in the course of time. . . . Out of uninterpreted sense-experience science cannot be distilled, no matter how industriously we gather and sort them.'[1] He believes that scientists arrive at their hypotheses by guesses and he quotes Einstein as saying that they are reached by intuition.[2] If so, there is strictly no method of discovery and the production and success of scientific theories remain logically and epistemologically inexplicable and mysterious. This is another lapse into psychologism, for only psychology could throw light upon the intuitive process of inspired guesswork.

However, once the hypothesis is put forward the business of testing can proceed and is done (as we are already familiar from Braithwaite) by deducing from the hypothesis consequences which can be tested experimentally or by observation. This is what is known as the hypothetico-deductive method.

In the absence of any better theory than has so far been given either of deductive reasoning or of observation and its role in confirmation, little light can be expected from this notion of scientific procedure, and J. O. Wisdom,[3] who lays great stress upon it, as the alternative to induction, does nothing to elucidate either source of obscurity. He merely quotes and summarizes Popper with approval, leaving unexplained the source of the hypotheses to be tested; and the origin of what Popper felicitiously calls 'the horizon of expectations' remains shrouded in miasma. Deductive reasoning is still, for these writers, purely analytic and tautological, so that it is no clearer in their account of it than in Braithwaite's how new empirical hypotheses of any kind can be derived from old by its means. At the same time, even were this difficulty overcome, we should still be left with the problem on our hands of the nature and legitimacy of the process of falsification and its persistent involvement with induction. But more than this—Wisdom brings out the fact, though he somehow thinks he can get round it, that the hypothetico-deductive method is itself entangled in the problem of justifying inductive procedure. For, as he admits, when testing has eliminated false hypotheses, we are left only with hitherto unfalsified hypotheses, and we are to assume

[1] *Op. cit.*, pp. 279 f. [2] *Op. cit.*, p. 32.
[3] *The Foundations of Inference in Natural Science* (London, 1952).

that it is rational to rely upon them in preference to those regularly or occasionally falsified. On what ground? Can it be other than that we expect hitherto unfalsified hypotheses to remain unfalsified in future and hitherto falsified hypotheses to be continually falsified? And this raises the problem of induction over again. Wisdom seeks to avoid this difficulty by refusing to postulate any belief as to the probability of continued falsification (or lack of it), and points out merely that in a chaotic universe no reliance either way could be placed on any hypothesis. Our gamble, therefore, is only on the possibility of a world sufficiently orderly to justify the expectation that hypotheses hitherto regularly falsified would always be false and that those hitherto unfalsified would therefore be preferable. Yet this again is precisely the gamble of induction and would, if it were successful, justify inductive argument equally well. It is simply the gamble on the possibility that (after all) the future will resemble the past, at least in some measure.

PART II

HISTORICAL

CHAPTER V

NON-EMPIRICAL ASPECTS OF SCIENTIFIC PROCEDURE

The kernel of the view we have been criticizing is that any reliable theory about the world can be derived only from sense-observation and that all sound science proceeds from observation to hypothesis and back (for confirmation) to observation. Theories evolved by pure deduction, recommended for purely logical reasons or on considerations simply of mathematical elegance and coherence, are non-empirical and must be recognized (it is alleged) as purely formal, giving no information about the real world. If they are treated as informative about matters of fact, they become 'unscientific', at best speculative, and at worst metaphysical. The cosmological theories of the Ancient Greeks are, for the most part, decried as theories of this kind, without factual backing—that, for the modern empirically minded philosopher, is why they are so palpably false—the heavenly bodies were held to revolve in circles because the circle was thought to be the most perfect geometrical figure; the earth was regarded as the centre of the universe because the abode of man must be that towards which all else is directed; reason dictates that heavy bodies must fall downward, hence it follows that the heaviest (the earthiest) must concentrate at the centre—and so forth. This was the prevailing character, so it is often said, of Ancient science, with some few exceptions in the case of more empirically minded thinkers, like Archimedes, who made lasting discoveries. Any such account is an historical travesty, which examination of the record readily exposes, but our immediate purpose is not to defend the Ancients but to examine the methods of modern scientists.

i. THE COPERNICAN REVOLUTION

The Copernican revolution is commonly regarded as the turning-point at which speculative theorizing gave way to observation and experiment as the source and support of scientific theory. The classical systems of Aristotle and Ptolemy and their Mediaeval adaptations are taken as typical of the first, while the work of Copernicus and his successors of the sixteenth and seventeenth centuries exemplify the second. This dichotomy is quite false. It would almost be less misleading to say that Aristotle and Ptolemy more faithfully followed observation than any of the scientists of 'the great instauration'. But in truth it is misleading to allege that scientists of any period give special precedence to observation. Its importance, as we shall presently see, is neither more nor less than that of theory; each without the other becomes scientifically barren, if not altogether meaningless.

(*a*) *Copernicus*

The hinge on which the Copernican revolution turned was the science of astronomy, the oldest (with the possible exception of medicine) and most advanced of the natural sciences at the time. The facts with which the science is concerned are the visible movements of the heavenly bodies, which, observed with the naked eye, appear as the diurnal transition of the sun, moon and stars across the heaven from east to west, the annual process of the sun and the planets eastward around the ecliptic, the phases of the moon, its successive positions in the sky relative to the stars and the sun, the periodic retrogressions of the planets, occasional eclipses, and the appearance and disappearance of comets. All these had been observed and recorded in some detail (if not with any specially great accuracy) by the Ancients; but, as T. S. Kuhn has remarked, they do not in themselves provide any structural information.[1] In Ancient times they had given rise to what Kuhn has aptly called the two-sphere cosmology. This is a conception of the universe as consisting of a central spherical earth, surrounded by a celestial sphere in which the heavenly bodies are embedded. No evidence of the source of movement is revealed by observation of the heavens. True it is that movements of terrestrial bodies are frequently observed not only visually but also tactually and kinaestheti-

[1] 'They tell us nothing about the composition of the heavenly bodies or their distance; they give no explicit information about the size, position or shape of the earth. Though the method of reporting the observations has disguised the fact, they do not even indicate that the celestial bodies really move. An observer can only be sure that the angular distance between a celestial body and the horizon changes continually. The change might as easily be caused by a motion of the horizon as by a motion of the heavenly body. Terms like sunset, sunrise, and diurnal motion of a star do not, strictly speaking, belong in a record of observation at all. They are parts of an interpretation of the data, and though this interpretation is so natural that it can scarcely be kept out of the vocabulary with which the observations are discussed, it does go beyond the content of the observations themselves.' *The Copernican Revolution* (New York, 1959), pp. 25–6.

cally (though this is by no means always the case.) But as no movement is *felt* in the earth, it is natural to conclude that it is stationary, and to impute the passage of the heavenly bodies to the revolution of the heavenly sphere about a stationary earth as centre. To this piece of 'observation' the ancients and mediaevals remained constantly faithful.

The sphericity of the earth is not apparent to the senses. In ancient times, it was postulated by the Pythagoreans (possibly by the Master himself) for purely mathematical reasons and without observational grounds, but it was held to be confirmed by the shape of the earth's shadow on the moon in eclipse. However, in the absence of a cosmological theory which suitably explained the lunar eclipse, the appearance cannot be so interpreted, and that it is the earth's shadow which obscures the moon at such times is not strictly a matter of observation at all.

The aberrant movements of the planets were explained by the Greeks as due to the revolution of transparent crystal spheres, into which the celestial region was divided, and which according to different theories were alleged to rotate in various ways. The eventual outcome was the Ptolemaic theory, with its complicated system of deferents and epicycles, by means of which it was possible to give a plausible, though never quite accurate, account of the movements of the heavenly bodies as observed. At this point we may neglect the details both of the theories preceding Ptolemy's, and of the complexities of the Ptolemaic system itself. It is sufficient to note that, within broad limits, it gave a tolerable explanation of the observed facts, and within a fair margin of error, could be used to predict the positions of the heavenly bodies. It was far from being mere arm-chair astronomy, and allowing for the difficulties of precise observation and measurements with the instruments available, it could claim considerable empirical support.

When Copernicus recommended the heliocentric system as an alternative to the Ptolemaic, no new empirical evidence was advanced and none had been discovered. The reasons for which he recommended it were without exception theoretical, and many of them of precisely the same kind as those offered by Aristotle and his followers: The earth revolves on its axis (Copernicus asserts) because that is the motion natural to its shape. The reasons advanced against its motion by Ptolemy are hollow and apply even more strongly against the diurnal revolution of the heavens. Heavy bodies fall directly toward the centre of the revolving earth because they 'retain the nature of that to which they belong' and will continue to move with the earth's revolution as well as back to their natural place (as Aristotle averred). 'Further we conceive immobility to be nobler and more divine than change and inconsistency, which latter is thus more appropriate to Earth than to the Universe'.[1]

These are not good scientific reasons, nor are they the ones for which the heliocentric system recommended itself either to Copernicus or to his successors. The real reasons were mathematical and aesthetic, but they were still not empirical. Copernicus was able to account for the irregularities of planetary motion without the help of major epicycles,[2] especially the apparent motions of the inferior planets (Mercury and Venus) were much more simply and naturally explained than by Ptolemy. Copernicus himself impresses on his reader the harmony and necessity of his system. In his prefatory letter to Pope Paul III he complains that 'the mathematicians are so unsure of the movements of the Sun and Moon that they cannot even explain or observe the constant length of the seasonal year', and that they use different principles of explanation for the movements of the different heavenly bodies. In his own system, on the other hand, he claims:

'I have at last discovered that, if the motions of the rest of the planets be brought into relation with the circulation of the Earth and be reckoned in proportion to the circles of each planet, not only do their phenomena presently ensue, but the orders and magnitudes of all stars and spheres, nay the heavens themselves, become so bound together that nothing in any part thereof could be moved from its place without producing confusion of all the other parts of the Universe as a whole.'

In the Ptolemaic system a change in the size of an epicycle or the addition of new ones to the system was always possible without affecting the rest. But in the Copernican the alteration of any orbit affects all calculations and each is determined by its relations to the others.

[1] *De Revolutionibus Orbium Coelestium*, I, Ch. 6.

[2] Ptolemy had made use of two kinds of epicycle. The major (larger) epicycle accounted for planetary retrogression and minor epicycles for eccentricities or elliptical-seeming orbits. A lucid explanation is given by Kuhn, *op. cit.*, pp. 67 ff.

This was the main appeal of Copernicus' system—its unity, consistency and coherence. For it gave a simpler and more elegant theory of the qualitative characteristics of planetary motion than did the Ptolemaic. But quantitatively it was no more successful than Ptolemy's. Copernicus' circular orbits no more fitted even the observational records then available than did Ptolemy's deferents and epicycles. In his attempts to 'fit the facts' Copernicus had to introduce minor epicycles of his own, eccentrics and similar complications which, so far as 'simplicity' was concerned, left little to chose between his heliocentric hypothesis and Ptolemy's geocentrism. To quote Kuhn: 'When Copernicus had finished adding circles, his cumbrous sun-centred system gave results as accurate as Ptolemy's, but it did not give more accurate results. Copernicus did not solve the problem of the planets.'[1] The indictment he made against 'the mathematicians' in the end applied also to himself.

Thus there were neither new facts to justify his innovations, nor better quantitative demonstrations of the old. Empirically the hypothesis was a failure. Nowhere is any attempt made to justify it on the ground that it agrees better with observation than its rivals. No argument appears in *De Revolutionibus* to the effect that what has been frequently experienced in the heavens must be universally true—for no such argument could be made appropriate. There is no trace here of 'induction'. Nevertheless, the heliocentric hypothesis was recognized by the greatest among Copernicus' immediate successors as scientifically superior. He recommended it on the score of its 'harmony', its coherence, its geometrical simplicity in principle, and, in the first instance, it appealed to those who adopted it for these reasons alone—for there were no other respects in which it improved upon what it sought to supersede.

(*b*) *Tycho Brahe*

If Copernicus was too conservative to depart from the tradition further than he did, the same is even more true of Tycho Brahe, meticulous and persistent observer though he was. He refined the methods of naked-eye observation and recorded the position of the planets more regularly and more accurately than ever before. The value of these observations to his successors, especially to Kepler, cannot be exaggerated, but instead of inclining Brahe to the

[1] *Op. cit.*, p. 169.

heliocentric hypothesis of which (since it is true) the observations provided the evidence, they left him still wedded to geocentrism.

The Tychonic system attempts to combine Copernicus' doctrine with Ptolemy's in a new way, making the planets revolve around the sun, but the sun and the moon revolve around the earth as the fixed centre of the universe. The particular interest, for us, of this hybrid cosmology was that its virtues were still those of Copernicus' hypothesis. It simplified the explanation of the apparent movement of the planets in the same way, since it preserved the same geometrical relations between the earth and the sun as the heliocentric system. This is simply illustrated by the following figure (borrowed from Kuhn).[1]

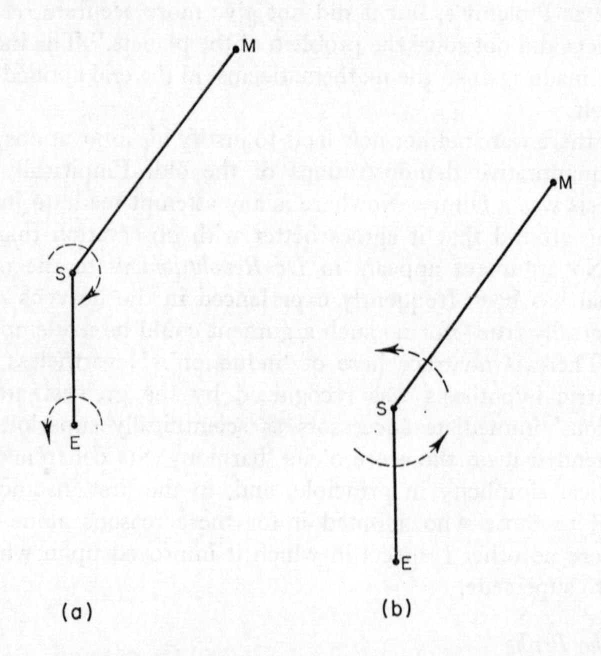

The geometrical equivalence of (*a*) the Tychonic and (*b*) the Copernican systems.

In (*a*) the sun S is carried eastward about the stationary earth E by the rigid arm ES. Simultaneously, the planet Mars, M, is carried westward about S by the steady rotation of the arm SM. Since ES rotates more rapidly than SM, the net motion of Mars is eastward except during the brief period when SM crosses over ES. In the

[1] *Op. cit.*, Fig. 38, p. 204.

second diagram (*b*), the same arms are shown rotating about the fixed sun S. The *relative positions* of E, S, and M are the same as those in (*a*), and they will stay the same while the arms in the two diagrams rotate. Notice particularly that in (*b*) the angle ESM must decrease as it does in (*a*) because ES rotates about the sun more rapidly than SM (from T. S. Kuhn, *The Copernican Revolution*, p. 204).

Again, the reasons for offering the hypothesis are theoretical and philosophical, not empirical, even though the empirical evidence was not only available but was actually accumulated by Brahe himself. The essential point (which we shall take up in the sequel) is that the observational data acquire significance only in the light of the theory by means of which they are interpreted. The interplay between theory and observation gives superiority to neither and gives temporal priority more often to theory than to empirical data.

Evidence of the mutability of the heavens which helped to undermine the Aristotelian astronomy appeared early in Tycho Brahe's career and his observation of comets and discussion of their parallax contributed still more evidence against the existence of crystal spheres. But none of this evidence was new. Comets are frequent phenomena which had been observed often enough before, and instruments, with which parallax could be measured with sufficient accuracy to demonstrate that they were not sublunary bodies, had been available for at least two millennia. Even new stars do not appear so infrequently that astrologers and other observers of the night skies would altogether have failed to notice their occurrence. And since none of these phenomena have any direct bearing on the movement of the earth, none give direct support to Copernicanism. They could all have been accommodated by resolute Ptolemaics, who were not, for consistency, compelled to postulate the existence of solid spheres. Once again, therefore, empirical evidence fails to be the decisive factor. Once again we may note that no astronomers (least of all Brahe) argued that because they had observed some novae and comets such objects must always or regularly have appeared—if any had so argued they would have argued with obvious fallacy. Appeal must be made to other factors to discover what finally convinces scientists of truth.

(*c*) *Kepler*

The astronomer who made theory and observation coalesce was Kepler, and his reasoning will repay more detailed examination at a later stage. Here we may note, however, that the reasons supporting his laws of planetary motion were mathematical and theoretical, not empirical. Kepler was a firm believer in the sun-centred universe; he therefore set out to purge Copernicus' system of its Ptolemaic survivals. Copernicus measured the eccentricity of the earth's orbit from the sun, but of all the other orbits from the centre of the earth's. Kepler corrected this anomaly, not by reference to observed data, but to make the heliocentric theory self-consistent. He found that he could accommodate the observed facts if he generalized the figure of the orbit from circle to ellipse (of which the circle is a special case) with the sun at one of the two foci. This gave him his first law. The second law, which ought on empiricist principles to have been dictated by the facts, historically was not.

'In its origin,' writes Kuhn, 'the Second Law is independent of any but the crudest sort of observation. It arises rather from Kepler's physical intuition that the planets are pushed around their orbits by rays of a moving force, the *anima motrix*, which emanates from the sun. . . . Therefore the number of rays which impinged on the planet would decrease as the distance between the planet and the sun increased. . . . Long before he began to work on ellipical orbits or stated the law of areas in its familiar form, Kepler had worked out this inverse-distance speed law to replace both the ancient law of uniform circular motion and the Ptolemaic variant which permitted uniform motion with respect to an equant point.'[1]

Once again, he rejected these alternatives not for empirical reasons, but to make the Copernican system more consistent, because he believed in the mathematical simplicity of natural phenomena, and because, as a neo-Platonist, he held the sun to be the most important and the most dignified of the heavenly bodies.[2] In fact Kepler had not at his disposal sufficient measurements to derive his second law from empirical data. He knew only that the planets moved fastest at perihelion and none of the other relevant facts at his disposal were sufficiently quantitative. But he sought mathematical simplicity and the second law was the result.

The same is true of the third law, that the ratio of the squares of

[1] *Op. cit.*, pp. 214 ff.

[2] Cf. *ibid.*, p. 214, and Kepler's own encomium on the sun.

the orbital periods of different planets is equal to that of the cubes of their distances from the sun. At the time this law had no practical use. It was not needed to determine either the orbital periods or the distances, both of which were already known. But it was important to Kepler as evidence of the mathematical regularity of God's creation.

The underlying presupposition of all Kepler's thinking is that the universe is fundamentally a harmonious rational structure mathematically ordered and that the discovery of this mathematical order in the observed facts is the only way to understand and to explain them. Thus the astronomical hypotheses were not, for him, just mathematical devices convenient as tools for calculation and prediction; they were the only true means of discovering the nature of the real world; and the best hypothesis was that which linked phenomena otherwise left unrelated in a single system of necessarily interconnected elements, so that it showed why the facts 'are as they are and not otherwise'.[1] This faith in the mathematical orderliness of nature he held in common with Copernicus before him (whose appeal to it we have observed in his dedicatory Preface) and with his contemporary, Galileo, whose insistence on it we shall shortly examine. The discovery of this order he saw as the aim of science, and the facts were important only because it was in the facts themselves that the order was to be discovered.[2]

Accurate observation and measurement were indispensable to Kepler's researches on planetary orbits, but they were not what prompted his conclusions. Without the conceptual scheme with which he worked he could not have made the use of them that he did. The value of his laws of planetary motion lay to a great extent in the agreement which they achieved between prediction and observation, and if this served as confirmation of the theory, it was *not* because it occurred frequently, but because it occurred at all.

(*d*) *Galileo*

Further support for the Copernican system came from what Galileo's telescope revealed in the heavens. Here indeed was new

[1] See *Mysterium Cosmographicum*, *Praefatio ad Lectorem*, *Opera* I, ed. Ch. Frisch (Frankfurt, 1858), p. 106.

[2] Cf. *Harmonia Mundi*, *Cap.* III, *Opera* V, pp. 226 f. Also II, p. 687, where Kepler protests against the dogmatism of 'other sects of philosophers who drag commonly experienced facts by the hair' to accommodate their own axioms.

data, some of it—the visually patent phases of Venus—amounted to crucial evidence. But, here again, what was observed directly was not planetary motion in orbits about the sun, nor was the evidence conclusive proof of the main contentions of the Copernicans,[1] for either the Ptolemaic or the Tychonic system might have accommodated the new facts, merely by introducing some further complications. Moreover, all that the telescope revealed was spots of light; the understanding of what they represented depended upon the astronomical theory adopted. Had these phenomena been visible at any earlier period there is little doubt that they would have been otherwise interpreted and explained. As it was, even in the seventeenth century their validity as evidence was highly suspect. Galileo's opponents impugned them as illusions.[2] It was difficult for others to repeat his observations and he had to insist that his own telescope should be used. There was no adequate theory of telescopic vision then or for long afterwards, and without that, what it revealed has simply to be taken on trust.[3]

Taken on trust it was not. Few, if any, of Galileo's contemporaries, who were not, like Kepler, already convinced Copernicans, would accept the evidence which the telescope offered. The tragedy of Galileo's life was largely due to his ingenuous expectation that his opponents would be convinced of the physical reality of a heliocentric universe by what his telescope revealed (though the arguments to which he resorted to persuade them were, as we shall see presently, far from restricted to this evidence). In actuality, that conviction was prerequisite to accepting the evidence—and Galileo was coërced into recantation.

The telescopic evidence was not in fact his most significant contribution to the scientific revolution, for though that began with astronomy, only half of its concern was with the theory of the heavens. The interdependence of Ptolemaic astronomy and Aristotelian physics and dynamics was the great obstacle to belief

[1] See Kuhn, *op. cit.*, p. 224.

[2] Cf. Francesco Sizi's polemical *Dianoia Astronomica*; see Giorgio di Santillana, *The Crime of Galileo* (London, 1958), p. 13 f. Santillana comments: 'Let us imagine the telescope in the hands of a Giambattista della Porta or some such known contemporary, and we shall understand how much its importance was a matter of the intellectual personality of its author' (*op. cit.*, p. 38).

[3] Cf. P. K. Feyerabend, *op. cit.*, p. 229, and Vasco Ronchi, 'Complexities, Advances, and Misconceptions in the Development of the Science of Vision: What is being Discovered?', in *Scientific Change* (London, 1963; ed. A. C. Crombie), pp. 549 ff.

in Copernicanism, which, by removing the sphere of the stars from contact with those of the planets (in order to eliminate stellar parallax), had deprived the whole Aristotelian universe of its source of motion, the *primum mobile*. By placing the earth in the heavens to circulate about the sun, it had disrupted the whole doctrine of natural motion, while at the same time it attributed movement to the earth, to explain which it provided no dynamic theory. Little wonder that the visual evidence made no impression on the Aristotelians. Kepler's doctrine of an *anima motrix* proceeding from the sun might suggest a way in which the planets could be pushed around their orbits, but this had no connexion with and gave no clue to the movement of terrestrial bodies. It was, however, just here that the Aristotelian doctrine had shown weakness, and this had induced Jean Buridan two centuries earlier to introduce the conception of impetus. Galileo was fully aware both of the cracks in the Aristotelian theory and of the shortcomings of Copernicanism. His major contribution was not the telescope but the beginnings of a new mechanic.

The significance of these historical facts lies on the one hand, in the resistance to Copernicanism on account of its disruptive effect upon the systematic explanation of a wide area of disparate phenomena which Aristotelian physics and Ptolemaic astronomy had offered; and, on the other, in the efforts of its protagonists to evolve a similarly comprehensive system. Not before Newton was this accomplished, but, in the meantime, the appeal of the new theory to men like Kepler and Galileo was not its empirical character but the promise which it held of elegant coherence and simplicity (a term the application of which to the astronomical problem of the Renaissance is obvious enough from its ordinary usage, but which as a precise technical term, needs further examination). To these aspects of the matter we shall return; what we are trying in this chapter to unveil is only the non-empirical and deductive character of the actual approach and argumentation of the pre-eminent scientists.

In the *Dialogue Concerning the Two Chief World Systems*, Galileo's arguments are hardly at all empirical, and are never such as empiricists allege inductive argument to be. They are almost invariably deductive in some form, either mathematical or other. It is in the mouth of Simplicio, the Peripatetic, representing his opponents, that he puts all the appeals to sense perception. Moreover, the 'experiments' to which frequent reference is made are experiments which, for the most part, have never been performed but are what we should call today *Gedankenexperimente*. Even when experimental evidence is, or could easily be made, available, Galileo tends to sweep it aside in preference to a deductive proof to demonstrate the *necessity* of the result. The appeal throughout is a Platonic appeal to reason and conceptual necessity as against the misleading and deceptive evidence of the senses. Let me give examples of these points.

At the very beginning of the *Dialogue*[1] Salviati is made to demonstrate the futility of Aristotelian (Peripatetic) proofs of the three-dimensionality of space, not by any appeal to sense, but by offering a geometrical proof of the same conclusion based on the fact that not more than three mutually perpendicular lines can be drawn through a single point. The only demur Simplicio can offer is that Aristotle held mathematical demonstrations not always to be necessary in 'physical (*naturali*) matters'. Again, a little further on,[2] Salviati, in the dialogue, claims to be able to calculate the distances of the planets and their speed in orbit by deduction from a purely speculative theory of their place of origin combined with the doctrine of accelerated motion which Galileo was attempting to develop. In fact the calculation he produces is in conflict with the astronomical data and with what Kepler had already ascertained. When Salviati is arguing in favour of the earth's rotation on its axis, the evidence is quoted of the way in which a stone will fall which is dropped from the top of a mast in a moving ship. Simplicio complains that Salviati has 'not made a hundred tests, nor even one', and Salviati replies 'Without experiment, I am sure the effect will happen as I tell you'.[3] He proceeds to prove deductively that a body moving with uniform velocity will continue in the same straight line and so gives a deductive argument to the effect that the stone must fall parallel to the mast, or what appears to be perpendicularly to the observer moving with the ship (because its horizontal movement with the ship, which is shared by the observer, is for that reason not apparent). This is a case which especially lends itself to empirical investigation, yet the only experimentation cited is on the wrong side, is obviously fictitious, and is *proved* to be erroneous by purely deductive argument.[4]

Likewise reference is made to the description, given by Sagredo,

[1] Stillman Drake's translation (Berkeley, 1962), pp. 12–14.
[2] *Op. cit.*, pp. 28–9. [3] *Op. cit.*, p. 145. [4] Cf. *Op. cit.*, pp. 141–9.

of the path traced by the point of a pen used for writing or drawing on board ship during a long voyage, as an 'experiment'; yet it is one the result of which, if performed, could never be recorded or observed (namely, that the line traced would not perceptibly diverge from a straight line or smooth curve, though the pen had actually drawn houses, landscapes and animals).[1] The result is reached with the help of what might be termed geometrical imagination, but could never be sensuously perceived or observed. That 'experiment', for Galileo, meant, as much as anything, 'proof' or 'demonstration' and in general '*Gedankenexperiment*', is shown by Salviati's remark 'If you want to present a more suitable experiment, you ought to say what would be observed (if not with one's actual eyes, at least with those of the mind). . . .'[2] Experiments with cannon shots to the east and to the west, like most of the others to which the interlocutors allude, are experiments adduced by the Peripatetic opponents of Copernicus, and when Salviati explains what their results would be he admits not only that they have not been performed, but that even if they had been, it would not have been possible to detect the alleged differences produced by the earth's motion.[3] If so, they would appear to disconfirm the Copernican hypothesis.

Nowhere in the *Dialogue* is any argument presented in the form (which we may adapt from Bertrand Russell): 'because the sun has always risen every day, as far as human memory and records penetrate, it will (most probably) rise tomorrow'. There is no sign of the traditional inductive arguments and no conceivable argument in this form could support the contention that the earth rotates on its axis, or that the heavens do not revolve around the earth as centre, or that the earth revolves annually about the sun.

So clear is Galileo in his own mind about this, that he frequently decries the evidence of the senses and of familiar observations, to which his opponents (in the persons of Simplicio and the authors whom he quotes) constantly appeal. It is the evidence of the senses which persuades Simplicio that heavy bodies always fall perpendicularly and it is Salviati who demonstrates deductively that this evidence is deceptive. Scipio Chiaramonti is quoted as contending against Copernicus that his theory subverts the criter-

[1] *Dialogue concerning the Two Chief World Systems*, p. 172.

[2] *Op. cit.*, p. 143.

[3] See *Op. cit.*, pp. 180 f.

ion of the sciences—the senses;[1] and Salviati replies, with support from Sagredo, that the senses constantly deceive us. He appeals to reason as the one sure source of conviction and truth.[2]

It is Chiaramonti who uses empirical observations—though in fact he misuses them—to demonstrate the sublunar status of new stars; and Galileo in the person of Salviati refutes him by geometrical demonstrations of the amount of parallax which must be observed in order to prove his case, evidence of which Chiaramonti does not produce.[3] 'How powerless', Salviati declares, 'are our senses to distinguish large distances from extremely large ones, even when the latter are in fact many thousands of times larger!'[4] His reliance is placed wholly upon reason and mathematics, as against purely empirical evidence, and his enthusiasm mounts in praise of those who, like Copernicus and Kepler, follow the dictates of their intellect in preference to sensuous appearances in affirming the earth's motion:

'Nor can I sufficiently admire the outstanding acumen of those who have taken hold of this opinion and accepted it as true; they have through sheer force of intellect done such violence to their own senses as to prefer what reason told them over that which sensible experience plainly showed to the contrary.'[5]

As we have said above, Galileo's most important contribution to the scientific revolution was his theory of inertial and accelerated motion—the beginnings of a new mechanic needed to develop the Copernican system into one more coherent than, and equally comprehensive with, the Aristotelian that it was to supersede. 'Against the physical principles of conventional cosmology, which were always brought out against him', di Santillana writes, 'he needed an equally solid set of principles—indeed, more solid—because *he did not appeal to ordinary experience and common sense as his opponents did*' (my italics).[6] In developing these new principles he almost invariably used geometrical and deductive arguments; and, though observation and experiment played no mean part in his researches, it is the way in which they were used that is most revealing and not the bare fact that they are (admittedly) often important. Leaving to a later chapter the

[1] *Op. cit.*, p. 248.
[2] *Op. cit.*, p. 256.
[3] *Op. cit.*, pp. 276–318.
[4] *Op. cit.*, p. 367.
[5] *Op. cit.*, p. 328.
[6] *The Crime of Galileo.*

relation of observation to theory, we may illustrate here, from Galileo's *Dialogues Concerning Two New Sciences*, his deductive method of procedure.

In the course of the third Dialogue the following laws are enunciated concerning the behaviour of falling bodies.[1]

(i) The spaces described by a body falling from rest with a uniform accelerated motion are to each other as the squares of the time-intervals employed in traversing these distances (p. 167).

(ii) The momenta or speeds of one and the same moving body (falling along an inclined plane) vary with the inclination of the plane (from the horizontal) (pp. 173–4).

(iii) If a body falls freely along smooth planes inclined at any angle whatsoever, but of the same height, the speeds with which it reaches the bottom are the same (p. 176).

(iv) The times of descent along planes having different inclinations but the same vertical height stand to one another in the same ratio as the lengths of the planes (pp. 179–80).

These laws are all, in effect, empirical in character for they state relationships between the distance fallen from rest, the time which it takes and the velocities attained. There is, in the first place, no suggestion that the propositions are purely formal consequences of arbitrary definitions without necessary applicability to actual examples of falling bodies. Yet, in the second place, there is no attempt to describe any actual experiments which confirm them or to list recorded results. The laws are enunciated as theorems or corollaries which are proved deductively *geometrico ordine*.

The proof of the first relies on an earlier theorem, derived purely geometrically, to the effect that a body moving from rest with uniform acceleration traverses any space in the same time as it would if it moved at a uniform velocity half the final velocity of the accelerated motion. He then shows that the ratio of velocities reached in uniformly accelerated motion is the same as that of the times. But in the case of uniform velocities the ratio of the spaces traversed is equal to the product of the ratio of the times and the ratio of the velocities. Therefore in the case of uniformly accelerated motion it is equal to the ratio of the square of the times.

The second proposition is one which specially lends itself to

[1] Page references are to the translation by Crew, de Salvio and Favaro (Evanston, 1950).

experimental test, and it is known that Galileo did perform experiments with inclined planes, though the methods of time measurement at his disposal were not nearly accurate enough to enable him to derive any exact law from them. But here there is no appeal to any actual experiment and everything is deduced from concepts. Heavy bodies tend to fall vertically towards the centre and no heavy body of its own natural motion tends to move upward. Hence a body at rest on a horizontal plane will have no tendency to motion and will offer no resistance to external forces setting it in motion (Galileo overlooks what we now call inertial mass). Accordingly the 'momentum' of the falling body is at a maximum when it falls vertically and diminishes as the inclination of the plane along which it falls diverges from the vertical. To prove his point Galileo describes a *Gedankenexperiment*. A body may be prevented from rolling down an incline by a weight attached to it and hanging vertically over a pulley. If the body were falling vertically the counteracting weight would have to be equal. When the plane is inclined from the vertical the force acting on the body still acts only along the vertical and so is proportional to the perpendicular distance traversed, which is less than the actual distance along the incline. As the two weights, however, are inextensibly attached to each other, they move (if at all) through equal times and distances. It follows that the one which falls vertically must be less than that which moves along the incline if they are to balance. Since their 'momenta' are represented by the distance each would move (if moving freely) in equal times, it follows that, if they are to move equal distances in equal times, their weights must be in inverse ratio to the 'momenta'. But the distances traversed freely in equal times would be in the ratio of the length of the inclined plane to its height, hence the ratio of the weight of the body moving perpendicularly to that moving on the incline will be the inverse of this. Consequently the 'momentum' of a falling body will vary with the inclination of the plane.

The third proposition is proved as follows. The acceleration of a falling body (if uniform) is in proportion to the time; the distances traversed, in proportion to the square of the times. The ratio of the force impelling a body down an inclined plane to that acting along the vertical has been shown to be the inverse of that of the length of the plane to its height. By marking off a length along the incline equal to the third proportional of these lengths, Galileo is able to prove, purely by equating ratios of times, distances, and velocities, that the final speed of a body falling through a vertical distance is equal to that of a similar body falling along any inclined plane from the same initial height. The proof is purely geometrical, and no appeal is made to observation.

The fourth proposition follows naturally from the last, for if the speeds of two moving bodies are equal the times taken will be proportional to the distances.

There is no need to multiply examples. Galileo is obviously developing a purely deductive system, not with the intention of checking its conclusions by observation, but in the hope of being able to develop from first principles a dynamic theory that would account for the earth's motion compatible with that of falling bodies on the earth. He did not succeed, but Newton, combining Kepler's results with Galileo's, did succeed.

Experiments were performed and may have prompted hypotheses or strengthened convictions. There is not the least doubt that Galileo made copious and careful observations, but his reasoning omits almost all reference to them and, as it is presented, is independent of them. The accounts popularly given of his experimental work are almost certainly apochryphal. As Kuhn points out, if bodies of different weights had been dropped from the Tower of Pisa and the times of their fall accurately measured, the heavier would be found to fall more rapidly. The experiment would have vindicated Galileo's opponents and would merely have been an obstacle to the development of his own theory.[1]

Galileo's own view of the empirical data was that the senses provided the phenomena to be explained. From observation we get the material of science, but we get it in a confused and relatively unintelligible form, which demands clarification and organization: i.e. explanation. This can be given only by determining the mathematical relationships between the sensible phenomena. Once that has been done, further deductions might be made which could be corroborated by experiment, and this is useful to prevent our being led astray by errors in the deduction, but the experiment is not the source of the explanation, nor is generalization of its result the purpose for which it is performed.

[1] See Kuhn, *op. cit.*, p. 95: 'To verify Galileo's law by observation demands special equipment; the unaided senses will not yield or confirm it. Galileo himself got the law not from observation, but by a chain of logical arguments. . . . Probably he did not perform the experiment at the tower of Pisa. That was performed by one of his critics, and the result supported Aristotle. The heavy body did hit the ground first.'

The world of nature does not reveal its true character to man's senses but, as Galileo says, it is 'a vast book' which 'cannot be read until we have learnt the language and become familiar with the characters in which it is written. It is written in mathematical language and the letters are triangles, circles and other geometrical figures, without which means it is humanly impossible to comprehend a single word.'[1] It is this rational, mathematical order which alone provides the explanation of sensible facts and its discovery is the scientist's principal aim. This belief was fundamental for Galileo as it was for Kepler and for Copernicus. Both facts provided by observation and mathematical deduction contribute equally towards its attainment, for without the former there would be nothing to explain and without the latter no explanation, but the so-called inductive method, as described in empiricist thought, plays no part in Galileo's procedure.

Einstein, in his Foreword to the University of California edition of the *Dialogue Concerning the Two Chief World Systems* sums up the matter thus:

'It has often been maintained that Galileo became the father of modern science by replacing the speculative, deductive method with the empirical, experimental method. I believe, however, that this interpretation would not stand close scrutiny. There is no empirical method without speculative concepts and systems; and there is no speculative thinking whose concepts do not reveal, on closer investigation, the empirical material from which they stem. To put into sharp contrast the empirical and the deductive attitude is misleading, and was entirely foreign to Galileo.'[2]

(*e*) *Newton*

The work of Newton forms the culmination and completion of the scientific revolution begun by Copernicus; and it is largely due to his pronouncements on the method of 'philosophy' (which was his compendious name for science and metaphysics regarded as a single discipline), as well as to much of his practice, that science after his day has been looked upon as essentically empirical. He professed to be primarily an experimentalist and to derive the laws of nature only inductively from observed phenomena; he vehemently renounced 'hypotheses' and demanded experimental

[1] *Il Saggiatore*. Cf. *Discoveries and Opinions of Galileo* (New York, 1957), pp. 237–8.

[2] *Op. cit.*, p. xvi.

demonstration even of conclusions he could deduce mathematically from accepted principles.[1] Nevertheless, the fundamental concepts of his entire natural philosophy are asserted, not merely without experimental support, but even despite the impossibility in principle of any empirical confirmation.

First and foremost among these is his first law of motion, the principle of inertia, which is the essential foundation of all modern mechanics. It states that 'every body continues in a state of rest or uniform motion in a right line unless compelled to change that state by forces impressed upon it'. That Newton was aware of the special status of this and his other two laws of motion is apparent from the fact that he ranks them as 'axioms' in his system. But until very recently they were taken almost universally (and with some reason) to be synthetic empirical propositions based upon the experimental findings of Galileo. Not without reason, because they state how bodies, under certain specified conditions, do and would always move, and they are therefore general statements about actual and possible events in the physical world. From them, along with other propositions stating initial conditions, observable consequences can be deduced; so they function (at least *prima facie*) as do all empirical laws of nature.

Moreover, the principle of inertia, in particular, is in direct contradiction to the Aristotelian principle that a body continues in a state of rest only if it is already in its natural place, and in a state of uniform (or any other sort of) motion only if compelled to do so by forces (or causes) exerted upon it; and this contrary principle was supported by empirical evidence at the common-sense level, was claimed by its proponents to be true on that account, and was presumably enunciated by Aristotle for that reason. Its falsity, therefore, could presumably be legitimately proved only by the production of counter evidence.

Further, Galileo had argued, without departing seriously from the Aristotelian position, that a smooth sphere, falling vertically towards the centre of the earth, accelerates at a certain rate, which is decreased if it rolls down an inclined plane, proportionately to the inclination of the plane from the vertical. Thus if the sphere by rolling down such an incline acquired a certain velocity, with which at the bottom it were propelled along a horizontal surface it would continue to move at the same velocity without any tendency to decelerate unless it were somehow impeded. For on a

[1] See *Opticks*, Bk I, Pt I, Prop. VI, Theor. V (New York, 1952), pp. 76–7.

horizontal plane it has no tendency to move downward or in any direction other than that in which it is already moving as a result of having rolled down the inclined plane. Here then is empirical confirmation, if only in a *Gedankenexperiment*; and though Galileo (who did perform experiments with inclined planes) had no instruments capable of measuring the required quantities with sufficient accuracy, they could in principle be measured and we do now have such instruments.

But no such experiment would serve as a test of the law of inertia. For, first, no flat runway can be constructed on the earth's surface and, on any allegedly 'horizontal plane' so constructed, the body would not be moving in a right line; secondly, if a flat runway could be constructed tangential to the earth's curvature, the body would not move uniformly unless its deceleration due to gravity were counteracted by some impressed force. Thirdly, it is impossible in actual fact to find or produce experimentally a body moving free from the influence of impressed forces—and this is *in principle* impossible for to do so it would be necessary to annihilate all other bodies in the universe, and if that could be done it would not then be possible to tell whether or how the sole remaining body was moving, or to form any intelligible concept of motion.

Accordingly Newton's first law of motion is an axiom, and it has even been held to be empty of empirical content; for the extent to which a body is moving free from impressed forces can be determined only by discovering the degree of approximation of its movement to uniform motion in a straight line, and that can be discovered only by considering the nature and mutual balance of the forces acting upon it.[1] The principle is, notwithstanding, the indispensable foundation not only of classical Newtonian physics, but equally, in the definition of a Galilean frame of reference, of the special theory of relativity.

The historical thesis that Newton based the first law on experimental evidence adduced by Galileo is upset by what we have already seen of Galileo's work; for we have found that Galileo's own argument was not inductive and drew no support from experi-

[1] Cf. Henri Poincaré, *Science and Hypothesis*, Pt III, Ch. VI; Sir Arthur Eddington, *The Nature of the Physical World* (Cambridge, 1928), pp. 123 f.; and Einstein and Infeld, *The Evolution of Physics* (Cambridge, 1938), pp. 8–9. This fact has recently been rediscovered by N. R. Hanson and Brian Ellis: see Hanson, 'Newton's First Law: A Philosopher's Door into Natural Philosophy' and Ellis, 'The Origin and Nature of Newton's Laws of Motion', in *Beyond the Edge of Certainty* (ed. R. G. Colodny).

mental evidence. As a matter of historical fact, the first statement of the principle was made, not by Galileo, but by Descartes. I have hitherto omitted all reference to Descartes and have given no account of his very considerable contribution to the development of seventeenth-century science, for the weighty reason that Descartes was notoriously a thorough-going rationalist who believed explicitly in the possibility of deducing the physical system of the world from self-evident first principles. To argue for non-empirical procedures in science from Descartes' work would therefore appear as egregious special pleading. But the first explicit statement of the principle of inertia appears in Descartes's *Le Monde*, written sixty years before the publication of Newton's *Principia*, and first printed in 1644 in the *Principles of Philosophy* (Pt II, XXXVII–XXXIX).

Descartes asserts that, just as a body retains its shape and other properties, so it must retain its state of motion (speed and direction), unless some external agency acts upon it. This follows from his principle of conservation of motion, which he claimed to deduce from the immutability of God. That again follows necessarily from God's perfection, from which, by the ontological proof, Descartes held God's existence to be indubitably established. Thus the total quantity of motion which God included in the universe is conserved (since He is immutable He keeps this quantity constant), and every change of motion that occurs is compensated by an equal and opposite motion (cf. Newton's third law). If this is indeed the *proximate* origin of Newton's laws (for the ideas can be traced back much further), clearer demonstration that, historically, they were not empirically derived is hardly needed.

In his application of these laws Newton's procedure is no more empirical than in his derivation of them. To give but one example he deduced the forces producing curvilinear motion by combining rectilinear motions and applying the parallelogram law.[1] He conceives the motion along the curve as one which is primarily rectilinear, but which is continually deflected from the straight line by discrete impulses repeated at very short intervals. This intermittent impulse is reduced to a continuous force by imagining the time interval between applications to be reduced progressively to zero. In this manner Newton proves that motion obeying Kepler's first law must be the result of a force acting upon the planet directed towards the focus of the ellipse at which the sun is

[1] Corollary I to Axioms, or Laws of Motion, *Principia*, Bk I.

situated.[1] What must be noticed is that no intermittent discrete impulse occurs and so none is observed, and the generalization from this imagined impulse to continuously acting force is not inductive generalization. Thus the final solution of the problem of planetary motion was not inductive, in the traditional sense, but was reached first through Kepler's successful attempt to make Copernicus' theory self-consistent—a theory itself reached by non-inductive methods—secondly by Galileo's mathematical deductions of the parabolic movement of projectiles, and finally by Newton's mathematical generalization of ideal motions based on non-empirical laws and postulates.

In point of fact, Newton's procedure was very much the same as that of Galileo, deductive and mathematical. He laid down axioms, which (as we now see) were not in the usual sense 'empirical' propositions, and from them he deduced laws having empirical application. Even his repudiation of hypotheses in the famous General Scholium is followed in the very next paragraph by the most flagrant example of unsupported hypothesizing: 'And now we might add something concerning a certain most subtle spirit which pervades and lies hid in all gross bodies; by the force and action of which spirit the particles of bodies attract one another at near distances, and cohere, if contiguous; . . .'[2] We have just been told that 'in this philosophy particular propositions are inferred from the phenomena, and afterwards rendered general by induction'. But from what 'phenomena' did Newton infer the existence of this 'subtle spirit', and by what 'induction' did he 'afterwards render general' his assertion of its pervasive character? He confesses that it 'lies hid' and so is unobservable. Likewise he steadfastly believed, with Francis Bacon, John Locke and many others of his contemporaries, in the corpuscular (or atomic) structure of matter[3] and used it in his theorizing (as he does here) without the least shred of empirical evidence—to say nothing of 'forces' and 'powers'[4] for the unobservability of which Hume was later most persuasively to argue.

[1] *Principia*, Bk I, Sect. II. Props I and II, and Sect. III, Prop. XI.

[2] *Principia* (ed. F. Cajori), p. 547.

[3] Cf. *Opticks*, Bk III, Pt III, Props IV-VII.

[4] Cf. *Principia*, Preface: 'But I consider philosophy rather than arts and write not concerning manual but natural powers, . . . and therefore I offer this work as the mathematical principles of philosophy, for the whole burden of philosophy seems to consist in this—from the phenomena of motions to investigate the forces of nature, and then from these forces to demonstrate the other phenomena. . . .'

That Newton was not wrong to think along these lines the future history of science proves clearly enough. These hypotheses, unsupported and merely speculative at the time, were to bear ample fruit a hundred years later. Nor, despite his overt pronouncements, was it really hypotheses as such that Newton deprecated but only wayward speculations that bore no clear relation to the conceptual structure which he was himself completing and which gave promise of ordering and comprehending the experienced phenomena more fully and more coherently than any hitherto developed. The famous dictum '*hypotheses non fingo*' is with little doubt a direct reference to the cause of gravitation. Concerning that, in the *Principia*, Newton offers no hypothesis. In this particular context it is not a general protest against the formulation of hypotheses, though in other contexts he makes it plain enough that he disapproves of speculations that have not been derived directly from observed phenomena and experiment. The method of derivation, however, is sometimes said to be induction and sometimes deduction, neither of which does Newton explain. In the *General Scholium* he tells us that 'whatever is not deduced from the phenomena is to be called an hypothesis', but in the fourth of the Rules of Reasoning which preface Book III of the *Principia* he speaks of 'propositions inferred by general induction from phenomena' as those to be regarded as accurate.

Though the precise meaning of these terms is left obscure some conjecture can be made of Newton's intentions from a passage in *Opticks*, Query 31, where he remarks:

'To tell us that every species of things is endow'd with an occult specifick quality by which it acts and produces manifest effects, is to tell us nothing: but to derive two or three principles of motion from phaenomena, and afterwards to tell us how the properties and actions of all corporeal things follow from those manifest principles, would be a very great step in philosophy. . . .'

Here we see first what is meant by an hypothesis. It is a theory without explanatory significance, one of which the phenomena give no evidence and which throws no light upon them. But how 'two or three principles of motion' are derived from phenomena is not explained. Certainly it could not be by the sort of induction described by Hume and later empiricists, for we have seen that Newton's own principles were not and could not have been derived in that way. What seems to be meant is a kind of deduction from concepts implied or presupposed in our interpretation of the phenomena. For example, if a body at rest requires a force to set it in motion, it will remain at rest unless and until a force is impressed upon it; *per contra*, if it is already in motion it will not come to rest unless forced to do so. Equally, if already in motion and if no force acts upon it, it is not conceivable what could cause its motion to change either in velocity or direction. Therefore, every body continues in its state of rest or of uniform motion in a straight line, unless compelled to change that state by forces impressed upon it.

The phenomena, here, are our common experiences of moving bodies and the deduction is from ideas by which we have come to interpret them. In Newton's case, these ideas were evolved in a long conceptual development dating back to Jean Buridan and the theory of impetus. This sort of thinking is what Newton seems to be calling indifferently 'deduction' or 'inference by general induction' from phenomena; though he also thought it possible to universalize propositions found true within a limited experience (perhaps by assuming the validity of the kind of quasi-deductive reasoning I have tried to exemplify). Finally, only by mathematical deduction did Newton ever attempt or think it possible to show, 'how the properties and actions of all corporeal things *follow* from those manifest principles' (my italics). For him deduction and induction seems to have been complementary processes of reasoning and we shall find in a later chapter that scientific thinking is typically of this character.

We may conclude that the Copernican revolution was not an empirical revolution at all but rather the reverse (though it would not be strictly true to insist on this reversal, for the earlier scientists were neither more nor less empirical in their methods). The Copernicans did not found their theories or their argument merely upon empirically discovered particulars, but upon rational grounds. Those who appealed to the evidence of the senses were their opponents, the Peripatetics. Cardinal Bellarmine wrote to Foscarini:

'... though it may appear to a voyager as if the shore were receding from the vessel on which he stands rather than the vessel from the shore, yet he knows this to be an illusion and is able to correct it because he understands clearly that it is the ship that is in movement. But as to the Sun and the Earth, a wise man has no need to correct his judgement, for his experience tells him plainly that the Earth is standing still and that his eyes are not deceived when they report that the Sun, Moon and Stars are in motion.'[1]

The view, so widely held, is false, that the work of Copernicus and the seventeenth-century astronomers marks the turning-point at which science became empirical, *a priori* deduction was abandoned and the modern mind, throwing off the shackles of the Church, set out on the path of progress. Both the changes and their causes were very different. It is for this reason that I have devoted so much space to the Copernican revolution here and it will be necessary to return to this vital period of scientific development to show that it does not differ in form from that of other periods of major scientific change. But unempirical procedures can be illustrated equally from more modern instances, a few of which we may examine without attempting any exhaustive historical survey.

ii. DALTON AND CHEMICAL COMBINATION

A good example, usually held to be a paradigm of experimental method, is the innovation made by John Dalton in chemical procedure, which was so fruitful of later advance. This, as is well known, was the introduction of the idea that the combination of chemical elements was due to bonds (later to be called valency bonds) between their particles or atoms in simple numerical proportion. There was at the time no empirical evidence for this idea and on the basis of the accepted theories none could have been found. It was only after Dalton's view was accepted that the evidence for it became so much as recognizable.

The atomic theory, in the sense of a belief in the particulate structure of matter, had been widely held, at least since the sixteenth century. At the end of the eighteenth, chemists believed in the existence of a cohesive force between the atomic particles of a substance which held them together. They also believed in the existence of an 'affinity' between the particles of different substances which accounted for solutions of one in another. For instance, a metal dissolved in an acid because its particles were attracted by those of the acid more strongly than to one another.

[1] Cf. G. di Santillana, *The Crime of Galileo*, p. 100.

Similarly the particles of salt had an affinity to those of water; and no distinction was made between the two cases. There was no clear conception of chemical combination, as opposed to physical mixture, or to solution, though when effervescence or heat was generated on mixing, chemical combination was presumed.

In these circumstances, no experimentalist was likely to look for or to recognize combinations in constant proportions by weight, and it was impossible to devise any clear test for the distinction of chemical compounds from solutions or from physical mixtures. J. B. Richter, who first discovered and tabulated regularities among combining proportions did not make this distinction clearly and was inspired in his researches by the wholly unempirical faith that chemistry was a branch of applied mathematics and that chemical reactions would therefore display mathematical relations.[1] Other chemists of the day were too impressed by the apparent exceptions, among which they included mixtures (of gases), solutions (like salt in water), alloys, and so forth, to give credence or importance to a few isolated cases of regularity in combining proportions. Proust believed that all chemical combinations occurred in fixed proportions but Berthollet was equally convinced of the contrary and the debate between them could not be settled because compounds and mixtures were indiscriminately confused.

Dalton approached the matter from an entirely different angle in an effort to deal with problems connected with meteorology of mixture and absorption between gases and moisture. Thinking he could solve these more easily if he could determine the relative weights of the atomic particles involved, he assumed, wrongly and independent of all empirical evidence, that in chemical combinations atoms could combine only in pairs.[2] The immediate consequence of this theory was that there *must be* a law of constant proportion, but it was only after the introduction of this notion of atomic combination that evidence for such a law could be recognized as such. Only if one assessed the weights of combining elements in proportional terms was the evidence apparent. As Kuhn remarks: 'Chemists stopped writing that the two oxides of,

[1] See J. R. Partington, *A Short History of Chemistry* (London, 1951), p. 162.

[2] He subsequently admitted the possibility that one atom of one element could combine with two, or more, of another, but even so his theory dictated his interpretation of the evidence. When forced to account for the sesquioxides, which appear to contain $1\frac{1}{2}$ atoms of oxygen to one of the combining element, he said 'Thou knowest thou canst not cut an atom . . . *they*, are two to three.' Cf. Partington, *op. cit.*, p. 169.

say, carbon contained 56 per cent and 72 per cent of oxygen by weight; instead they wrote that one weight of carbon would combine with either 1·3 or with 2·6 weights of oxygen. When results of the old manipulations were recorded in this way, a 2 : 1 ratio leapt to the eye. . . .'[1]

What is of importance in this story is that the theory preceded the evidence and was proposed on purely theoretical grounds. There was no evidence prior to this date even for the particulate theory of matter, which had simply been reintroduced at the Renaissance because of the general reaction against Aristotle and the increasing interest in Lucretius and Epicurus. An atomic theory was simply assumed, and later affinity between particles likewise, without any empirical foundation. So Dalton's innovation was quite unempirical, but it opened up the possibility of discovering empirical evidence which supported the earlier beliefs, though they, in the form they had hitherto been held, not only failed to stimulate scientists to look for the evidence, but obscured it so that when it was actually before them they failed to observe it.

There are numerous other cases in the history of science of similar situations, which one need only scan the pages of science histories to discover. We shall have occasion to use other such examples when we discuss the relation of theory to observation in a later chapter. One or two typical and significant cases are all that need be cited here to serve our present purpose.

iii. THE CONSERVATION OF MASS AND ENERGY

Emile Meyerson has shown that, not only the principle of inertia, but equally those of conservation are incapable of empirical demonstration, and that in the history of science they have been accepted prior to empirical evidence simply because they appealed to theorists as rationally credible and this again he attributed to a kind of *a priori* demand for an identity throughout time of the essentially real.[2] This is really only another form of expression of the demand for logical completeness, mathematical elegance and systematic coherence. Meyerson traces the belief in the conservation principles back as far as the Ancients, and shows, with great weight of historical evidence, the continuity of development of the ideas concerned.

[1] T. S. Kuhn, *The Structure of Scientific Revolutions* (Chicago, 1962), p. 133.
[2] Cf. *Identity and Reality* (London, 1964), Chs III, IV and V.

The most significant example is, perhaps, the principle of conservation of energy into which, nowadays, the principle of mass conservation has been absorbed. 'The certainty', says Meyerson, 'with which the principle of the conservation of energy appears invested far exceeds what the experimental data would seem to admit.'[1] As he makes clear, no experimental demonstration of the principle is actually possible, if only because we cannot be sure that we know all the forms that energy can take, much less that we can trace and measure all its transformations. If we could, it is by no means certain that we should find the quantity unchanged. Moreover, the principle applies only to closed or isolated systems and it is impossible in practice to produce a physical system which is wholly isolated.

When Joule attempted to discover whether there was a constant ratio between the heat generated by the electro-magnetic machine and 'the mechanical power gained or lost',[2] his experimental results were discrepant to a considerable degree (varying within a range of 33 per cent.) Yet instead of coming to a negative conclusion, he merely took the average and proclaimed that to be the constant ratio. In short, he assumed the principle of conservation regardless of the experimental results, asserting that 'we might reason *a priori* that such absolute destruction of living force ($mv^2/2$) cannot possibly take place . . .'.[3] Other scientists did not condemn this departure from empirical faithfulness, but shared the conviction in the validity of the law for reasons equally unempirical.

In our own day the same faith in its rational validity was manifested even more spectacularly, when the foremost physicists refused to abandon it in the face of palpable evidence revealed by the anomalies of the beta-ray spectrum from spontaneous radioactive decay. To preserve the principle of conservation Pauli literally invented the neutrino, a particle without mass or charge which was virtually impossible to detect. The subsequent experimental confirmation of its existence would almost certainly never have occurred if it had not first been deduced in order to preserve a law in defiance of empirical evidence—a law which had never been, and probably never could be, directly confirmed experimentally.

[1] *Identity and Reality*, p. 198.

[2] J. P. Joule 'On the Calorific Effect of Magneto-Electricity and on the Mechanical Value of Heat', *Philosophical Magazine*, Vol. XXIII, 1843.

[3] *Scientific Papers* (London, 1884–7), Vol. I, pp. 268 f.

The observations of F. Reines and C. L. Cowan, which are taken to confirm Pauli's postulation of the neutrino's existence, are as indirect as can be. No neutrino is ever observed. It is inferred from accepted theory that, if a neutrino is absorbed by a proton to form a neutron, a positron will be emitted, and if this is captured by an electron, mutual annihilation will give rise to an emission of gamma-rays, which should be detectable. An experimental arrangement is set up to bring about these interactions and all that is observed is an effect in liquid scintillator detectors which is interpreted as the emission of the necessary gamma-rays. That the source and explanation of the rays are what has been deduced might appear to the layman to be a precarious assumption, but for the scientist the theoretical assurance outweighs any such possible doubts.

'Why', asks Fermi, 'accept this concept of the neutrino? It cannot be observed in a Wilson chamber or a bubble chamber—nor has it ever been directly detected by another means, prior to the effect discovered in 1956 by Cowan and Reines. Besides, such a particle seems both unlikely and unsettling. So why accept the neutrino? Because if you do, the continuous beta-ray spectrum will be explained, and the energy principle will remain intact. What, indeed, could be a better reason?'[1]

In short, the paramount consideration is the maintenance of a conceptual scheme in which all the facts can be explained without conflict or contradiction; and if the facts observed conflict with what is already theoretically demanded, it is necessary to invent new facts to bring the empirical data into line. If the theoretical demands are based upon a sufficiently comprehensive range of scientific experience the invention will, as likely as not, be vindicated by the observation of new phenomena, or by the re-evaluation of old ones that were observed earlier but misinterpreted.[2]

[1] Quoted by N. R. Hanson, *The Concept of the Positron* (Cambridge, 1963), p. 48. Cf. also E. Fermi, *Elementary Particles* (Yale, 1951) and 'Versuch einer Theorie der β-Strahlen', *Z. Phys.* LXXXVIII, 1934.

[2] Millikan observed the positron in 1929 but did not recognize it as such. The phenomenon which he recorded was only rightly interpreted in 1932 after the existence of the positron had been demonstrated by Dirac from theoretical considerations and later observed by Anderson, Blackett and Occhialini. See below pp. 189–196.

iv. RELATIVITY

The firm faith in the mathematical unity and harmony of the physical world and its discoverability purely by thought and imagination is as strong in Einstein (though he is no pure rationalist) as in the seventeenth-century scientists. He repeatedly affirms it:

'Certain it is that a conviction akin to religious feeling, of the rationality or intelligibility of the world lies behind all scientific work of a higher order.'[1]

'All these endeavours are based on the conviction that existence should have a completely harmonious structure. Today we have less ground than ever before for allowing ourselves to be forced away from this wonderful belief.'[2]

The means of discovery is the exercise of pure thought:

'The structure of the system is the work of pure reason; the empirical contents and their mutual relations must find their representation in the conclusion of the theory. In the possibility of such a representation lies the sole value and justification of the whole system, and especially of the concepts and fundamental principles which underlie it. These latter, by the way, are free inventions of human intellect. . . .'[3]

And generalization from experience is of no use:

'Once this formulation of general principles is successfully accomplished, inference follows on inference, often revealing relations which extend far beyond the province of reality from which the principles were originally drawn. But as long as the principles capable of serving as starting-points for the deduction remain undiscovered, the individual fact is of no use to the theorist; indeed he cannot even do anything with isolated empirical generalizations of more or less wide application.'[4]

[1] *The World as I See It* (London, 1935), p. 131.
[2] *Ibid.*, p. 141. [3] *Ibid.*, p. 134.
[4] *Ibid.*, p. 128.

But quotations from great scientists are not by themselves the best evidence for our thesis. We must seek corroboration of Einstein's pronouncements on method in the actual practice of the science. This is not difficult in his own case, for the theory of relativity was the fruit of reflection upon concepts and their mutual consistency and was by no means a mere generalization from observed facts. This is as true of the theory of relativity, and in much the same way, as it was of the Copernican theory. What stimulated Einstein's reflections were contradictions and conflicts which proliferated as classical physics developed.[1] Not even the Michelson–Morley experiment was essential to its development. As a matter of historical fact, Einstein himself admitted, in a communication with Professor Michael Polanyi, that he had thought out the essentials of the theory before he had heard of the Michelson–Morley result.[2] Even when he did come to know of it, the thinking by which he developed the theory was independent of it.

Methodologically, the relation of the Michelson–Morley experiment to the theory of relativity is not very different from that of Galileo's experiments with inclined planes to Newton's laws. They provide supporting empirical evidence only if the laws are presupposed in their interpretation. So the Michelson–Morley experiment proves the impossibility of measuring velocity relative to the aether precisely because the simultaneity of distant events is indeterminable. But, as Eddington has pointed out, no experiment is required to prove that.'[3] Distant simultaneity can be measured only with synchronized clocks. In systems moving relatively to one another clocks can be satisfactorily synchronized only by means of light signals, the times of which have to be adjusted to allow for the time of transit. But that can be done only if the velocity of light is known, which again can be measured only with synchronized clocks. Consequently, any definition of simultaneity of distant events will be viciously circular.

This logical obstacle was overlooked by Michelson and Morley, who assumed both the absolute character of time (i.e. ubiquitous simultaneity) and the value of the velocity of light relative to the

[1] See below, pp. 215, 223 f., 359, and also Max Born, *Einstein's Theory of Relativity* (New York, 1965), pp. 218 f.

[2] See M. Polanyi, *Personal Knowledge* (London, 1958), p. 10, n. 2: Einstein is quoted as saying: 'The Michelson–Morley experiment has no role in the foundation of the theory.'

[3] Cf. *The Philosophy of Physical Science* (Cambridge, 1939), pp. 38–9.

absolute frame of the aether. But for these assumptions, their attempt to measure the velocity of the aether wind would have had no theoretical basis. If there were an absolute frame of reference and the velocity of light were defined as c in relation to it, the motion of the earth through the aether should be detectable, because there would be a difference between the apparent velocity of light relative to the earth (*a*) in the direction of the earth's motion and (*b*) perpendicular to it. The interferometer in effect measures these two velocities, and their difference should therefore be revealed in a shift of the interference fringes when the apparatus is rotated through an angle of 90°.

The time (t_1) taken for light to travel a distance l in the direction of the earth's motion and to return to its point of origin should be

$$t_1 = \frac{l}{c+v} + \frac{l}{c-v} = \frac{2lc}{(c+v)(c-v)} = \frac{2lc}{c^2 - v^2} = \frac{2l}{c}\left(\frac{1}{1 - \frac{v^2}{c^2}}\right).$$

In the direction perpendicular to the earth's motion

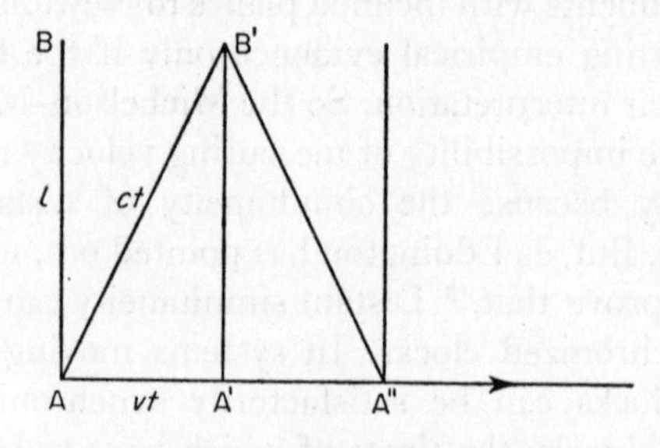

during the time (t) taken by light to travel from A to B the light source would have moved; so the path of the ray will actually be AB′, AB having moved to A′B′ in the time t. If its velocity is v, the distance AA′ will be vt and AB′ will be ct. By the Pythagorean theorem the relation between these lengths is

$$ct = \sqrt{l^2 + v^2t^2}$$

and

$$t = \frac{l}{c}\left(\frac{1}{\sqrt{1 - \frac{v^2}{c^2}}}\right),$$

and the time of the double journey AB′A″ will be

$$t_2 = \frac{2l}{c}\left(\frac{1}{\sqrt{1-\frac{v^2}{c^2}}}\right)$$

Hence

$$t_1 - t_2 = \frac{2l}{c}\left(\frac{1}{1-\frac{v^2}{c^2}} - \frac{1}{\sqrt{1-\frac{v^2}{c^2}}}\right)$$

If it is assumed that we know l and c and that $t_1 - t_2$ is calculable from the shift of the interference fringes, v can easily be found.

Determinations of the velocity of light prior to the Michelson–Morley experiment had shown no variation with differing orientations of the earth in its diurnal and annual rotations, and it was therefore assumed that the earth's velocity through the aether was, for experimental purposes, negligible. The value of c given in these experiments, however, was taken as only approximate, and the refinement produced by the sensitivity of the Michelson–Morley apparatus was so great that it was expected to reveal the hitherto negligible discrepancies. But the reasoning relating to the new experiment makes it necessary to assume the value of c given as relative to a frame at rest in the aether. The only available value, however, was the one previously determined relative to the earth, apart from that given by Maxwell's theory, in which c represents the ratio of the electromagnetic to the electrostatic unit of charge, which is stated without explicit reference to any frame.[1]

In the calculation of the time (t_2) taken by the light beam directed perpendicularly to the earth's movement, vt is assumed to be proportional to ct by the Pythagorean theorem; but unless c is defined relatively to the absolute frame of the aether, ct can be given only relative to the earth, and relative to the earth v is zero. c could be determined relatively to the aether only if clocks could be synchronized at a distance in the absolute frame, which is impossible independently of c. This means that c cannot be defined relatively to the aether and vt can never be calculated in terms of ct on the assumption that c is so defined. The quantity c as measured by the apparatus can only be relative to the earth, and the difference $t_1 - t_2$ must therefore be 0, which is what the result of the experiment reveals.

[1] Cf. G. Temple, 'From Relative to Absolute' in ***Turning Points in Physics***, ed. A. C. Crombie (Amsterdam, 1959), p. 74.

It is as if an experimenter wished to calculate the velocity of a current by sending a motor boat equal distances in perpendicular directions and noting the difference between the times taken. He could do this successfully if he knew the speed of the boat independently of the experiment and could measure distances from a fixed point. But if he were on a floating platform, could measure distances only relative to the time taken by the boat to traverse them, and calculated the speed of the boat only in relation to these distances, his final result could only be zero.

In 1881 Michelson and Morley could hardly have been expected to think in this way, for the influence of Newton's *Principia* still dominated physical science, and we shall see later that scientists habitually reason (and also observe) in terms of the dominant conceptual scheme until compelled to do otherwise by contradictions which that conceptual scheme forces upon them.

The function of the experiment was precisely that it did bring to light such contradictions. The null result was implicit (as we have just seen) in the concepts used, and the empirical evidence revealed, more than anything else, the illegitimacy of the assumptions on which the experiment was based. The invariance of c in all frames of reference was deduced by Einstein without direct appeal to this or any other experiment—though his thinking was by no means altogether independent of experience, for it did develop implications of the conception of the electromagnetic field, an idea which Faraday had introduced to explain experimental findings.

In his autobiographical notes, in Schilpp's volume, *Albert Einstein: Philosopher-Scientist*, Einstein explains how he developed the main ideas of relativity theory by reflection upon the inconsistencies in classical physics and the conflict between classical mechanics and electrodynamics. That Maxwell's theory maintained the dualism of electromagnetic energy and kinetic energy troubled him, and the fact that, while Maxwell treated the magnetic field of a moving electrical charge as inertia, inertial mass was omitted from the expression for electromagnetic forces but retained in the expression for gravitational force. Because Maxwell regarded the bearer of the field as matter rather than empty space (which was treated as a special form of dielectric), the possibility was implied that the carrier of the field, and so presumably also the aether, could have velocity. In the theory of Lorentz only matter (in the form of atoms) was the seat of electric charge, and the field was situated in empty space between the charged particles. But it was the position and velocity of the charges that created the field, which again exerted forces upon the charged particles, determining their motion according to Newton's laws (which were independent of Maxwell's).

Einstein saw that if somehow the field laws could be made to account for kinetic energy and the mass-point could be represented as a singularity in the field, the laws of motion should be derivable from the field equations, and he sought a more unified theory. In all this he makes not the least reference to the Michelson–Morley experiment, but declares that as time went on he despaired of discovering the new laws by efforts based upon the known facts, and became convinced that what was needed was a new universal theoretical principle similar to the theorem in thermodynamics that the laws of nature did not permit the construction of a *perpetuum mobile*.[1] What put him in mind of such a principle was a paradox which had occurred to him first at the age of sixteen. If one could pursue a beam of light with the velocity *c*; one should observe a spatially oscillating electromagnetic field *at rest*. But there can be no such thing, for the existence of such a field results from the motion of a charge both as observed by Oersted and Faraday and according to Maxwell's theory. Thus he realized 'from the very beginning' that, in all circumstances, what is observed must be in accordance with the same laws of nature, whatever the observer's state of motion; for unless this were so no observer would be able to tell in what state of motion he was.[2]

Here is a pure piece of deduction from the intrinsic nature of the facts as understood in terms of theory. And Einstein continues in the same vein to reason that, as in physics spatial co-ordinates presupposed some reference-body in some definite state of motion, and involved some method of measurement with rigid rods, the validity of such measurements, and consequently of Euclidean geometry, would depend upon the physical relations of the measuring instruments. The assumption that rods remain rigid in all circumstances (he points out) is arbitrary. And so also is the assumption that the measurements of time in different inertial systems must agree. We commonly acquiesce in this last assumption only because, for practical purposes, light signals

[1] *Op. cit.*, p. 53. [2] *Ibid.*

serve as a measure of simultaneity owing to their very high velocity.

The paradox which he had discovered, therefore, arose from the assumed relations in classical physics between spatial co-ordinates and the time of events in the transition from one inertial system to another, according to which there was an inconsistency between (i) the constancy of the velocity of light and (ii) the independence of the choice of inertial system of the laws of nature (including (i)).[1] To render these two requirements mutually consistent the Lorentz transformations must be substituted for Galilean transformations and then the desired theorem can be enunciated:

The laws of physics are invariant, with respect ot the Lorentz transformations, for transition from one inertial system to any other.

The result is reached entirely by reasoning from the implications in classical physics of the concepts, inertial system, spatial co-ordinates, time-measurement, etc., and of their use in framing the laws of mechanics and of electrodynamics.

The development further of the general theory of relativity emphasizes still more the unempirical character of Einstein's procedure. He begins from the acknowledged equality of inertial and gravitational mass, with a reference, it is true, to Eötvös' measurements, but without basing any argument upon them. It is the concepts with which he works, proceeding rather from the question raised by Mach seeking a theoretical basis for (or else the elimination of) the special status of inertial systems with respect to physical laws. He notes the independence of gravitational acceleration on the nature of the falling body, and postulates an equivalence between a gravitational field in an inertial system and the introduction into a space free of gravitation of an accelerated frame of reference.

The elaboration of the theory leads to the introduction of Gaussian co-ordinates and the notion of space curvature, and the equations required by the new law of gravity are such that he can say of them:

[1] *Op. cit.*, p. 56–7. Einstein adds: '*trotzdem beide einzeln durch die Erfahrung gestüzt sind*' (though both are separately supported by experience). There is a hint here of question-begging, for, as he himself has shown, the 'experience' of the choice of inertial systems depends on the *assumption* that the laws of physics are independent of it; and the constancy of *c* in Maxwell's theory is not simply a matter of experience, but arises out of the mathematical correlation of the electromagnetic to the electro-static unit of charge.

'No collection of empirical facts, however inclusive, can ever lead to the setting up of such complicated equations. A theory can be tested by experience, but there is no way from experience to the setting up of a theory. Equations of such complexity as are the equations of the gravitational field can be found only through the discovery of a logically simple mathematical condition which determines the equations completely, or very nearly so.'[1]

Here is no inductive generalization, and, if the admission that theories can be tested by experience suggests a hypothetico-deductive procedure, it still leaves much to be explained that the logical theory of that method requires, and much to be said about the nature of the experience which affords the test. These are matters for investigation in what follows.

The historical examples that we have examined in this chapter are typical enough of scientific thinking, and appeal to them has been made simply to emphasize the negative point that the essential character of natural science is not its dependence upon and generalization from empirical data. But this is, at best, only half the story. Observation is not to be despised even if also it is not to be worshipped. The true relationship of observation to theory is more fully revealed in researches some of which we shall scrutinize in the next chapter.

[1] *Op. cit.*, pp. 88–9. Einstein goes even further than I do in his insistence upon the unempirical character of physics. He says: 'Physics is an attempt to grasp reality as something conceptual, which is thought independently of its observed nature (*Wahrgenommenwerden*).' *Ibid.*, pp. 80–1. (I have departed very slightly from Schilpp's translation.)

CHAPTER VI

'DEDUCTION FROM PHENOMENA'

In the *General Scholium* at the end of the *Principia* Newton insists that theories must be 'deduced from the phenomena'.[1] The phrase has been questioned by modern logicians who hold that from empirical facts neither general theories nor other particular facts can be derived by deduction. This follows, of course, from a logical theory that makes all deductive reasoning analytic and ultimately tautological, and all inference to factual matters inductive. But doubts have been expressed about the alleged logical character of both these types of reasoning, and the question whether they are really separable and mutually independent can legitimately be raised.

The aim of the last chapter was by no means to argue that scientific method is pure *a priori* deduction or to defend the notion that observation is of no importance. Einstein's allegation that scientific theories are free inventions of the human intellect is certainly an over-statement. The foregoing examples were intended simply as illustrations falsifying the dogma that all scientific laws are nothing but generalizations from a host of observed similar instances, discovered by induction and confirmed by reference to new observations of the same kind. Though that is a somewhat simplified statement of the dogma, we have already learned from the theories criticized in Part I that it is no straw man.

In the natural sciences all theory has essential reference to observed facts, but it is clear from the actual procedure of scientists that the relation between the two is not what has been alleged in most of the recent writings on philosophy of science. We have seen that theoretical conclusions of far-reaching empirical importance are sometimes reached by methods which, according to past and present empiricist doctrines, are unempirical. Our next step will be to examine a selection of particular episodes in the history of science to try to decide what the actual relation was

[1] 'In this philosophy particular propositions are inferred from the phenomena . . . for whatever is not deduced from the phenomena is to be called an hypothesis. . . .'

in each case of observation to theory, and what part they played in the process of discovery. This should help us to judge by what means a new hypothesis is reached and how it is confirmed and established. Examination of these examples, including one taken from the work of Newton himself, may throw light on what he had in mind when he spoke of 'deduction from the phenomena', may modify our conceptions of induction and deduction, and may open the way to a more satisfactory theory of scientific method.

In discussing the methods actually used by scientists, recent philosophies of science have drawn a distinction between what they call 'the context of discovery' and 'the context of justification'. Some writers have maintained that what belongs to the first falls outside the proper sphere of the logician, and that its explanation, if it is to be sought, must be provided by the psychologist. The logician, they say, is concerned only with the context of justification, with the form and validity of arguments by which scientific conclusions are supported and hypotheses confirmed. Others take the view that, while both types of theoretical situation are to be investigated, one must be careful to distinguish the type of reasoning, if any, which leads to discovery from that by which a satisfactory theory once discovered is justified and confirmed. This distinction of 'contexts' is clearly associated with the view of scientific method as hypothetico-deductive, according to which the hypotheses entertained are conceived by some means that is independent of their inductive confirmation, or (if one rejects induction) of the process of falsification. Science then becomes a congeries of hypotheses from which empirical falsification eliminates some and leaves others unrefuted, though none of them is ever verified. If investigation of the actual modes of argument and reasoning that scientists have used should reveal that they are not what the theory of hypothetico-deductive method would have us believe, the distinction between method of discovery and method of justification may break down. Further, denial that the process of discovery is subject to logical analysis, on the ground that it is psychologically guided and not 'logically determined',[1] seems to stem from the doctrine that formal deduction is purely analytic, while inductive reasoning succeeds only in forming generalizations from already known facts. It would follow that the discovery of any fresh knowledge must originate in

[1] Cf. Carl G. Hempel, 'Studies in the Logic of Confirmation', *Mind*, LIV, 1945, pp. 1–26.

some other way; and as it is believed that there is no other valid logical procedure, discovery must be explicable, if at all, only psychologically. If, however, we find scientists using a form of reasoning that is neither deductive nor inductive according to the accepted canons, our notions of discovery and justification may well be radically altered.

The examples which follow have been selected more or less at random, each as a case of outstanding and revolutionary advance in the science concerned. Some attempt has been made to range over different sciences, for it is sometimes said that there is no one scientific method but that scientists in different branches adopt such methods as they discover empirically to prove fruitful. If the methods pursued in different sciences are different in character (as indeed there are likely to be different techniques involved in dealing with diverse subject matters), this fact should be detectable in the ensuing descriptions. Despite any such differences, however, we shall find that there is a pattern or structure of argument relating evidence to conclusions which is common to all these instances, while the diversity of their subject matter and technical methods is sufficient to warrant our accepting this general pattern as characteristic of science as such.

CASE HISTORIES

i. KEPLER'S DETERMINATION OF THE ORBIT OF MARS

'The greatest piece of retroductive reasoning ever performed' was Peirce's description of Kepler's determination of the martian orbit. But for all his efforts, Peirce never made very clear precisely what sort of logical process was involved in retroduction. He was convinced that it involved reasoning and inference and was no mere summarizing of observed data, but equally that it was not deductive reasoning. In effect, our object in this chapter is to examine cases of retroduction, and in what follows (Part III) to try to give an account of its logical character. What we shall find will be compatible with Peirce's insights that the reasoning which science employs is neither deductive nor inductive as these words are usually understood by logicians—but it remains to be shown whether very much or any reasoning is either.

The background to Kepler's resolution of the problem of the

martian orbit is twofold. First, there were his own theoretical convictions of the truth of the Copernican system and of the mathematical coherence of the structure of the universe. Secondly, there were the copious records of astronomical observations accumulated by Tycho Brahe and made with instruments more precise than any that had hitherto been devised. From these records, by themselves, as a mere collection of data, nothing significant was evident. Despite his possession of them, Tycho retained geocentrism and knew no more about the orbit of Mars than did Copernicus or Ptolemy. He still believed it to be circular and his assistant, Longomontanus could make little sense of the observational data. As Kepler wrote in 1600:

'Tycho possesses the best observations, and thus, as it were, the material for the building of a new edifice; he also has collaborators and everything one could wish for. He only lacks the architect who would put all this to use according to his own design. For although he has truly auspicious talents and real architectonic skill, yet the multitude of the phenomena and the fact that the truth is deeply hidden in the particular details are obstacles to progress.'[1]

The stages of Kepler's research were briefly the following:

(i) As already mentioned he removed the hub of the solar system from the centre of the earth's orbit to the sun. This was no induction, nor abduction, nor was it in any traditional or current sense the result of deductive argument. A consistent Copernican simply could not regard the centre of the earth's orbit as of primary significance once the earth had ceased to be the focus of the system. The earth was a planet, and a planet was no longer merely a heavenly wanderer, but a satellite of the sun (Kepler was the first to use the term). Measurements and calculations, therefore, to be soundly based must be made from the sun and not from a purely imaginary point associated with the earth to which, in the past, the fictitious rotation of the heavens had been referred.

The fact that the shift from the centre of the earth's eccentric to the sun was originally prompted by Kepler's seemingly e-logical desire to make the planetary orbits fit between the regular solids does not invalidate this reasoning, first because the idea is not strictly prompted by an e-logical desire, and, secondly, because the reasoning is in principle the same in any case. Even the theory

[1] Max Caspar, *Johannes Kepler* (Stuttgart, 1950), p. 117.

of the *Mysterium Cosmographicum* was an attempt to find a systematic, mathematical inter-relation between known facts such that it would explain (as he said) why they were as they were and not otherwise. If the theory of the regular solids could do this adequately the orbits must fit. If they were heliocentric orbits, they were more likely to fit if measured from the true sun than if measured from the so-called mean sun. The move was, therefore, towards consistency in the construction of an explanatory system, and has the same validity whether made to facilitate the cosmographic theory or to remove incongruencies in application of the heliocentric hypothesis. That the former is false is irrelevant, for it could be true only if it adequately fitted the facts, and then only if the facts were accurately represented. But if planetary orbits were measured from the centre of the earth's orbit the facts would not be accurately represented; if from the sun, they would.[1]

(ii) Once the sun is taken as the central body its significance in relation to the motion of the planets is altogether different from what it had been in the Ptolemaic system. Aristotle and Ptolemy believed that the heavens moved by their own natural motion in which the earth had no part, causal or other. Copernicus simply changed the geometrical relationships, but this change immediately raised the question of the source of the earth's motion—a difficulty which constituted one of the main obstacles to the general acceptance of the theory in the sixteenth century. A follower of Copernicus had to find a theory of dynamics which accounted for the new geometrical relationships, and the revolutionary importance of the work of Kepler and Galileo was no more than its contribution to this development, finally completed by Newton. The credit for the definitive discovery must go to Kepler, who became convinced that the force which moved the planets emanated from the sun.

This conclusion was not derived from any empirical data but simply by developing the implications of the new conceptual scheme, which made the sun the focal point and thus the body to which the revolutions of the planets must be referred. As his investigations proceeded, this physical insight proved to be an unerring guide to Kepler, enabling him to interpret the empirical

[1] Note that here the accurate representation of the facts depends on their interpretation in the light of the (heliocentric) theory. Cf. *Mysterium Cosmographicum*, Cap. XV.

data most fruitfully, despite the fact that his actual conception of the nature of the force concerned was false.

The essential point here is the domination in Kepler's thought of the principle of organization governing the heliocentric system. Throughout the course of his researches this influence repeatedly makes itself felt. What is inconsistent with it is rejected, what does not fit is adjusted, and what is incongrous is corrected. The next move in Kepler's advance is a direct result of the same regulative control.

(iii) That the line of the apsides of the orbits of Mars and of the earth are slightly inclined one to the other was known to Copernicus, but because, despite his revolutionary innovation, his thinking was still earth-bound (under the lingering influence of Ptolemy), Copernicus alleged that the orbit of Mars oscillated about that of the earth. Relative to the earth (regarded as stationary), of course, it does, but relative to the sun this oscillation is simply the effect of mutual inclination of the orbital planes. To Kepler, whose thinking was heliocentric, the suggestion was monstrous that the martian orbit should oscillate, as he says, 'according not to the laws of motion of its own eccentric, but of the earth's orbit with which it has nothing to do'.[1] The inclination of the orbital planes appeared to him not simply as a geometrical fact, for the sun's position was in his view a *cause* of the planetary movements as well as a mere spatial centre. His aim was to discover why the facts are what they are and not otherwise. If the sun is at the centre of the system, there must be a reason for this; and if the sun were the source and physical determinant of the movement of the planets that would be a sufficient reason. But, again, if the sun is the physical cause of the planetary motion, the orbital planes must intersect at the centre of the sun. Once more, the reasoning is governed by the plan and structure of the conceptual system.

The confirmation came from the Tychonic records, which showed that the orbital planes maintained a constant angle of 1°50′ and Kepler comments in triumph: 'Whereby I congratulate myself the more heartily that the observations are found to support me, as in the case of many other preconceived opinions.'[2] The remark is significant, for it underlines the priority of the conceptual scheme to the observations. The theory was 'preconceived' and

[1] *Astronomia Nova,* Cap. XIV, *Opera* III, p. 234.
[2] *Ibid.*

without it the observations meant nothing. Tycho who had made them over the years could not see that they threw light, one way or the other, on any teaching of Ptolemy or Copernicus. It was only as fitted into the system which Kepler was perfecting that the demonstration emerged from the observed planetary positions that the angle between the orbital planes did not vary.

(iv) The sun, however, is not actually in the centre of the planetary orbits, which Kepler, like his predecessors, knew to be eccentric. He still believed them to be circular so that to obtain accurate measurements the centre of the circle had to be determined and the position of the sun related to it. But if the sun is the cause of planetary motion why do the planets revolve round a different geometrical centre? Kepler resolved this problem by postulating a second force lodged in the planet itself—a 'laziness' or inertia, which resisted the push of the sun's *vis motrix*. This second physical intuition was nearer the truth and prepared the way for Newton's account of the combined effect of gravity and inertia. In Kepler's theory the functions of the two forces were, however, reversed, the sun accounting for the tangential motion and inertia for the retardation which produced the eccentricity. For he came to regard this inertia as the consequence of magnetic attraction between the planet and the sun.[1]

For such physical conceptions there was no direct observational support and they follow more obviously upon the earlier theories. Ptolemy had dealt in eccentrics with their equant points, to which he related the angular velocity of the planet. Copernicus had abandoned the equant but had retained the eccentrics and uniform motion, relying on epicycles to account for eccentricities. Kepler abandoned the epicycles and also the uniform motion, resorting once again to the equant in order to reconcile the appearance of uniform motion with actual retardation and acceleration. All these theoretical jugglings were designed, like their precursors since the time of Plato, to 'save the appearances', to provide a schema into which the phenomena could be fitted and which was a self-consistent mathematical unity.

(v) Before considering the first attempt which Kepler made with the help of these ideas to construct the orbit of Mars, let us

[1] It was the sort of attraction which, he maintains in the introduction to *Astronomia Nova*, is exerted between all bodies whatsoever, anticipating, without empirical evidence, the theory of universal gravitation. On the basis of this he gives the correct explanation of the tides, which Galileo despite all his ingenuity failed to recognize.

jump ahead in the chronological sequence of steps, to note the abiding results that followed from the progress outlined thus far. His first failure (to which we shall return) led him to digress in order to determine precisely the earth's orbit round the sun. Copernicus had alleged that the earth's orbit pulsated, because he believed in uniform motion and could not otherwise account for the fact that the sun's apparent motion is slower between the vernal and the autumnal equinox. This hypothesis seemed merely fantastic to Kepler, being completely incompatible with his own physical conceptions. He adopted the explanation that the earth moved more slowly at aphelion than at perihelion. He then formulated the 'law' that the velocity of the planet varied inversely as its distance from the sun—which seemed to follow from the view that the force which moved it emanated from the sun. That the law was wrong Kepler was aware, for he considered the intensity of the force to vary analogously to that of the sun's light, and that he knew varied as the inverse *square* of the distance. However, the eccentricity of the earth's orbit being very small the error made little difference in its case; so Kepler disregarded it.

Nevertheless, accepting the law, he next assayed to find a means of calculating the earth's position at stated times. He divided the orbit (still taken as circular) into 360 equal segments and then assumed that, for any group of contiguous segments, the measure of time taken to traverse the portion of the orbit they covered was the sum of the distances from the sun (calculated severally for each arc). He then argued that the sum of these distances (i.e. of the contiguous segments) is equal to the area swept out by the radius (or line joining the sun to the planet), and so he arrived at his second law of planetary motion: that, as the planet revolves, this line sweeps out equal areas in equal times.

(vi) Contributory to this achievement was a brilliantly inventive series of calculations to determine precisely the orbit of the earth, its eccentricity and the times of transit in different parts of the orbit.[1] It is an episode in the history of scientific discovery well suited to illustrate the relation between observation and theory, and we can profitably begin our analysis here. For this purpose let us interrupt our summary of Kepler's procedure to

[1] The method was to take the position of Mars at opposition as one fixed point and the sun as another and then calculate the position of the earth by triangulation.

reflect upon its epistemological characteristics as we have traced it so far.

The need to work out all these facts itself arose only because of the adoption of the new heliocentric conceptual scheme. For observations of Mars and the other planets are taken from the earth, and if the earth moves it is essential to know precisely how it moves in order to interpret the observations correctly. Therefore its orbit and rate of motion around it must be correctly known. Any observations would be meaningless and worthless except as interpreted in terms of the theoretical system, first by reference to the earth and its motion, secondly in the light of the earth's relation to the other heavenly bodies concerned. This relationship, again, is dependent upon the presumed heliocentric arrangement and movements of the planets, so that without the theoretical organization it is impossible to know the precise relevance of any particular observations, or just what is being observed. And it is equally impossible to tell how the observations bear upon accepted theory, either as additional information or as evidence confirming or falsifying any aspect of the theory.

Further, the device hit upon by Kepler as the means of accomplishing the task was available only to one who adopted the heliocentric system as a cosmological (not merely a mathematical) theory. Only if the planets really were moving round the sun in roughly concentric orbits could one argue fruitfully from calculated results of the observation of the earth by an imaginary observer on Mars. Moreover the accurate calculation of the sidereal period of Mars depended on the knowledge that the earth moved round the sun and not the sun round the earth,[1] and this must be known in order to effect the series of triangulations made by Kepler to fix positions of the earth relative to the sun and Mars at successive oppositions.

The empirical facts thus reveal themselves in part as the product of the development of the theoretical system, in part as the interpretation of observations in the light of that system. Let us consider this matter further and try to decide what sort of reasoning has been used.

First we must note that Kepler does not begin *in vacuo*. He had

[1] The sidereal period would be directly observable from the earth in the latter case; but in the former it has to be calculated from the synodic by using the relation of the observed period of the earth, thus: $\frac{1}{P\oplus} - \frac{1}{Psid.} = \frac{1}{Psyn.}$

the theories of Ptolemy and Copernicus, as well as that of Tycho to work upon and develop. These again did not float loose from observation. They were different schemata, each developed from its predecessor, into which the interpreted observations were fitted and in which they were ordered and structured. In the process of organizing the data incompatibilities and inconsistencies arose. For example, the various observed positions of Mars in relation to the ecliptic leads the geocentrist to allege that the planet's orbit oscillates about the earth. But Copernicus, who held this view, advocated at the same time a heliocentrism with which it is inconsistent. Consequently, Kepler corrects the inconsistency by viewing the orbits as fixed in mutually inclined planes which intersect at the sun's centre.

The persistent relation of theory and observation is that the former is a structure or organized system of concepts by reference to which the latter can be consistently correlated. The theory is not derived from the observations but from the earlier theory—an already constructed system of ordered facts. Thus it is equally not concocted or 'deduced' in sheer abstraction and isolation from observation, for the observations are already encapsulated or ingredient in the preceding theory.

The shift of the centre of the solar system to the sun, for instance, was not prompted by new observations or any different from those known to Copernicus or even to Ptolemy. It was not made because what Tycho had observed differed from what Copernicus or his predecessors knew. Copernicus made the first move in order to correlate the known data more coherently, and Kepler made the second to effect more satisfactorily and more completely what Copernicus had only very partially achieved. Kepler's physical theory, once more, was not inspired by new empirical discoveries, it was the pure result of thinking in accordance with the intellectual demands of a conceptual system. It was not a generalization from observed facts but the development of the implication of a theoretical schema so that the observed data, when they became available, could be intelligibly, consistently and coherently interpreted by its means. The observations by which the inclination of the orbital planes was confirmed had themselves to be fitted into the theoretical system before their significance could be recognized. It was only if the right question were asked that the observations could be seen as an answer to it. The question was: What is the angle between the orbital planes as projected mathematically from the various observed positions of the planet Mars in relation to the sun and to the earth? It is only when we conceive the orbits as the heliocentric theory requires—as in fixed planes intersecting in the true sun's centre—that the question of the angle between them arises. Thus it could not sensibly have been asked by Tycho or Copernicus, who did not conceive the orbital planes in that way; and if it had been it could not have been correctly answered by anybody who took the centre to be the mean sun. Yet until it was asked, the observations gave no relevant information and the position and relation of the orbital planes could never have been derived from them, or confirmed by them. On the other hand, as we have seen, the hypothesis tested was not just a hunch or an isolated flash of insight. It was the conclusion of a reasoned argument based on the requirements of a celestial system in which the sun produces the force which impels the planets and is, consequently, situated at a focus around which they rotate.

The hypothesis that the planet does not move round its orbit with uniform velocity and that this is due to a second force exerted by the planet itself is prompted by observation, but it is no mere generalization from particular cases. What is observed is successive positions of a planet relative to the sun and the earth. From these the apparent rate of motion of the planet across the heaven can be calculated; but this apparent motion can be variously interpreted, as really uniform but eccentric, as uniform and epicyclic, as non-uniform and circular, or as non-uniform and eccentric (either circular or elliptical). The choice depends on the general theoretical schema adopted and the extent to which the observed facts when ordered in accordance with it constitute a self-consistent structure.

Finally, the type of reasoning involved has been neither inductive generalization from similar cases nor deduction from arbitrarily chosen axioms. The system of the heavenly bodies and the mutual relations of their motions are presumed, but the way in which this system is organized is modified and adjusted to make it self-consistent and self-sustaining. This requires a determination or choice of organizing principle (e.g. the identification of geometrical centre). Once this has been done, certain consequences follow from the nature of the system. What kind of implication is involved here requires closer investigation which we must defer to a later chapter. It seems to be in some sense deductive, yet as it is synthetic, current theories of deduction cannot account for it; in some sense it is *a posteriori*, depending on our common experience and that of the specific facts with which the astronomical theory is dealing; and in some sense it is a dialectical development of a system already set up, a development stimulated by an internal conflict or discrepancy, to the correction of which both *a priori* and *a posteriori* movements of thought contribute.

We may leave the matter thus for the moment and return to the examination of Kepler's investigation.

(vii) Kepler's first attempt to construct the martian orbit was based upon the supposition that it was circular. He sought to calculate its radius, the direction of the line of the apsides and the position on this line of the sun, the centre of the orbit, and the equant point. Selecting four observations of Mars in opposition he proceeded by successive approximations until he arrived at a result which fitted the majority of observations to within two minutes of arc. It was then necessary to discover whether other observed positions of the planet coincided with the orbit thus constructed; but he found in two cases a discrepancy of 8′, more than the accuracy of Tycho's observations allowed as tolerable.

It was this initial failure that set Kepler thinking along new lines and led to the conclusions already discussed. It is significant first because the two discrepant observations prompted the abandonment of an otherwise encouraging and attractive theoretical construction, and secondly because they stimulated the continued progress of the investigation and the modification of theoretical ideas which were subsequently made. 'Just those eight minutes pointed the way to a complete reformation of astronomy.'[1]

(viii) He then, after digressing as we have noted into celestial physics and the recalculation of the orbit of the earth, attempted, with the help of his inverse-distance speed law and more exact knowledge of the earth's movement, to reconstitute the orbit of Mars, still assuming it to be a circle. That assumption, after all, had at that time the status of an established dogma. Nobody had hitherto doubted it, and behind it was the authority not only of Aristotle and Ptolemy but also of Copernicus. Not even ardent Copernicans like Rheticus and Galileo ever called it in question. But by now Kepler's suspicions were roused and his respect for tradition was being eroded away by the innovations he had already been forced to make. Yet before he could abandon the hypothesis

[1] *Astronomia Nova*, II, Cap. XIX, *Opera* III, p. 258.

of circular orbits he had to show quite definitely that it was wrong, so he made one final attempt, with the most painstaking labour to fit the data of observation to a circular path. Their recalcitrance brought him to what was, in a sense, the triumphant negative conclusion 'that the planet's orbit very definitely diverges from the circular form'.[1]

(ix) But he had also discovered something positive; the orbit receded from the circle at both sides. What shape was it? He decided it was some kind of ovoid but thought it was broader at perihelion and narrower at aphelion, the influence of the circle perhaps, still having some hold.

(x) With this awkward, assymetrical figure he struggled for a year, trying every possible device to make it answer his purpose. He reintroduced an epicycle turning in the opposite direction to that of the orbital movement as a result of the imputed 'laziness' or 'reluctance' of the planet in its submission to the forward-sweeping *anima motrix* of the sun. He temporarily abandoned his newly discovered law of planetary motion. He made more than seven thousand computations of distances of the planet from the sun to determine different points on the orbit—all to no avail.

But again, amidst the manifold of data a new pattern began to emerge, and he lamented that if only the figure were an ellipse he could solve his problem. In fact he had for some time been using an ellipse as a mathematical approximation to assist his calculations, remarking that 'it differs little' and considering 'what would follow if the orbit were a perfect ellipse'. Yet he persisted with the ovoid to the verge of despair.

Nevertheless, there was an inconsistency between the physical theory by which Kepler sought to support his ovoid orbit and its assymetrical egg-like shape. For the sun's impelling force varied inversely as the distance, but the inertial resistance of the planet was uniform. The differing curvatures of the alleged orbit could not therefore be properly accounted for, and Kepler came to doubt the appropriateness of his figure.[2]

(xi) Accordingly, he made a new start, calculating afresh a number of radical distances. Again he found the path to be like an ellipse and he proceeded to calculate the dimensions of the *lunulae* between the circle and the receding portions of the oval shape. He found the greatest width to be 0·00429 of the diameter.

[1] *Op. cit.*, Cap. XLIV.

[2] Cf. N. R. Hanson, *Patterns of Discovery*, p. 81.

Then, almost accidentally, he noticed that the angle subtended by the distance between the sun and the centre of the orbit at the midpoint of the *lunula* is 5°18′, and that its secant is 1·00429. This afforded an equation linking the sun's distance from the planet at any point in its orbit with the angle subtended at that point by the sun's distance from the centre.

(xii) Kepler did not immediately notice that this equation defined an ellipse, and in his first attempt to reconstruct the orbit with its help he made a mathematical error which resulted in the wrong shape (the notorious *via buccosa*). He then actually set the equation aside for no less a reason than that he now wished to begin anew, trying out the ellipse as the precise shape of the orbit, only to discover that this re-established his despised equation:

'Indeed, the greatest concern was this, that, considering and searching almost to distraction, I was not able to discover why the planet, to which the variation [*libratio*] LE in the diameter LK may be attributed with so much probability and such near agreement of observed distances, prefers to go in the elliptical path indicated by the equations. O how ridiculous of me! As if the variation in diameter could not point to an elliptical path. So that notation in no small measure persuaded me that the variation and the ellipse go together; as will become clear in the next chapter, where it will also be demonstrated concurrently that no figure remains for the planet's orbit other than a perfect ellipse, reasons derived from physical principles corroborating the empirical observations and the new hypothesis brought forward in this chapter.'[1]

Thus we come at last to the first law of planetary motion: The orbit of the planet is an ellipse, with the sun at one focus.

We may begin our analysis with the 8′ of arc, the famous discrepancy between theory and observation generally heralded as the turning point (or 'watershed' in Koestler's image) of the revolution of modern science. The data had been provided by Tycho (plus a few observations of Kepler's own). The task was to organize them into a system coherent in itself and with the rest of the astronomical structure which Copernicus had introduced.

To do this Kepler had to select observations of the planet's position in the heaven and calculate from these, as he had done for the earth, the distances from the sun. This gave him points on

[1] *Astronomia Nova*, IV, Cap. LVIII, *Opera* III, p. 400.

the orbit, which he had to construct so as to incorporate all observed positions. He had to calculate other positions at other times and check them with the actual observations, as well as calculate sun-Mars distances from other observations and find whether they were included in the orbit constructed. The recorded observations were no more than points of departure for calculations made with the help of geometrical constructions (e.g. triangulation), which gave the real elements to be included in the orbital curve. Thus before they could be significantly useful the observations had to be incorporated into some theoretical structure; and before the new structure which was subsequently developed could be checked, other possible observations had to be deduced from it, and the whole complex had to remain self-consistent. If an assumed orbit fitted to certain of the calculated data had implications which conflicted with other data, it was only because if fitted to the latter it would have other and divergent implications. So Kepler's first attempt to fit a circular orbit to four selected positions of Mars failed by 8′ to fit other observed positions, and had he fitted it to these it would have failed by some margin to fit those with which he had started. Thus the theoretical requirements were not met and the figure of the orbit had to be changed.

Changing the figure, however, disrupted the system—at least, *prima facie*—for how could a system of circles accommodate an ovoid orbit? Yet this inconsistency was only apparent. The structural principle of the system was its common centralization on the sun. This implied a causal connexion between the sun and the orbits; that again, could be made to account for the non-uniform rate of rotation of the planets, and so for an assymetrical path.

Nevertheless, the structure of the system was so important that circular orbits could not be lightly abandoned and Kepler tries every possible avenue before he is satisfied that the circle must be given up. Even when he does give it up he remains under its spell and at first produces the modification by means of an epicycle.

But now he is faced with a new inconsistency internal to the theory. The departure from the circular path is attributed to two forces one of which varies consistently with distance and the other of which is uniform, yet the alleged effect is non-uniform and does not vary consistently with distance. Add to this the continued failure to bring into line the measurements and calculations derived from the observations, and no alternative offers but a new modification of the theoretical structure.

The new figure, however, is already at hand and has, in fact, been used as a mathematical prop in the effort to arrive at a formula for the ovoid. The ellipse had already been incorporated in the figures which Kepler constructed, and the sun was always placed at one focus. It is, therefore, not surprising that detailed investigation of its properties should turn up numbers which served as clues to the final formula.

All that then remained to be done was to fit the details of the observed positions to the new mathematical structure, and when all are found to agree the harmony of the whole system is restored.

What must be stressed is that Kepler's performance throughout is architectonic. He is building a structure, using the Tychonic observations as scaffolding, the calculated distances as bricks, the mathematical formulae as mortar and the geometrical construction as a blue-print. It is a structure, the function of which is to organize the facts inferred (or developed) from the observations, and it is only to the extent that they are successfully incorporated into the structure that the proper significance of the facts is brought to light.

Further, the blue-print is suggested to him by, and is an elaboration of detail within, the larger plan of the Copernican solar system, the requirements of which from time to time dictate its details. His task differed from that of the builder who has the blue-print complete from the start; it was more like that of the architect who has to construct the plan. But the scientist has to do this not just to satisfy his aesthetic fancy, even though limited, as the architect is, by the requirements of the terrain and the structural materials. The scientist must adapt his plan to his materials or rather to seek the plan in the materials.

Kepler's problem was not to discover regularities of occurrence revealed by Tycho's data—for they revealed none. There was no question of arguing that because Mars had been observed repeatedly in certain positions in the heavens, it would be constantly or continually observed there. It was known (or at least assumed) that the planet followed a regular closed orbit within the system as set out in the general theory (Ptolemaic, Copernican, or Tychonic, whichever was adopted). The question was the shape of the orbit. This was implicit, or embodied, in the observations, for they were observations *of a body pursuing that orbit*. By fitting together the facts, as indicated by, and calculable from, the observations, Kepler had to construct the blue-print. It was like solving a jig-saw puzzle by drawing the picture which the pieces when assembled constitute, at the same time as, and as a means to, discovering how they fit together.

Three factors had to coalesce, the hypothesized orbital shape, the physical theory which accounted for the motion, and the observed positions. Each was used as a clue to the other two, and whenever discrepancies appeared between any of these factors, mutual adjustments had to be made. The final result was successful only when observational details fitted together (among themselves), theoretical elements were mutually consistent and theory and observation supported one another. All cross references had to cohere. Thus Kepler himself says, as he celebrates his triumph, the reasons derived from physical principles 'conspire' (*conspirant*) with the experienced observations and the hypothesis of the form of the orbit.

The method of discovery, therefore, is one of construction, but a construction which is in some sense 'deduced' from the facts, while at the same time the facts are 'deduced' by applying the hypothesis to the observations and making relevant calculations. Conflicts and contradictions in the course of this process reveal errors, which must be corrected by mutual adjustment of the elements (accuracy of observation must be checked, interpretation of observation must be reconsidered, and hypotheses must be modified to suit), and the conclusion is considered adequate only when a complete system has been evolved in which all the contributory factors and elements fit harmoniously together.

Three characteristics of the procedure must be specially noted:

(i) Observational data are throughout dominated and given significance by the theoretical scheme in the light of which they are interpreted, and they constitute 'facts' only in conjunction with that theoretical scheme so far as the 'facts' are reached by mathematical calculation. The use of a calculus here is obviously necessary, but it plays a different role from that alleged by Braithwaite. It is not used in order to deduce 'lower level hypotheses' from 'higher level hypotheses', but to deduce facts from observations by reference to a theoretical construct. This relationship clearly demands detailed analysis which must be deferred to a later chapter.

(ii) Throughout the course of his research, the scientist is fitting the evidence at his disposal together into systematic relationships so that each piece of evidence supports the others, not by repeating them but by cohering with them. The calculated positions of a planet in its orbit are not repetitions of conjunctions from which a generalization can be made (e.g. because AB frequently, therefore AB always), they are points whose mutual relation constitutes a geometrical figure; and they rank as evidence for the accuracy of the figure only so far as they fit together.

(iii) The entire course of Kepler's progress is marked by the mutual interplay of observation and theory, what we may in some sense term induction and deduction, neither sufficing alone and independently of the other. What any observation (or set of observations) signifies depends on the way in which what is observed is illuminated by the theory into which it is incorporated. At the same time the attempt to incorporate the processed[1] data into the theoretical scheme reveals discrepancies (when they exist) in the proposed system, which demand modifications of the system in order to make it self-consistent. From such new theoretical developments it is then possible to deduce new 'facts' which must agree with further observations.

The obvious dependence of observation on theory as well as that of theory on observation demands for its comprehension a close investigation of the epistemological character of observation itself (a theory of perception), to which attention is given in Chapter VIII below.

ii. HARVEY'S DISCOVERY OF THE CIRCULATION OF THE BLOOD

In astronomy the observer, especially without a telescope, sees only points of light on the dark background of the sky. Even with the aid of a telescope, whatever is known of the heavens beyond the solar system and much within it is the fruit of calculation and interpretation from the merest indications in sensuous appearance. But if we turn to sciences whose subject-matter is nearer at hand, the facts in all their detail seem open to our inspection, and observation, we might think, would be less tortuously involved with speculation and deduction. Especially in physiology the availability of the facts to direct experience seems obvious, so an

[1] Processed in the sense that the 'actual fact' depends on calculation.

outstanding example of discovery in that field should repay examination.

William Harvey, seven years Kepler's junior, is generally regarded as the Newton of modern physiology. He was not by any means the first to practise dissection or to observe the anatomy of the body at first hand, for, though such direct observation was largely neglected in the Middle Ages, the practice was very ancient, had been revived by Vesalius and was continued by Fabricius and Colombo. Nevertheless, the slavish deference to the authority of Galen, for whom Harvey did not lack respect, was finally discredited by his discoveries; though the difficulty with which it was overcome is illustrated by the expression of his own fears of the public reaction to his findings:

'But what remains to be said upon the quantity and source of the blood which thus passes, is of so novel and unheard-of character, that I not only fear injury to myself from the envy of a few, but I tremble lest I have mankind at large for my enemies, so much doth wont and custom, that become as another nature, and doctrine once sown and that hath struck deep root and respect for antiquity, influence all men.'[1]

Nevertheless, this doctrine which had struck deep roots was the background and context within which Harvey's own teaching developed and we shall best understand the process of his reasoning if we begin with a brief summary of the theory which he rendered obsolete.

Galen had taught that the blood was formed in the liver from ingested material passed to it from the stomach. From there, mixed with 'natural spirits', it was distributed to the rest of the body through the veins, to provide nutriment. Then with a sort of ebb and flow motion it returned and passed to the heart, where some of it percolated through invisible pores in the septum of the heart from the right into the left ventricle and there was mixed with air drawn from the lungs. The heart was the source of heat in the body and here the blood was 'concocted', mixed with 'vital spirits', and then distributed to other organs of the body through the arteries to enable them to perform their vital functions. This movement of the blood in the arteries also went back and forth, the main activity of the heart being in diastole, when, by its

[1] *De Motu Cordis*, Ch. VIII.

dilatation, it drew in air from the lungs as well as blood from the various vessels. A third portion of the blood was sent to the brain, where it percolated the *rete mirabile*, a fine net of vessels (which, as it happens, is present in *ruminants* but not in men). Here it generated 'animal spirits' which passed pure and unmixed into the nerves and were distributed through them to the body at large.

Before Harvey, certain new discoveries were already throwing some doubts upon the Galenic doctrine, however faithfully scholars adhered to it. First Vesalius called in question the permeability of the septum; next Colombo discovered the 'lesser circulation' from the right ventricle of the heart through the lungs to the left ventricle; then Fabricius revealed the presence of the valves in the veins, though he mistook their function and thought that they served to prevent the blood from collecting, under the influence of gravity, in the extremities. It was at this stage that Harvey made his entry and it was against this theoretical backdrop that he performed his transformation scene. Like Kepler, therefore, he did not begin *in vacuo*, but made his experiments and observations with the earlier theories in mind.

Harvey begins by cataloguing the contradictions and absurdities involved in the accepted theories of the day. The function of the heart and the pulsing of the arteries was supposed to supplement the work of the lungs in ventilating and cooling the blood; but in that case, Harvey contends, one would expect the structure and movements of the heart and of the lungs to be similar, but they are not. Moreover, if the arteries draw in air and expel 'fuliginous vapours' one would expect them to contain some proportion of both, yet when opened, as even Galen certifies, they are found to contain only blood.

The current theory, in keeping with the foregoing view, held that in diastole the heart and arteries drew in air through the skin. Yet if the body is immersed in water or oil and the passage of air is obstructed, the pulse is not diminished. Moreover their action is the same in whales and other cetaceans, though they live perpetually under water.

Note that Harvey 'deduces' consequences from the accepted theory. If the expansion of the ventricles and the arteries were like that of a bellows (an analogy he often uses) certain consequences follow of necessity: air must enter through the pores or other orifices. If air is sucked in, obstruction of the passages must make the diastole more difficult or impossible (as with a bellows). These results are the logical consequences of the systematic relations presumed between the structures and events. The observed facts, however, are otherwise; and the consequences of the theory contradict one another. Nevertheless, the accepted theory is not divorced from fact. It is not a purely theoretical invention. There is a diastole and a systole of the heart and arteries, and the blood does percolate the spongy tissues of the lungs. The criticized theory is one way of ordering the observed facts, but it is an order which does not consistently accommodate other features of those same facts; e.g. that the arteries are never found to contain air, but only blood, never emit air when pierced, but always blood with some force, in contrast to the windpipe which emits and draws in only air.

Further, if the arteries, as was alleged, carried warming spirits in the vital blood, they could not at the same time perform a cooling function. If blood flowed through them in one direction, air and fuliginous vapours could not flow through them at the same time in the opposite direction. Especially the *arteria venalis* had attributed to it several different and mutually incompatible functions, that of passing air from the lungs to the left ventricle of the heart, of passing 'fuliginous vapour' from the heart to the lungs, and also of distributing spirituous blood to the lungs.

A blatent contradiction, which evokes a profane exclamation from Harvey, is occasioned by the incompatible functions attributed to the valves of the heart. They are shown as part and parcel of the contradictory and impossible theory as a whole. The left ventricle was supposed, in diastole, to draw air from the lungs, and, at the same time blood from the right sinuses of the heart, to manufacture spirits and send the spirituous blood into the aorta at the same time as it drew 'fuliginous vapour' from that vessel and sent it to the lungs. 'How,' asks Harvey, 'and by what means is the separation effected?' (between the vapours and the fresh air, between the air and the blood). How can the semilunars prevent the regress of spirits from the aorta if they allow the vapours to pass? The mitral valves were supposed to prevent the return of air to the lungs, yet not to obstruct the passage of fuliginous vapours escaping from the heart. How can spirituous blood pass to the lungs without hindrance from these valves, as they are explicitly alleged to prevent retrogression of the air (which is supposed to pass to the heart by the same route). 'Good God!

how should the mitral valves prevent regurgitation of air and not of blood?' Harvey exclaims.

Both ventricles contract and both dilate simultaneously. How is it then possible for either action to force blood through the imperceptible (imputed) pores of the septum? These pores had never been discovered and the substance of the septum was found to be more compact and dense than any other bodily tissue, apart from bone and sinew. Yet blood was thought to be drawn through it, while air was provided with open tubes for its passage. Again, the septum has its own separate supply of blood vessels (which would hardly be needed if the blood traversed it through the imagined pores) and, in any case there is a far easier and more open way for blood to pass from the right to the left ventricle, through the pulmonary veins.

If these theories are accepted other known anatomical facts conflict among themselves. The structure and action of the ventricles are alike; why should their functions differ? On the other hand the structure and rigidity of the trachea are altogether unlike those of the pulmonary vein, so how can their functions be similar? If air passes from the lungs into the heart, how is it that air artificially forced in through the trachea, expands the lungs but does not penetrate the pulmonary veins and arteries?

The theories taught are a mass of confusion and contradiction and Harvey concludes his introduction with the declaration: '. . . it is plain that what has heretofore been said concerning the motion and function of the heart and arteries must appear obscure, or inconsistent or even impossible to him who carefully considers the whole subject'.

It is because of its incoherence that the prevalent theory requires revision and reform, and the process of carrying out the reformation is the marshalling of evidence, its correlation and interconnexion so as to form so impressive a system of mutually supporting and corroborative facts,[1] that the conclusion cannot, in the face of it, be denied.

The reasoning throughout is obviously not inductive and equally, if regarded as strictly deductive according to the accepted formal rules, much of it would be invalid. Yet it is cogent. It does

[1] I use 'corroboration' always to mean support of a thesis by different but interrelated evidence from diverse sources, not by repetition of similar instances of a general rule.

not follow deductively that, if structures are dissimilar, functions must be, or *vice versa*. One might, however, say that an inductive inference to the correlation of structure with function could be based on frequent observation of such concomitance. But here the argument does not move in that way. Whatever might commonly have been observed, Galen and his followers alleged diverse functions for similarly structured organs and similar functions for organs of different types. In fact, the actual functions were not known, so no regular concomitance of structure and function could have been observed. Harvey, on the other hand, seeks to 'deduce' the function from the structure. Air tubes, if they are to function satisfactorily, must be rigid like the bronchial tubes; if liable to collapse they will not permit free passage of the air. But the pulmonary vein is relatively soft and flaccid, so it cannot function properly for gases. Moreover, Harvey's intuition in this matter is sound, for in living matter structure is the product, and nothing other than the expression and manifestation, of function.[1] This fact, of course, might conceivably be learned inductively, but it could also be grasped (as Aristotle holds a universal is seen in its particular exemplifications), by one familiar with biological phenomena, who draws out the implications of activity in a material medium constituted by self-adjustive processes.

Again, if open passages exist from one ventricle to the other through which air is thought to pass freely, it does not follow deductively that no pores exist in a solid partition between the ventricles. But it is reasonable to suppose that a denser fluid will not flow more freely through minute pores than through open channels. Or, if blood flows in one direction through a certain passage, it is not altogether inconceivable that air might pass at the same time in the opposite direction. But once the mechanism by which these movements are governed is considered this contrary motion becomes altogether unthinkable. In diastole the heart, like a bellows, might draw in air and in systole emit blood, but it could not do both at once; nor could it do them in succession if the direction of passage of either is controlled by valves which allow fluids to pass only one way. Then too, it becomes absurd to suggest that such valves could obstruct the more volatile fluid while allowing the more viscous to pass. It would hardly be sensible to maintain that we know this only from repeated experience of the behaviour of fluids, for it is implicit in the very

[1] Cf. J. S. Haldane, *Organism and Environment* (New Haven, 1917).

ideas, 'volatile' and 'viscous'. The cogency of Harvey's argument, however, as will appear more fully later, arises from the number and variety of considerations each and all of which lead the mind to the same conclusion, and the fact that in their variety they support and supplement one another, fit together coherently and form an interconnected system of ideas. The conclusion here is negative because the arguments prove inconsistency and incoherence in the system alleged by Galen; but each does so in a different way and (in Kepler's terminology) they all 'conspire' to the same outcome.

The positive argument, to which we must now turn displays the same structure and is built up by Harvey, marshalling his evidence with consummate generalship, in the most meticulous and systematic manner.

(i) First, Harvey describes the action of the heart as observed in the vivisection of animals. It is at times activated, and at times relaxed and 'at rest'. When active it is contracted and narrow, hard and erect, when relaxed limp and flaccid and also more distended. When contracted it is paler, when distended of a deeper red. Harvey notes that, as in the case of other muscles (e.g. of the forearm), action is accompanied by contraction, hardening and tension, relaxation by softening. So, he concludes, the activity of the heart is in systole when it contracts, not in diastole; and in this activity it expels blood, for then it is paler, while in relaxation it is distended, redder, and so contains more blood. The latter point he confirms by piercing the ventricle and observing the forcible emission of blood during contraction.

This conclusion is not inductive. It is not derived from frequently repeated observations (though, of course, these could be made), but from the structure of the evidence—consistency, plus colour, plus observed emission of blood, pointing to activity in systole, relaxation in diastole, and ejection of blood through activity. Attention drawn to the analogical behaviour of other muscles might count as inductive reasoning, but here it is used simply as corroborative make-weight. Harvey does appeal to different observations in different species (fishes, frogs, snakes, etc., as well as mammals) but he does so not to amass similar instances but to illustrate through variety what is more apparent in one case than in another (e.g. the elongated conical shape of the heart in fishes and reptiles displays better the contraction in systole and also shows the change in colour). The multiplicity of examples is significant because of their differences as much as of their similarities, so that they become mutually supplementary.

(ii) When the heart is in systole the arteries are dilated by the blood forced into them by the heart from the left venticle, while the right supplies the arterial vein (pulmonary artery). When the left ventricle ceases to contract the pulse in the arteries ceases. If an artery is punctured, blood is forcibly ejected whenever the ventricle contracts. He concludes, therefore, that the diastole of the arteries corresponds to the systole of the heart and *vice versa*, so that the arteries are distended like a sac or bladder, not like bellows, and the pulse is due to the impulse of the blood occurring simultaneously with the systole of the heart. He then adduces corroborative evidence from the behaviour of a tumour caused by aneurism. Once more, we have a system of facts so interrelated as to enforce the conclusion. Undoubtedly, the facts must be observed, but their significance lies in their interconnexions. It is these, not the frequency of their occurrence, that lead to the disproof of the theory that the action of the heart in diastole ('like a cupping glass') draws in the blood from the veins and arteries, and that by their dilatation the arteries suck in air like bellows. Incidentally, when so stated it is apparent that this arrangement is impossible, because suction by the heart would collapse the vessels and not dilate them.

(iii) The auricles are found to contract before the ventricles so that there are two simultaneous movements of each pair—not four successive movements. On the approach of death the auricles continue to contract after the motion of the ventricles has either become very intermittent or ceased altogether, the movement of the latter apparently being stimulated by the ingress of blood from the former and so depending on its quantity. If the tip of the heart is cut blood is seen to flow out at each pulsation of the auricles. Analogous phenomena are seen in the hearts of fishes, snails, and some insects, as well as chicken and human embryos.

Hence we conclude that the contraction of the auricles forces blood into the ventricles, and contraction of the ventricles forces it into the arteries and the *vena arteriosa*. Once more, the structure of the facts enforces the conclusion, not their frequent recurrence, and the comparison of different species merely proves that what is observed in one is not a special or exceptional arrangement but is the common structure and interrelation of processes in widely differing circumstances, such that as it occurs it must result in a passage of blood through the heart in a definite manner and by means of a definite mechanism.

(iv) The next step is to establish the existence of the lesser circulation through the lungs, ignorance of which was the source of Galen's perplexity as to how the blood passed from the right to the left ventricle. Here an appeal is made to comparative anatomy and physiology. Harvey protests that dissection of the human cadaver alone, however many times it produces the same result, is misleading because the same error is always involved and generalization from it is quite unwarranted. Examination of other species while alive gives varied examples all differently illustrating the same fact—that blood is transferred from the veins to the auricle, or an analogous organ, from that to the ventricle and thence to the arteries. In fishes which have no lungs this is patent to observation and in amphibians and reptiles even though they have lungs the connection is evident. Further, in the human foetus, as the lungs are not in use, the four vessels of the heart (*vena arteriosa*, *arteria venalis*, *aorta* and *vena cava*) are connected otherwise than in the adult, there being a passage from the *vena cava* to the *arteria venosa*. In this passage there is a membrane which serves in the foetus as a sort of valve regulating the direction of the blood flow, but which becomes in the adult an obstruction preventing its passage. Also the *vena arteriosa* has an extra branch, in the foetus, connecting it with the aorta, which atrophies and disappears after birth.

The traditional explanation of these facts had been that the passages in the foetus exist for the nutrition of the lungs, yet in the adult where the lungs are in greater need of nourishment the passages do not exist. It was also thought that they were needed in the foetus because its heart did not beat, but Harvey produces evidence, from incubated eggs and embryos removed from the uterus, that the heart moves in them just as it does in adults, and he quotes Aristotle in support.

Hence he concludes that these passages serve to enable the blood to pass from right to left directly when the lungs are not in use, but are blocked in the adult so as to force the blood through the lungs when they are functioning.

The purpose of this detour and the function of the lungs Harvey does not closely investigate here, as being too large a digression. But the conclusion that the blood circulates through the lungs from the right ventricle to the left auricle is now inescapable, the systematic arrangement of organs and functions having been demonstrated. The heart has been shown to force out blood by its contraction, the valves have been shown to make a certain direction of propulsion unavoidable, the arrangement of the vessels being such as has been described, the action of the heart must pump the blood into the *arteria venalis* whence it will percolate through the parenchyma of the lungs and re-enter the heart on the left side through the *vena arteriosa*. That this is possible Harvey argues from analogy with the percolation of fluids through the liver and the kidneys.

Thus,

'that the blood is continually passing from the right to the left ventricle, from the *vena cava* into the *aorta*, through the porous structure of the lungs, plainly appears from this, that since the blood is incessantly sent from the right ventricle into the lungs by the pulmonary artery, and in like manner is incessantly drawn from the lungs into the left ventricle, as appears from what precedes and the position of the valves, *it cannot do otherwise* than pass through continually' (my italics).

'This', he says, 'is obvious both to sense and reason.'[1]

(v) We now come to the climax of the entire proof, the quantitive demonstration that clinches the whole argument. In this phase the form of the argument is more deductive in character, yet it does not change in essential structure and continually incorporates observed facts.

The capacity of the heart is estimated—not precisely but within reasonable limits—and the estimate least favourable to the thesis being defended is used. The number of pulses per hour (or half-hour) is observed, and by simple multiplication it is shown that more blood must pass through the heart in that short span than is contained in the whole body. It is far more than could be manufactured in the time from ingested nutriment and more than is required for nutrition of the tissues. It can thus be explained only by the return of the blood from the arteries to the veins and thence to the heart. For we have seen that it passes continually from the *vena cava* to the aorta through the heart, and if the supply is to be maintained it can only be by circulation. Corroborating evidence is drawn from the fact that all the blood rapidly drains

[1] *De Motu Cordis,* Ch. VII.

away through an opened artery, both veins and arteries being emptied. The implication is obvious that the venous blood, if it is to be drained away through the arteries, must have passed to them through the heart, proving the circulatory movement.

(vi) Further experiment shows that a vein ligatured below the heart rapidly empties between the ligature and the heart. If an artery is ligatured it becomes distended and purple between the heart and the ligature. If a tourniquet is applied to the arm above the elbow and tightened sufficiently to restrict the artery, the hand remains white and the pulse below the ligature can no longer be felt. When the ligature is loosened somewhat the blood returns to the hand and the pulse revives, but the lower arm and hand become swollen and flushed. This is explicable from the fact that the blood is forced through the artery, overcoming the slackened stricture of the ligature, but its return through the veins is now prevented. Hence, restriction of the flow in the artery keeps both artery and veins in the forearm empty, but when this flow is restored not only does the pulse in the wrist return but the veins are also filled, proving the passage of the blood from the arteries to the veins at the extremities. It is further demonstrated that veins obstructed as above emit blood when punctured with some force whereas when the obstruction is removed the flow is relatively slow and weak. Further, if the ligature is retained with moderate pressure almost all the blood in the body escapes through a vein punctured below the obstruction, within about half an hour, including that contained in the arteries. One can easily estimate the rate of flow in a vein by this means and discover once more that the quantity passing through a limb in a given time is far more than could be supplied from an outside source without continual circulation.

It is to be noted that most of these observations had been made before. The facts, though they had not been submitted to quantitative analysis, were for the most part well known through the practice of blood-letting. But before Harvey, they had not been interpreted in terms of circulation and so their interrelation had not been investigated. It is Harvey's theory that gives them special significance and converts them into relevant evidence.

(vii) Harvey goes on to demonstrate how the valves in the veins permit passage only towards the heart and not away from it. Their existence is revealed in dissections, and also, in the living body, if the arm is ligatured as previously described 'knots' can be observed at intervals in the veins above where the valves occur. If the blood be pressed with a finger-tip upward from below one of these valves and pressure is retained at the inferior point from which the blood was pressed upward, it does not return through the valve, though the vein remains distended above it. Nor can the blood be forced back through the valve by applying pressure above. On the other hand, when the pressure at the lower point is removed, the emptied vein is immediately refilled.

Again it is possible to calculate the amount of blood flowing through a vein in this manner in a given time and it is found to be much more than could be manufactured in that time by ingested food.

So Harvey concludes:

'Since all things, both argument and ocular demonstration, show that the blood passes through the lungs and heart by the action of the auricles and ventricles, and is sent for distribution to all parts of the body, where it makes its way into the veins and pores of the flesh, and then flows by the veins from the circumference on every side to the centre, from the lesser to the greater veins, and is by them finally discharged into the *vena cava* and right auricle of the heart, and this in such a quantity or in such a flux and reflux thither by the arteries, hither by the veins, as cannot possibly be supplied by the ingesta, and is much greater than can be required for mere purposes of nutrition; it is necessary to conclude that the blood in the animal body is impelled in a circle, and is in a state of ceaseless motion; that this is the act or function which the heart performs by means of its pulse; and that it is the sole and only end of the motion and contraction of the heart.'[1]

He finds this conclusion, not merely probable, but 'necessary', for he clearly considers the supporting argument to be so compelling as to be apodeictic. Yet, as we have seen, it is not formally deductive and derives much of its force from 'ocular demonstration', which according to current logical and philosophical doctrines could give only a probable result. Obviously these doctrines are wide of the mark and altogether inadequate to account for the structure of this most typically scientific thinking.

What then is this structure? We have observed that it contains a deductive aspect or element. 'If-then' arguments abound within

[1] *De Motu Cordis,* Ch. XIV.

it, but the necessity of the nexus springs not from mere definition of terms nor simply a relationship of truth values like that which validates material implication in symbolic logic. Given a certain structural arrangement of (observed) facts, certain relationships are excluded and others required. As we saw, the arrangement alleged by Galen would require certain facts which observation does not find: air in the arteries (which is never emitted when they are pierced), the passage of blood through the pulmonary vein in the opposite direction to that of air, past valves which would obstruct it (and were alleged to obstruct the return to the lungs of air). Thus, certain relationships are both required and excluded by the theory which is seen to be self-contradictory.

A cavity filled with liquid, if contracted by muscular action *must* force the liquid out through any available passage, thus the systole of the heart *must* propel the blood into the arteries (passage into the veins being obstructed by valves). The successive contraction, first of the auricles and then of the ventricles requires propulsion of the blood from the former to the latter and thence to the arteries. The formation of the valves being what it is, they *must* obstruct flow in one direction and permit it in the other. Further, if the beat of the heart forces blood into the arteries, the dilatation of the arteries is bound to be caused by blood pressure and not by a bellows-like expansion drawing in air or liquid.

The direct connexion of the right ventricle to the left in organisms without lungs and the special provision in the foetus of passages to effect this connexion indicates the route taken by the blood, and the blockage of the special passages or their atrophy in the adult makes it inevitable that the blood which formerly passed through them should go through the lungs. The interesting feature of this conclusion is that it could be made with cogency, though the actual percolation of blood through the lungs had not been directly observed and its possibility is postulated only on the analogy of other organs like the kidneys and the liver. Yet the conclusion cannot be denied given the facts (*a*) that the heart continually pumps blood from the right ventricle into the pulmonary artery and draws it into the left auricle from the pulmonary vein and (*b*) that the position and structure of the valves would not permit its flow in any other way. Hence, Harvey declares 'it cannot do otherwise'; and this is a conclusion required by reason on the strength of the observations.

The quantitative considerations combined with ocular demonstration of the directions of flow in the arm make it impossible to account for the rate of flow except in terms of circulation, for if the known quantity of blood in the body passes through the heart (or any limb) within the space of half an hour, the flow could be maintained only by constant circulation; for this quantity of blood cannot be supplied at that rate from any extraneous source, nor has it any external destination.

So we have an assemblage of facts and relations, derived indeed for the most part from observation, but leading to an inescapable conclusion, not as a generalization from repeated occurrences, but as the consequence of the interlocking implications inherent in the system. The facts are observed to be so related among themselves that no other conclusion is possible. At no point is it suggested that as the blood is observed to circulate in *n* different organisms, therefore the blood always circulates; but only that in the observed arrangement of organs and activities, the observed movement of the blood, in the observed (and calculated) quantities, *must* be one of circulation.

Contemporary logicians would probably maintain that they can account for Harvey's reasoning in terms of their own principles. Such statements as have been made above (that a cavity filled with liquid, if contracted must force the liquid out) they will regard as inductive generalizations from common experience, even though (in this example, for instance) a sort of geometrical necessity is involved. They could deny the necessity by saying that it is purely a matter of experience that liquids are not compressible and that fluids tend to flow in the direction of least pressure. That these are merely matters of experience may be disputed, but if it is we shall be told that they can be regarded as other than empirical only if, by making non-compressibility and fluidity implicit in the definition of the term 'liquid', we reduce our statements to tautologies. Harvey naturally assumed numbers of such generally accepted hypotheses, and, they will say, his 'deductions' are really no more than the application to these hypotheses of formal logic by incorporating them into a deductive system in the manner already explained.

But this account of the matter is unsatisfactory for several reasons, some of which have already been aired. First, we have found no sign of the alleged inductive procedure anywhere in Harvey's work. Those instances most nearly similar to it were all cases in which different phenomena were cited as throwing light from new angles upon the fact under investigation. They were not cases in which the same phenomenon was repeated and a probability estimate reached that it would be met with again. If Harvey does not use the inductive method to establish hypotheses that he is himself seeking to prove, why should we assume that he accepts as premises other hypotheses that are established inductively? To reply that this is the normal way in which common sense opinions become accepted is to beg the whole question over again of the legitimacy of inductive reasoning. And if (as we have earlier maintained) it should not be legitimate, to base scientific reasoning on premises inductively established would be to invalidate the whole scientific procedure.

Secondly, beginning from the various phenomena anatomical, physiological and zoological to which he draws attention, Harvey reaches a conclusion which goes beyond them and is synthetic with respect to them. That the blood circulates is neither synonymous with nor contained in any of the evidential material that he describes. It is an inference from a body of facts from which it follows only when they are seen in their mutual systematic relations. It does not follow (by any sort of logic) merely from the fact that the heart's action is in systole, nor merely from the fact that there are one-way valves in the veins, nor merely from the structure of the valves of the heart, nor from any single fact among those that Harvey adduces; but it does follow inescapably from all of them taken together, not just as a list or collection, but in systematic arrangement and interconnexion. As the conclusion is in this way synthetic, the accepted theory of deductive inference could not account for it, even if the deductive system did contain empirical statements as initial formulae, or as axioms, or as a result of substitution of empirical terms for variables.

Nowhere does Harvey accept a current theory and use it as a major premise, then, specifying initial conditions, introduce a factual minor to reach his conclusion. When he does use an accepted theory as a premise, it is in order to disprove it by *reductio ad absurdum*. Where his argument is deductive, as when he calculates the volume of blood passing through the heart in half an hour and finds it to be as great as that contained in the whole body, he uses this result as a premise for his final demonstration, by an inference that is deductive in a different way, that the blood 'is impelled in a circle'. The amount of blood in the body, the capacity of the heart and the rate of its beat, as well as the rate and quantity of ingestion, are all discovered empirically. From the second and third of these items the volume of blood passing through the heart in a given time is calculable. The relation of this sum to the first and the last items establishes the conclusion—*not* by inductive generalization. Continuous flow of a liquid through a vessel can be maintained only by continuous supply, either from within the system or from without, or else by circulation. Is this an empirical generalization? If not, is it a tautology? Neither seems plausible. It is not an arbitrary postulate, but it is in no way uncertain.

The identification and disentanglement of deductive and inductive elements in this type of reasoning must be left until later; but it is plain that no calculus is here being employed (as in Braithwaite's model), and the necessity of the conclusion is not simply due to a train of mutually entailing tautologies. Nor is there any process of deducing from one hypothesis another which is then subjected to observation for testing. Harvey does not argue 'If the blood circulates, so-and-so should follow' and then investigate to see whether so-and-so is the case. His procedure throughout is: The organs and movements are found to be thus and thus, therefore the blood must circulate. Once this conclusion has been reached it can, of course, be submitted to further tests. But no further tests are needed beyond the evidence Harvey produces to establish the conclusion. Once so established it demands no further confirmation.

iii. NEWTON'S EXPERIMENTS

It may well be the case that when Newton spoke of deduction from phenomena, he was not himself very clear about the method to which the phrase was supposed to refer. Nevertheless, his reasoning in the *Opticks* has characteristics which make the application of the phrase plausible. He was always scrupulous in his efforts to stick to the facts and to rest his theories on observation without wild and unfounded speculation, but in spite of his protestations, he does not refrain from the formulation of hypotheses as long as he can derive them rationally from the observed phenomena. The special interest of Newton's procedure in the *Opticks* lies in the fact that the experimental work which he describes in great detail is obviously part of the process of discovery, and yet at the same time is used by Newton as a means of justification (or what he calls 'proof') of his hypotheses. An examination of his procedure will show that, particularly in the form in which it is set out, it is very difficult to assimilate to the traditional and current accounts of inductive method; and even when it appears to be hypothetico-deductive, the relation between the hypothesis and the deduction is not obviously that which the theory as presented by Karl Popper and J. O. Wisdom would require.

Newton begins the *Opticks* in his usual way, by stating definitions and axioms, the obvious purposes of which are to give precise meanings to terms and to construct a framework within which the observed facts can be organized and his subsequent reasoning can proceed. Some of the Axioms obviously depend upon observation (e.g. Axioms I, II, IV, V) and in some cases elaborate geometrical demonstrations are given to deduce consequences from them. For instance, Axiom V states that there is a constant ratio between the sine of incidence and the sine of refraction. This could, of course, be stipulated arbitrarily, but Newton unquestionably derived it from observation and measurement. One may presume that after finding the angles of incidence and refraction in a number of actual cases and finding the proportion between them to be constant he generalized the finding in the form of this axiom. He then proceeds at once to demonstrate geometrically how to determine the direction of a ray after refraction, when the proportion is known for the colour of light and the medium concerned. This procedure could be held to conform pretty closely to what Braithwaite[1] describes. An hypothesis is reached by inductive generalization, then by the use of a deductive system (in this case Euclidean geometry) consequences are deduced from the hypothesis which are lower level hypotheses (e.g. that in such-and-such a case red light falling on the surface of stagnant water at such-and-such an angle will be deflected thus and thus). These can then be tested by observation to give further support to the hypothesis.

It is to be observed, however, that Newton does not use the results of such deductions to help confirm his 'axioms'. He uses them to reach new conclusions from experiments designed to demonstrate quite other laws in ways which we shall presently examine. He does not confess to having discovered his axioms by induction and is content to state them as axiomatic. The conclu-

[1] *Op. cit.*, Chs I–III.

sions of proofs which follow, for instance that by which he shows how rays converging after refraction or reflection will produce a picture of the object from which they originated, are left as established results, not treated as hypotheses requiring further confirmation. Taken at its face value, this procedure might be described as 'deduction from phenomena'. The phenomenon in the example cited is the observed constant relation of the angle of incidence to that of refraction and the conclusions reached by geometrical demonstration are deduced from it.

But when we turn to the description of the experiments which Newton performed we find clearer and less ambivalent cases of what we are seeking. Proposition I, Theorem I of the first part of Book I states that 'lights which differ in colour, differ also in degree of refrangibility', and two experiments follow, each designed to show in a different way that this is the case. In the first, each half of a strip of paper is differently coloured (red and blue) and the paper is viewed through a prism (adequate precaution being taken to eliminate reflected light). It is found that, as the prism is moved so as to raise the image by refraction, the blue half is displaced more than the red, and likewise if the prism is so moved as to lower the image.

The next proposition is that the light of the sun is a mixture of rays differently refrangible, and is demonstrated by a series of eight experiments. Some are designed to show that the rays falling with equal incidence upon a prism are differently refracted, others to show that this difference is not due to dispersal or splitting of the rays. Some are designed to deflect the whole beam, some to deal with successively refracted portions of it. Yet others are so devised as to show the relation between reflection and refraction of the same rays. The only mention of repetition in any of these experiments is made when it is necessary to ensure accuracy of measurement or to exclude fortuitous factors, like flaws in the glass of the prisms.[1] Then Newton repeats to check quantities, or changes his prism or lens to eliminate chance effects produced by inessential differences in the apparatus. There is no repetition of experiments in order to accumulate similar evidence and no argument of the form 'AB constantly, therefore AB always'. In Experiment 5 a series of prisms is used to refract the sun's image successively a number of times, but the purpose is to discover whether the elongated spectrum, which Newton takes as evidence

[1] See Experiment 3, Bk I, Pt I.

of variable refrangibility, is ever broadened by further refraction, which could be evidence for dispersion. The fact that no broadening of the image is observed, and that passage through successive prisms results only in a change in the inclination of the image from the perpendicular, is taken to show that the effect is not due to dispersion. The experiment thus provides corroborative evidence in favour of the hypothesis being demonstrated.

The point to be emphasized is that the elegantly devised and varied experiments that Newton here describes provide a body of quite different but mutually corroborating observations, each of which, when interpreted in the light of the theoretical framework established by the definitions and axioms, supports the proposition to be proved. It is to be noticed again that Newton speaks of 'proof by experiments' and not of confirmation to any degree of probability.

The observations, to be useful and relevant, have to be interpreted. In Experiment 3[1] all that is directly observed is the position, movement and dimensions of the spectrum of the sun's light proceeding from a circular aperture ¼ inch in diameter, as a prism is rotated in the beam 18½ feet from the wall on which the spectral image is projected. The prism is fixed so that the refraction of the light on each side of the refracting angle is equal to that on the other side. The angle subtended by the breadth of the image at the prism is calculated (adjusting the measurement to allow for the dimensions of the aperture through which the light is admitted) and found to be equal to the apparent diameter of the sun (½ a degree). From the fact that the angles of refraction on each side of the prism are equal it is deduced that if all the incident rays were equally refrangible the sun's image projected on the wall would be circular and would subtend the same angle at the prism both vertically and horizontally (i.e. ½ a degree). But the observed image is much longer than it is broad (10¼ ins × 2⅛ ins) and its length subtends an angle at the prism of approximately 2½ degrees. From this it follows that some rays of light have been refracted more than others, a conclusion which has not directly been perceived by sight but must be derived with the help of geometrical construction and (of course) measurement.

That the sun's image would be circular if all the rays were equally refrangible is deduced from axioms and geometrical considerations. It cannot be observed because homogeneal (i.e.

[1] Bk I, Pt I.

monochromatic) rays are not available, at this stage, to pass through the prism. The circular image produced by the sun's light when not refracted through the prism gives no support to any hypothesis about what would occur after refraction. So, at this stage, this hypothesis is submitted to no empirical test, but is taken as established by deduction.

The fact that the refraction of homogeneal rays does not produce an elongated image (Proposition V, Theorem IV) is later demonstrated by Experiment 12; but this cannot be done until it is established that the light of any one point of the spectrum is homogeneal and the proof of this depends upon Experiment 3 where the monochromatic circle is taken as deductively necessary. So Experiment 12 gives no more than corroborative evidence contributing to a systematic body of theory all of which supports the hypothesis under investigation. Newton, in fact, states that the essential part of Proposition V, Theorem IV, has already been established by Experiment 5.[1] The purpose of this demonstration can therefore be little more than corroboration of what has already been proved.

Still more significant is the step by which Newton ensures that the light projected in this experiment is homogeneal. This follows from the results of Experiments 3 and 5. He argues that rays which are equally refrangible will form an image of the sun's disc which is circular (for reasons already noted). Therefore, if the image is elongated with semicircular ends (as is observed) this must be because the unequally refracted rays form a succession of overlapping circles, each placed according to its degree of refrangibility. At one end will be a complete circle of violet light, at the other of red, and intermediate between them a succession of overlapping circles of intermediate colours. He then argues that, as the breadth of the image is unaffected by refraction (the angle is always in the same plane as the angle of incidence), if the beam is narrowed, the diameter of the several circles will be decreased while the positions of their centres will remain unchanged. Accordingly, he reasons, the overlap between them will be decreased and the light will be proportionately more homogeneal at each point. To increase the effect to a maximum he recommends using a triangular slit as the source of the light with a base of $\frac{1}{10}$ inch and a height of an inch or more.[2] This will give a spectrum in the shape of a quadrilateral

[1] See below.

[2] See illustration of Experiment 11.

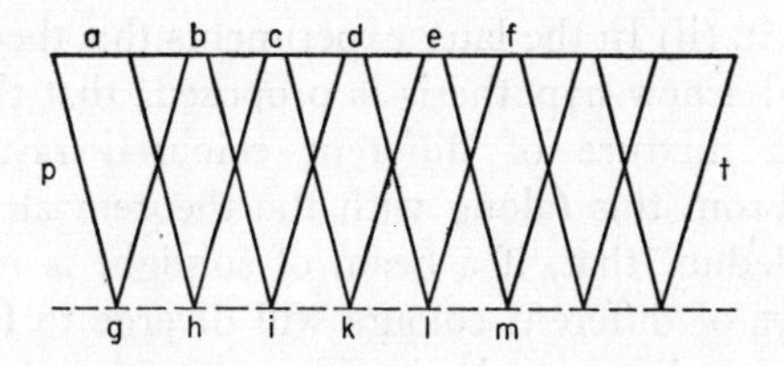

made up of overlapping triangles and having two sides parallel, the longer one corresponding to the base of the aperture and the shorter to its apex. On this shorter side the colours will be more completely separated and the overlap will be greatest along the longer.

It is to be noticed that once again the result of the experiment is not and cannot be directly observed. No monochromatic circles are visible. The gradation of colours along the spectrum is continuous, so however much its breadth is decreased there will be some degree of overlap except in the extreme case in which the imagined circles become points—and thus imperceptible. Or, in the case of the triangles, complete freedom from overlap is to be found only at the apex of each triangle on the side of the image where the illumination decreases to zero, only at the point of disappearance. The argument is no more than deduction from phenomena, for if in practice the aperture were reduced in size, a point would be reached at which the image would not be further reduced because of the effects of diffraction. The deduced result, therefore, is not observable.[1] Nevertheless, up to this point (as Newton is fully aware), the proportion of mixture is progressively reduced with the breadth of the image, but this is not observed, only deduced; and when the overlap is reduced to a certain sufficient degree he is able to experiment with light near enough, for practical purposes, to homogeneal (or monochromatic).

Defenders of the accepted logical theories might argue that the reasoning could plausibly be cast in a form which agrees with their schemata. (i) In the first experiments the hypothesis enunciated is that rays of different colours are differently refrangible. From this a lower level hypothesis is deduced to the effect that, if passed through a prism or lens, blue light will be displaced more than red light. The experiment submits this hypothesis to test

[1] In *Opticks*, Bk I, Pt I, Experiment 11, Newton uses a magnifying glass to concentrate the rays and so obtains an effect from an arrangement other than that from which the above 'deduction' was made.

and confirms it. (ii) In the later experiments this theorem is taken as proved and a new hypothesis is proposed: that the sun's light consists of a mixture of different coloured rays, differently refrangible. From this (along with the theorem already demonstrated) we deduce that, if a beam of sunlight is refracted by a prism the rays of different colours will diverge to form a many-coloured elongated image. Again this is tested and confirmed by experiment.

In reply it is to be noted, first, that Newton does not present his argument in this form. Secondly, if he did, it would still lack three features essential for conformity to currently accepted theories of scientific explanation. (i) The lower level hypothesis is not *inductively* established. The experiment, in each case, confirms it, but no resort is made or is needed to repeated observations in each of which a new confirmation is obtained. (ii) The higher level hypothesis is not derived from any inductive investigation. It is not maintained, or shown, that different coloured rays have in the past frequently been refracted to different extents, or that sunlight has in the past frequently been found to contain different coloured rays in mixture. Nor (iii) are these hypotheses purely theoretical constructs derived deductively from the definitions and axioms of the system.

It does not seem plausible to argue that it is only 'in the context of justification' that Newton cites the experimental result to confirm the hypothesis. He does not seem to argue that if the hypothesis is true the image will be elongated and multi-coloured and then experiment to see if this is the case. He observes the image and then accounts for it by deducing from the facts that the rays must have been variously refracted. Again, this form of argument is not merely suited to 'the context of discovery', for it is used by way of confirmation and proof of the hypothesis. Discovery and justification go hand in hand and one is not complete without the other. What Newton is doing is constructing a system of facts and geometrical relations which at once provides explanation of the phenomena and confirmation of hypotheses. Both are incorporated into the system in such a way that each elucidates and supports the other. The hypothesis makes clear what the fact is that is being observed (diverse refrangibility), and the observation so interpreted confirms the hypothesis. The argument is in a way circular, but only in the sense that each part of the system necessitates the rest. It is not viciously circular in the sense that to reach a conclusion one must first assume or include it in the premises. The axioms which serve as premises, do not contain the conclusion, yet unless the phenomenon (the elongated image produced after refraction) is understood as caused by refraction of the rays of sunlight along divergent paths, the conclusion does not follow.

Further the conclusion is not offered as probable but as apodeictic. Newton never suggests that because in a number of observations the spectrum of the sun is found to be elongated, its light probably consists of differently refrangible rays. He takes a single case, carefully checked for accuracy and the elimination of possible interfering effects, and from the facts that the lines followed by the rays are straight and after passing through the prism are inclined at an angle of (approximately) $2\frac{1}{2}$ degrees in the vertical plane, while those in the horizontal plane are inclined only $\frac{1}{2}$ a degree, like the rays by which the beam of sunlight is bounded before reaching the prism, he argues that the elongated image *must* be the result of the divergence of the colours after refraction. The theoretical structure requires this conclusion and without it the facts cannot be accommodated. Equally, then, the facts being what they are, the conclusion is necessary.

This sort of argument is surely what Newton means by deduction from phenomena. It is from the observed facts as interpreted in the light of the axiom system that the hypothesis is deduced and once having been deduced in this way the experiments are taken as establishing its truth.

The hypothesis that the homogeneal rays would form a circular image after refraction may be taken as (in the first instance) purely theoretical and deducible directly from the axioms, including those of geometry. Likewise the reduction of overlap by narrowing the beam is deduced. The former is used as a premise from which to conclude that the spectrum must consist of overlapping monochromatic circles, but no direct attempt is made to confirm that hypothesis by experiment. It is used to devise a means of isolating the monochromatic rays in the spectrum sufficiently to perform experiments with them demonstrating the facts that they are not spread or scattered by refraction (Proposition V, Theorem IV). The results of these experiments are then taken to corroborate those of Experiments 3 and 5, in which the earlier hypothesis is assumed.

That the elongated spectrum consists of overlapping monochromatic circles is a lower level hypothesis which is taken as established before it is submitted to experimental test. It has been deduced from the phenomena observed in the first five experiments, in conjunction with the definitions and axioms of geometry and of the theoretical system being constructed in the *Opticks*. Moreover, what explains the facts, as well as the lower level hypotheses is this deduction. The spectrum appears as an elongated succession of colours *because* each colour is produced by homogeneal rays which are differently refrangible and are thus separated by refraction into overlapping images of the source of white light. The higher level hypothesis does explain the lower and the lower the phenomenon, but Newton does not enunciate the hypotheses first, deducing the second from the first and then go on to confirm them by experiment. He experiments first and then derives the hypotheses by deduction from the phenomena, with the help of the axioms and definitions. But the deduction is not a series of formal entailments according to conventional rules of transformation, but requires a different logical theory.

This procedure is more clearly apparent in Part III of Book II. The proof of the first Proposition, that 'those superfices of transparent bodies reflect the greatest quantity of light which have the greatest refracting power', could be held to conform to Braithwaite's pattern—a deduction from axioms confirmed by observed examples. But the second Proposition is remarkable. It is really two propositions rather than one and no demonstration is offered of the second half. It states first, that 'The least parts of almost all natural bodies are in some measure transparent', and this is 'proved' by drawing attention to the fact that most substances, if pared sufficiently thin, will be seen to be somewhat transparent when held before a small hole through which a bright light is entering a dark room. This looks like a sort of induction but it has the odd appearance of a *non sequitur*. For it does not follow from the facts as described that the least parts of bodies (supposing they are particulate) are transparent. The light might pass between them, through interstices. But Newton here is not assuming that there are any interstices, for in Proposition III he sets out to prove that there must be, and uses this second Proposition as a premise. If there were no interstices and the bodies were transparent, their least parts must be transparent, and if that is so they could appear opaque only from some other cause. This is stated in the second half of Proposition II: 'the opacity of those bodies ariseth from the multitude of reflexions caused in their internal parts'. No proof is offered of this statement, for it follows from the facts given above. This, then, is no inductive inference. Objects, if very thin, are observed to be transparent. This must be due, if they are solid, to the transparency of their least parts. But when they are more bulky they are opaque; therefore this must be because of a multitude of internal reflections. These reflections are not and cannot be observed and no attempt is made at empirical confirmation. Their occurrence has been deduced from the phenomena (of transparency when the body is very thin and opacity when it is thicker).

The deduction continues in Proposition III. If internal reflections are the cause of opacity, there must be internal interfaces between substances of differing refractive density; for it has been proved, by Proposition I, that where no such interfaces exist there is no reflection. Therefore, 'Between the parts of opake and colour'd bodies are many spaces, either empty, or replenish'd with mediums of other densities'. The empirical support given to this proposition is both indirect, seeking to show that if these postulated interstices are filled up the substances become more transparent, and more direct, showing that if transparent substances are abraded or powdered they lose their transparency. Neither of these observed phenomena, however, is itself the fact to be proved. At best, they bear some analogy to it, but the actual observations listed are not favourable, or negative, instances of the general proposition. Reference is also made, at the end of the supporting argument, to the 23rd Observation in Part I in which very thin transparent substances were found to reflect light more strongly than the same substances when thicker, and this again affords no instance of the proposition to be proved.

This is not an inductive argument. It is simply the assembling of evidence from various sources, each of which in its own way points to the probability of the proposition to be proved. But the proposition itself is proved deductively from the earlier propositions, and the empirical evidence supports it just so far as from that evidence one might deduce the proposition—for example, if thinner transparent objects reflect more strongly, and a mass of discontinuous small particles is more opaque than one of more uniform consistency, it would follow that opacity results from multiple reflections between minute bodies. Take this along with Propositions I and II and you conclude to Proposition III.

Once again the deduction does not proceed according to purely formal rules of transformation. The conclusion follows from the nature of the systematic relationships between the facts which is not just the result of their incorporation into a formal deductive system derived from the definitions and Axioms. The transparency of thin bodies and the opacity of thick ones is not deduced, it is observed; but the cause of these properties, in the presence or absence of multiple internal reflections among minute disparate parts, is deduced; and it could not be deduced from the Axioms alone without the observed facts. Yet the facts have not been incorporated into the system merely as additional Axioms or by substitution of empirical terms for variables. Even if they could be, we should still be at a loss to account for the synthetic character of the conclusion, a synthetic proposition giving information about opaque bodies and their internal constitution which is not contained in any of the premises.

In like fashion, drawing upon the results of experiments and observations described earlier, Newton goes on to deduce the sizes of the least parts of bodies from their colours. He shows that the colours depend on the sizes of least parts and the interstices between them, as a consequence of the propensity of transparent bodies to reflect rays of one colour and transmit those of another according to their several thicknesses (Propositions V, VI, and VII). He does not, of course, claim precision, and says only that the sizes of the least parts may be 'conjectured'. But he arrives at his conjectures by a mathematical calculation and states the sizes of corpuscles in fractions of an inch.[1]

But perhaps the most spectacular of all deductions from phenomena is that by which, in Proposition XII, he concludes to the famous 'fits of easy transmission' and 'fits of easy reflection' of rays of light. This he does from the observations he has made of the alternate reflection and transmission of the same colour at the same angle of incidence by different thicknesses of the same substance (Observations 5, 9, 12 and 15, in Part I), and from the minute numerical correlations he has made of the thicknesses with the transmission or reflection of the various colours (in Part II). Because the alternate reflection and transmission depends on the thickness of the plate, he concludes that both surfaces are involved; because it requires both it must occur at the second (for otherwise both would not be necessary). Yet it must also depend on some effect upon the light produced at its passage through the

[1] Cf. Proposition VII.

first surface and propagated to the second 'because otherwise at the second it would not depend upon the first'. And this effect is intermittent, as is shown by the alternation of reflection and refraction at regular intervals. So, after tentatively proposing a speculative explanation for the sake of 'those that are averse to assenting to any new discoveries, but such as they can explain by an hypothesis', he concludes to the alternate disposition of rays to be easily reflected and easily transmitted. The special significance of this conclusion is that it establishes a systematic relation between the state of the rays and the thickness of the medium which suggests a periodicity in the character and behaviour of either or both, and this important result is deduced from the observations of coloured rings of reflected light, their mutual distances and that of the surfaces of the reflecting bodies.

Further detail of the arguments need not be given. What is needed is some understanding of their logical structure and character. What has been said is sufficient to show that it is not that of 'inductive inference' as usually described, nor yet that of formal linear deduction according to the current theory, nor is it a mixture of these of the form suggested by Braithwaite. Again it does not answer precisely to the hypothetico-deductive pattern, for the deduction is not drawn from the hypothesis as premise, and the conclusion is not offered for experimental testing. The experimental results are used as premises, and the conclusion, which is taken to be finally established by deduction from them, is the hypothesis or law.

As in the previous cases we have examined, we find here again a marshalling of evidence so correlated and interconnected that the conclusions drawn from it are necessitated by the interlocking parts of the system. Newton's definitions and axioms are not just initial propositions of a formal deductive system; they set up a framework of fact, some of it the fruit of observation and measurement, within which inferences can be made by pure geometrical deduction, and into which new observations can be fitted. The system enables us to interpret the new observations (the bending of light rays as refraction and reflection) and to infer, or 'deduce', from them new information about the properties of the light (whether homogeneal or heterogeneal, whether variable in refrangibility or not, how refrangibility is related to reflexibility, and so on). 'Deduction from phenomena' becomes possible just so far as the phenomena can be incorporated into such a system and one set of systematic relations points to and necessitates another. The 'rules' by which this necessitation follows are the principles of order or organization of the system (e.g. the principles of Euclidean geometry and the Axioms stating the geometrical relations of angles of refraction and reflection with the angle of incidence). But the system is not a purely formal deductive system interpreting a calculus in which empirical terms are being substituted for variables. It is difficult to see how Newton's deduction of multiple internal reflections as the explanation of opacity could be interpreted in this way.

But, it may be countered, Newton's conclusions are false. How then can they be necessitated by the observed facts? The answer to this is twofold. The conclusions are necessary within the system erected by Newton and on the grounds of the limited observations he was able to make. And they are not wholly false, but can be seen in the light of later discoveries to be at least partially true. When more elaborate observations had been made and a more comprehensive system constructed—by stages which involved intermediary systems of electrodynamics, wave and quantum mechanics—both the limitations and the insights of Newton's theories could be seen and understood. Newton's conclusions, though for long disregarded in favour of Young's and Clerk-Maxwell's, are now seen to bear essential relation to contemporary discoveries. As they stand they are a stepping-stone to the wave theory, and Newton himself hints at the possibility of waves of some sort in the medium, as well as some form of periodicity in the rays of light. In his Preface to the 1952 Dover edition of the *Opticks* Bernard Cohen writes:

'Newton's magnificent experiments, described in the text in Book Two, provided conclusive evidence that some kind of periodicity is associated with the several colours into which he divided visible light. Such periodicity can in no way be accounted for by the mechanical action of corpuscles moving in right lines and Newton was, therefore, forced into the position of having to postulate some kind of waves accompanying the corpuscles: these were the famous aether waves. . . . He suggested that the alternate "fits of easy reflection" and "fits of easy refraction" arise from the action of the aether waves which overtake the particles of light and put them into one or the other state.'[1]

[1] *Opticks* (New York, 1952), p. xl. Cf. Queries 18, 19, 28 and 29.

At the same time Newton's theory is a precursor of contemporary quantum conceptions of light, for his 'rays' follow trajectories orthogonal to surfaces corresponding to wave fronts in accordance with Jacobi's equation. As Louis de Broglie says,

'The intuition of a Newton or a Biot did not completely deceive them by making them suspect that the properties of light rays were closely related to those of the trajectories of material points in mechanics. . . . The great theorems of analytic mechanics, above all the theory of Jacobi, explain the true meaning of the laws of geometrical optics, but the optics of waves in its turn indicates the path to be followed to broaden classical mechanics and teaches us then that classical mechanics, just as geometrical optics, is only an approximation, often valid, but whose domain of application is none the less limited.'[1]

It is clear then that Newton's conclusions are not wholly false but are true within limits and it is only in the expanded conceptual system of modern physics that both the extent of their truth and their shortcomings are brought to light. Within the limited system that Newton himself set up they are true, are deduced with rigour, and may be accepted on the evidence as cogent.

iv. LAVOISIER AND COMBUSTION

In his *Opuscules Physique et Chymique*, Lavoisier investigates the role of the 'elastic fluid' fixed in metallic calxes and other substances. Guerlac[2] maintains that his interest in effervescence had been stimulated by his acquaintance with the work by Stephen Hales, though it was only later that he became more fully aware of the researches of other English pneumatic chemists. At all events he became persuaded that 'air' played a greater part in many chemical reactions than was at the time commonly believed. The researches recorded in the *Opuscules* are clearly directed towards the establishment of this fact, and the manner of his procedure is singularly enlightening to the student of scientific method. It is an *example par excellence* of experimental science and reveals several features which appear to conform to the familiar theories of induction and of hypothetico-deductive

[1] L. de Broglie, *The Revolution in Physics* (London, 1945), p. 49.
[2] *Lavoisier, The Crucial Year* (New York, 1961).

method. Nevertheless, they illustrate even better the shortcomings of these doctrines.

When he wrote *Opuscules* he had become aware of the work not only of Hales but also of Black, McBride, Cavendish and Priestley, and the first part of the book is an historical survey of pneumatic chemistry beginning with Paracelsus and discussing the work of all the major chemists since Von Helmont and Boyle. He then proceeds to describe experiments on the reactions of calcareous substances with acids and the resultant 'elastic fluid'. But it is the third part of the work to which I invite attention, dealing with the reduction of minium (the red calx of lead) and his method of demonstrating the fact that some part of atmospheric air becomes fixed in the calcination of the metals and is involved in combustion.

First he reduces the red oxide with charcoal by placing the substances in a crucible which is enclosed in an inverted glass receiver over water. The water is drawn up by suction to an assigned level in the receiver which is marked, and the substances are heated with a burning glass. He then repeats the experiment using an iron retort, into which the minium is placed with powdered charcoal and heated in a furnace, and the gas given off is collected in a tall container inverted in a trough of water. The apparatus is so arranged that the volume of air in the receiver can be measured by comparing the levels of the water, and the amounts of minium, carbon and the lead produced are carefully weighed, before and after each experiment.

The first experiment produced 14 cubic inches of 'elastic fluid' and $\frac{1}{32}$ cubic inches of lead, but not all the calx was reduced. The experiment was frequently repeated, as Lavoisier tells us, 'with different proportions', in order to discover what proportion of calx to carbon gave the best result, for in some cases he found the reduction more difficult with the consequent introduction of errors in the weights and volumes.[1] Decisive as this experiment seemed to be, he was not satisfied with it, because of limitations imposed by the apparatus (the narrowness of the focus of the burning glass, the excessive heat which sometimes broke the container), and he proceeds to the second method with the iron retort. This produced 560 cu. in. of gas and $\frac{3}{4}$ cu. in. of lead. None of the minium remained unreduced. Frequent repetitions all gave the same result.

A discrepancy between the weight of materials at the end of the

[1] See Lavoisier, *Oeuvres*, Vol. I (Paris, 1864), p. 600.

experiment and at the beginning was accounted for by the presence in and expulsion from the calx of water vapour, as moisture was observed in the neck of the retort. So a modification of the experiment was devised to enable Lavoisier to collect this moisture and weigh it, to find that it satisfactorily accounted for the deficiency.

From these experiments he concludes that a considerable quantity of elastic fluid is contained fixed in the calx and is given off during reduction. But does it possibly originate from the carbon? To test this hypothesis he heats a quantity of charcoal to a high temperature in a gun barrel, substituted for the retort in the previous experiment, but can produce no fixed air. He then tries the same process with minium alone with results similar to those of the former reduction with charcoal, but giving less gas proportionately to the amount of minium used. At this stage, however, Lavoisier overlooks this difference and does not test the resultant gas (and so fails to discover the difference in the properties of gas given off during reduction with charcoal and without). But he does mention the need to make such tests, and at a later stage he compares what results from reduction with charcoal and what is obtained from 'effervescent mixtures'. Meanwhile, he concludes that the elastic fluid is discharged not only by the minium and not by the charcoal alone, but requires both, and he has demonstrated, at least, the presence of some gas in the metallic calx, which is intimately involved in the chemical reaction.

In two instances Lavoisier records the fact that he repeated the experiment frequently, but in one case it is with different proportions of the substances and the results are not the same at each attempt, some being less satisfactory than others. This is especially instructive. Had Lavoisier been using induction by simple enumeration he should have considered these results disconfirmatory, but so far from regarding them in this light, he uses them to determine the proportions that 'constantly succeeded best' (*constamment le mieux reussi*). But by what right could he prejudge which result would be the best? The point is that he had set out to find (*a*) whether a gas was emitted and (*b*) if so in what quantities and proportions to the substances used. He was convinced that these proportions were definite and sought quantities by which they could most successfully be fixed. His repetitions of the experiment, therefore, were designed only to secure accuracy of measurement, not to seek a constant conjunction of events. He rejects the unsatisfactory cases because too much charcoal made the reduction difficult to complete, and so made the measurements uncertain. The aim is to make the experiment precise. When that is achieved the result is accepted, not because it occurs a number of times (or not only for that reason), but because the circumstances are so related to one another that the conclusion is inescapable. In this case it is the disposal of the apparatus that ensures that if there is a gas fixed in the calx which is released on reduction with charcoal, its volume *must* be revealed and measured by the difference between the levels of the water in the receiver. Estimated from an assumed specific gravity (Lavoisier takes it as similar to air), its weight can then be deduced and, by weighing the solid substances before and after treatment, the proportions can be worked out. Repetition will increase accuracy but the main fact is established by the structure of the experiment.

In the second case of repetition the result is always the same and so this does conform to the requirements of induction by enumeration. But Lavoisier does not argue that because the same results recur they always will, but that, since they have been maintained in several repetitions, the quantities must be the right ones.[1] Moreover, his point is not simply to establish the fact that a gas is always emitted during the reduction of the calx, but that an elastic fluid (gas) taken from the atmosphere is combined with the metal in the calx and for that purpose these experiments do not suffice. As he himself says when the whole series of experiments have been described: 'Though the experiments I have reported may appear, in this respect, of such a nature as to leave no doubt, it must be confessed, nevertheless, that one only reaches conviction in physics so far as one arrives at the same point by different routes.'[2] And like the other scientists whose procedure we have examined he sets out to assemble corroborative evidence from different quarters.

He turns next to the opposite process. If elastic fluid can be released from calx, we should be able to find evidence that it is absorbed during calcination. He uses apparatus similar to that of the first experiment, placing lead in a stone crucible on a stand

[1] It is, of course, assumed that as the same results are obtained several times with these quantities they always will be, but this is far from being the central issue.

[2] See Lavoisier, *Oeuvres*, Vol. I (Paris, 1864), p. 614.

in a trough of water and enclosing it in an inverted glass receiver. With a burning glass he then attempts to calcine the lead. Similar attempts, with suitably varied apparatus, are made with tin, with an alloy of tin and lead, and with iron filings in distilled water. The variation of the apparatus, again, is made in pursuit of better and more accurate results. In each case the air in the receiver is reduced in volume, some calx is procured but the calcination cannot be continued as long (so he asserts) as in open air, and ceases after a relatively short time, despite continued heating of the metal. The amount of air absorbed, in each case, is found to be proportional to the gain in weight of the calx. On this occasion Lavoisier tests the residual air and finds it supports combustion only indifferently and produces slight turbidity in lime water. But these results do not yet suggest any firm conclusion. Nevertheless, important progress has been made. It is clear that some part of the air is fixed in the calx and, more important, accounts for its increase in weight over the metal. It is also clear that the amount of calcination obtained is limited by the supply of air enclosed in the receiver and Lavoisier infers that if the vessel were exhausted of air the calcination would not occur (or if it did, would occasion no increase in weight).[1] This is new evidence from a new source pointing in the direction of the truth of his hypothesis.

The next step of the demonstration is to submit to tests the gas given off from metallic reduction and to compare it with that obtained from effervescence. Lavoisier uses the now familiar tests for carbon dioxide and finds that the gas obtained in both cases reacts similarly. As effervescence was recognized as chemical reaction, the force of this result is to identify the elastic fluid produced in metallic reduction as chemically involved, and not simply as an assisting principle, a medium or a menstruum, the view hitherto held by most other scientists. Beyond that it does not contribute very much at this stage, though it becomes of great importance later when Lavoisier has found new evidence enabling him to identify the gas which actually combines with metals in calcination.

More significant here are his experiments with phosphorus. Using, once more, apparatus similar to that in the first reduction experiment, he burns 8 grains of phosphorus in a crucible confined

[1] Cf. Lavoisier, *Oeuvres*, Vol. I, p. 621. At a later stage he confirms this inference experimentally so far as it applies to the combustion of phosphorus and sulphur.

in a receiver inverted over water. Between $\frac{1}{5}$ and $\frac{1}{6}$ of the contained air is absorbed. When mercury is used instead of water the absorption is slightly less but the solid product does not deliquesce as in the first attempt. When the quantity of phosphorus is increased but not the available air, the amount burnt does not increase, though the remainder is sublimated by the heat of the burning-glass without combustion. Once more, Lavoisier repeats the experiment a number of times and records the same result each time. Then, after failing to ignite the surplus phosphorus in the confined air, he allows the apparatus to cool, opens the receiver and lets in more air and is able to reignite the phosphorus, which is again extinguished after a short period, if the receiver is replaced. This process is repeated several times.

Again the repetition serves a purpose other than that of accumulating like instances. It serves not only as a check on accuracy of measurements, but, in this instance, it is accompanied by a variation in the experiment (the opening of the container to admit new air), and though this is done several times and the result is similar each time it is no mere repetition, for it shows that as the supply of air is successively renewed so the amount of phosphorus burned can be proportionately increased.

Even so, Lavoisier is not finally convinced that the experiment proves some part of the air to be combined with the phosphorus in combustion. He asserts that further experiment is needed to show that a combination of any kind really has been formed.

First he repeats his former experiment and finds the precise increase in weight of the phosphorus consumed in forming phosphoric acid. He then proceeds to show that this augmentation in weight cannot be due to the absorption of water. He does so by a most ingenious device. He collects the flowers of phosphorus produced from the combustion and dissolves them in distilled water with which he fills a bottle. This is weighed. It is then emptied and cleaned, refilled with pure distilled water, and re-weighed. The difference in weight is found to be 3 drachms 27½ grains. But the amount of phosphorus burned was found to be only 2 drachms 10 grains. Hence there must be in the solution some other substance besides the phosphorus and the water to account for the remaining 1 drachm 17 grains. The demonstration is conclusive.

Finally Lavoisier attempts to burn phosphorus, and then sulphur, and to explode gunpowder, in an exhausted container, and finds this impossible. So he has shown (*a*) that elastic fluid (similar to that obtained from effervescence) is given off during metallic reduction; (*b*) that some part of the air is absorbed during calcination which accounts for the augmentation in weight of the calx; (*c*) that calcination is incomplete in a restricted supply of air; (*d*) that 'air' is absorbed in considerable quantity during the combustion of phosphorus; (*e*) that no combustion occurs *in vacuo*.

This is all evidence converging towards the final conclusion that in combustion the burning substance combines with some special variety of gas taken from the atmosphere. But we have not yet reached the point at which we can say, with Harvey, 'it cannot be otherwise'. To bring Lavoisier to that point he had still to perform his famous experiments with mercury and its red oxide, and test the gases which remained after calcination and which emerged during reduction.

The first of these experiments are described in the Memoir to the Royal Academy of April 26, 1775 (published in August 1778). They were conducted in essentially the same way as those on minium and with similar apparatus. First he reduced the mercuric oxide with charcoal in a retort, collecting the emitted gas over water. It responded positively to the tests for 'mephitic' or 'fixed air' (CO_2), and so assured him that the substance was a true calx of mercury. The experiment also reconfirmed the hypothesis that the gas produced by metallic reduction with charcoal is 'fixed air', but, as Lavoisier here asserts with confidence, this air is 'not a simple being, but is, in some degree, a combination' of gas issuing from the metal and something derived from the charcoal. To discover what gas, if any, is united with the metal in the calx it is necessary to reduce the latter without charcoal. This he does with the same apparatus and records that it was more difficult and required a greater degree of heat.

The reverse process was achieved in the experiment described in the *Traité Élémentaire de Chimie* (published 1789),[1] in which mercury was heated in a retort to near boiling point until a quantity of oxide was formed on its surface. The retort was connected to a bell-jar inverted over mercury in which the volume of air before and after the experiment was measured. The calx was then removed and the earlier reduction experiment performed upon it. The weights of the substances were taken

[1] Lavoisier, *Oeuvres*, Vol. I, pp. 36 ff.

before and after each experiment and the volumes of gas emitted or absorbed. It was found that on calcination the mercury absorbed precisely the same amount of gas as the calx emitted when reduced. The gain in weight in calcination corresponded to the calculated weight of the air absorbed, and *vice versa* on reduction. The 'air' remaining in the receiver after calcination would not support combustion or respiration (as in the case of minium) and that which was collected from the reduction supported combustion better than atmospheric air, was highly 'respirable', was (unlike fixed air) insoluble in water and produced no precipitate in lime water.

The case is now proved. Metals on calcination combine with the respirable part of the atmosphere, which is not the fixed air of earlier experiments. When that is given off in metallic reductions it is, in part at least, due to the use of charcoal. Further tests with combustible materials showed later that the new gas, which Priestley had already prepared in similar fashion, was involved in all combustion, and the chemist was well on his way to the discovery of the composition of the atmosphere, as well as of water.

What has revealed these facts has been the way in which the evidence from various sources dovetailed together to form a close-knit system. The results of the experiments on phosphorus and sulphur, with those on minium, and these again with those on mercury; the relation of the tests of gases produced when metallic calx is reduced with charcoal and when it is reduced without charcoal. The agreement and corroboration of the quantities, of weight gained with gas absorbed or weight lost with gas exuded. Now one might well declare that it could not be otherwise. Let us seek for clues to the source of this assurance and cogency, and a satisfactory theory of scientific method.

The order of this exposition of Lavoisier's researches is probably not the same as the chronological order in which he performed the experiments. Those on phosphorus may (as Guerlac believes) have preceded those on minium. But in all this, discovery is so dependent on justification and justification so closely bound up with discovery that they are obviously only two sides of the same coin. What suggests the theory is what the experiments (his own or others') reveal; what constitutes the discovery is the mutual confirmation of experimental results. We may, therefore, without misgiving, look to this feature of his procedure for the principles of scientific method.

First, we have noted aspects of Lavoisier's procedure which give some support to the idea of induction by simple enumeration, but which also show that this is not the nature of his thinking. Clearly he often presupposes that certain conjunctions which have been commonly observed are general and invariable—for instance, that the commonly performed preparation of metallic calxes proceeds more easily in the open air than in the closed vessels with which he was experimenting. This he simply takes for granted. It is not part of what he is trying to demonstrate (though it bears upon it), for what he seeks to establish is the quantitative relation between the available air and the augmentation in weight of the calx, which will finally clinch his argument for chemical combination, and (as we shall see anon) his refutation of the Phlogiston Theory.

The occurrence of constant conjunctions contributed only indirectly to that purpose which was achieved only when a new conceptual system had been evolved. The function actually served by his observation of frequent conjunctions was different. His study of the earlier results of other chemists as well as his own observations impressed upon him the constant association of effervescence with various chemical changes, but this does not lead at once to a generalized conclusion. It stimulates interest and curiosity and sparks the demand for explanation. It is a preliminary phase of scientific investigation—the first fruit of exploration set in movement (as we have already seen and shall shortly more fully discover) by conflicts and contradictions in earlier theories.

The use of hypothetico-deductive method is apparent in Lavoisier's reasoning on the combustion of phosphorus. He conceives the alternative hypothesis that the increase of weight is due to the absorption of moisture from the air. 'This opinion was probable and presented itself with an appearance of truth apt to seduce.'[1] Notice, however, that it does not present itself *ex nihilo*. It is suggested by the experiment in which the phosphorus burned over water showed a slightly greater absorption of air than when burned over mercury. If the hypothesis is true, three consequences should follow:

(i) that it should be possible to prolong the process of combustion by restoring to the air confined under the receiver a quantity of water equal to that absorbed in the previous experiment;

[1] Lavoisier, *Oeuvres*, Vol. I, p. 645.

(ii) that there should then be no further diminution in the quantity of air in which the phosphorus burns;

(iii) that the diminution of about $\frac{1}{5}$ of its volume suffered by the confined air when the phosphorus is burned in it should be restored if a similar quantity of water vapour is introduced.

These three consequences deduced from the hypothesis are then tested by experiment. But the deduction is not purely analytic, the conclusions are not tautologically related to the premises, and they give new factual information. If they did not they could not be empirically tested.

The argument seems to run somewhat as follows: if water vapour supports combustion, burning phosphorus will probably absorb it. A greater supply of water vapour should then support combustion for longer time. This is obvious enough yet it does give us some new information. Combustion in a confined space has been found to reduce the volume of air contained in it by one-fifth; but it does not follow analytically that if the combustion is supported by water vapour this reduction must be due to its absorption. Need dry air be less voluminous than saturated air? If it is true that the addition of water vapour increases a volume of air and that combustion reduces it, the rest should follow. But Lavoisier, in his description of the ensuing experiment altogether overlooks the effect of the water vapour on the volume. He places two crucibles under his receiver, one containing water and the other phosphorus, and he uses the burning glass first to boil the water—but he reports no augmentation of the volume of air as a result. He then ignites the phosphorus and reports that everything happened as before: the same quantity of air was absorbed and the combustion ceased. The sole difference was in the liquid state of the products, which in the previous experiment (performed over mercury) had been dry and flocculent. Yet Lavoisier concludes that this experiment (along with others) disproves the hypothesis.[1] The final experiment, in which he compares the weight of the dissolved products of the combustion with an equal weight of water and finds that the excess is greater than the weight of the phosphorus burned, is, as we saw, highly ingenious. Yet he has not used any calculus, or anything equivalent, to deduce the consequences of the hypothesis that he wishes to test, and the validity of his reasoning does not depend solely on the

[1] Cf. Lavoisier, *Oeuvres*, pp. 647 ff.

laws of arithmetic. The conclusions follow from the real or assumed structure of the chemical and physical relations between the substances concerned, and it is only their quantitative aspect which submits to the arithmetical calculus.

This indeed seems to be the operative factor throughout. How does one know that the volume of air in the bell-jar changes? Obviously by observing the change in the level of the water or mercury. But this will not hold good if the bell-jar is not air-tight and what is 'observed' depends on a number of tacit assumptions of the same kind. How do we know that there is any connexion between this change in volume and the reaction taking place in the retort or the crucible? Because they are constantly conjoined? Hardly so, for they are not conjoined if the retort is not connected with the receiver nor the crucible included in the container, even though the reaction may still take place. That combustion takes something from the air or that metallic reduction exhales something into it cannot be directly seen. But if the apparatus is suitably ordered we can infer that changes in the level of the liquid indicate changes in the volume of air and that these are produced by the chemical reaction taking place in a retort. And unless we could draw freely upon a whole body of previously established fact, we should not expect that any reaction would take place at all. To urge once again that this body of established fact involved inductive inference would only be to raise the question afresh why one set of facts should be established in one way and another set of facts in an entirely different way. For the inference actually made is neither inductive nor formally analytic.

Observation, however evident its object, is not scientific observation unless it is implicitly informed by prior knowledge. Unless a great deal is known and understood about what is going on, very little, if anything relevant, can be observed. Inference from past experience as well as from the presented and designed arrangement of the data is constantly in progress and conditions every observed result. It is appreciation of the correlation of the facts, and deduction from the phenomena, which enable the scientist to reach his conclusions and this process is only a continuation of that which is already at work in observation. We need not, therefore, be surprised to find that no scientific discovery is made *in vacuo*. If what is observed itself always depends on what is already known, it is inevitable that every investigation should begin from and draw copiously upon theories already in existence

—even if these are not formulated as theories and are just assumed as common knowledge. It is, then, perfectly natural to find Lavoisier beginning with a review of discoveries already made and opinions already aired. He draws upon these freely even when he disagrees with them and, as we shall see later, is in debt to them even when he rejects them as untenable. Equally it is to be expected that the scientist's object will be to seek evidences which mutually support one another, though different in kind and found in different settings. It is the structure of the body of fact and its principles of organization which secure his conclusion. If the facts are not thus organized the theory is not scientific.

iv. DARWIN'S DEFENCE OF THE EVOLUTION HYPOTHESIS

'The whole of this volume is one long argument', writes Darwin in the last chapter of *The Origin of the Species*, and it is an argument famous for the wealth of observed factual information which is brought to bear upon the hypothesis it is designed to establish. It should therefore be an excellent illustration of the bearing of observation upon theory in scientific reasoning. But Darwin's undertaking is in several notable ways different from the examples we have so far considered. He is not attempting to establish the pattern of movement of a single object (as Kepler does with the orbit of Mars), or the character of a single process (as Harvey does in physiology); he is not examining special effects like the refraction and reflection of light (as Newton is doing in the *Opticks*). He is striving to establish the truth of an hypothesis which affects the scientific approach to a whole body of sciences, ranging from paleontology to psychology, and which was destined to have repercussions upon the entire philosophical outlook of the succeeding age. Further, Darwin's hypothesis is in large measure speculative. Evidence for it is, in the nature of the case, virtually impossible to observe directly, and it can therefore be established, if at all, only by elaborate inference; for it concerns processes which have taken place in the very distant past and have continued over long ages. For this reason they could not be observed actually occurring and it can only be by inference that the observed objects and events alleged to result from them can be recognized as their effects. Our object will be to investigate the form of this reasoning as it is made and stated by Darwin and to compare it with the theories of scientific inference criticized in Part I above. Finally, Darwin's treatise is not, as he says, just one long argument, but a constellation of arguments from different though related scientific fields, each bearing from a special angle upon the central issue.

The general ground plan of Darwin's exposition is as follows:

(i) (*a*) He notes the occurrence and extent of variation in inherited characteristics and speculates as to its cause. (*b*) He then draws attention to the effects, on breeds of domestic animals, of selection by man, either deliberate for the production of animals with special capacities, or unconscious merely to improve the stock.

(ii) (*a*) He draws attention to the difficulty of distinguishing between varieties within larger species and species within a genus; and to the facts of greater variation of species within larger genera and of greater numbers of varieties in the larger species. (*b*) Again he speculates on the possible causes of such variation.

(iii) He presents the famous deduction from the geometrical increase of living progeny among living beings and the consequent physical impossibility of their all surviving, to the inevitable struggle for existence. He elaborates the arguments by describing the numerous and complex conditions of survival and the way they operate.

(iv) From this he concludes to the survival of the fittest and deduces the occurrence and mode of operation of natural selection.

(v) He then considers at length objections to, and difficulties of, the theory of evolution by spontaneous variation and natural selection and suggests how they may be met.

(vi) In the course of this treatment of objections he reviews and marshalls one body of evidence drawn from the geological record and another from the geographical distribution of species.

(vii) Finally, he discusses the evidence of common ancestry provided by morphology, embryology and the classification of living forms into groups under groups.

Our purpose does not require, nor does space permit, a detailed repetition of all the minute exemplifications and arguments that Darwin presents. I shall select only a few main phases of the reasoning and consider its form and character. No other example of scientific reasoning could be more confidently expected to illustrate the traditional doctrine of inductive inference, yet nowhere throughout Darwin's discussion is there any argument obviously drawing a general conclusion from the mere repetition of conjoined phenomena. He argues from analogy but the structure of the reasoning is not that usually accredited as inductive. Men select their domestic birds and animals for breeding, choosing as parents those with the special characters they desire. This results in the development of well-marked breeds or varieties, which, if found in nature, would be classed by many naturalists as distinct species. This is matter of experience. So, Darwin argues, some similar selective agency in nature could produce well-defined species by concentrating on those characters most favourable to survival. This is no argument: 'AB frequently, therefore AB always'; but more accurately 'AB in one set of circumstances, therefore possibly A′B′ in another set of circumstances'. But more important, the basis of the inference is not frequent repetition of a conjunction but the fact of the conjunction's occurring at all. Human selection is the cause of variety in domestic breeds, therefore the cause of wild varieties and species may also be selection. If anything, the principle of the argument is 'Same effect, same cause (and *vice versa*)' and this goes well beyond any assumption of universality from constant conjunction. Of course, it is (at least *prima facie*) inductive reasoning to argue that, because it is commonly found to be the case that parents produce offspring which take after them, yet do not resemble them exactly, therefore we may expect this to happen universally. But this is not Darwin's point. He presupposes this universal fact quite explicitly; and combines it with a second presupposed inductive generalization that selection for a particular character over a number of successive generations will progressively modify the phenotype. Together these assumptions form a sort of major premise. The minor is then that in domestication man selects for particular characters, and it follows that the source of domestic varieties is human selection (deliberate or unconscious). The reasoning is syllogistic and deductive. But the crux of Darwin's argument is that, because this is known to be the case among domestic animals, there is no reason to suppose that among wild animals the cause of specific differences is not similar. This is analogical reasoning.

The argument from analogy is then buttressed and underpinned by a remarkable piece of deduction which is still rooted in and dependent upon experienced facts and could well be described as deduction from phenomena. The deductive reasoning runs:

(i) Living creatures reproduce in geometric progression.
(ii) After several generations their numbers will become exceedingly large.
(iii) The available supplies of food and living space are limited.
(iv) Therefore the multiplication of progeny must rapidly exhaust both.
(v) The average population of species tends to remain stable or to change very slowly.
(vi) Therefore many of the offspring of each generation must die before they can reproduce.
(vii) There is consequent competition for survival.
(viii) Any individuals which vary from their parents so as to have an advantage for survival will be favoured in the struggle for existence.
(ix) There is a natural selection of better adapted forms.

Propositions i, iii, and v are empirical in the sense that they are known from observation—though it might almost be claimed that iii could be presumed *a priori* under terrestrial conditions. Propositions ii and iv follow from i and iii by simple arithmetic. vi follows deductively either (and independently) from i and iv, or from i and v. vii follows deductively from ii and iv and from ii, v and vi. On the assumption that what is meant by 'advantage for survival' is that individuals having advantage are more likely to prevail, viii is tautological, and ix follows from vii and viii by deduction.

The facts, or phenomena, from which the conclusion is deduced are the observed geometric multiplication of progeny, the relative stability of populations and the limitation of natural resources. From these the struggle for existence and natural selection follow deductively on the one further assumption of what Darwin describes as spontaneous variation.

This is Darwin's main hypothesis—that the origin of species is the struggle for existence and the survival of the fittest. As a corollary it follows that the cause of adaptation is natural selection and accumulation of the more advantageous variations. He proceeds in large measure by a form of hypothetico-deductive method, deducing from these hypotheses what we should expect to find in nature. In some cases he shows that such predictions are, by and large, fulfilled; but much the greater part of the book is devoted to explaining the apparent failure of expectations which should follow from the theory. In the former class of cases the hypothetico-deductive method is exemplified—and I make no attempt to deny the use of arguments of this form in science, though reservations about the nature of deduction may still be entertained. But in cases where Darwin defends his theory against objections and faces up to the numerous phenomena which present difficulties for his view, the paradigm of hypothetico-deductive procedure is violated.

Hypothetico-deductive argument is exemplified wherever Darwin writes 'we should expect to find, and we do find . . .', and in such passages as the following:

'The existence of closely allied or representative species in any two areas, implied, on the theory of descent with modification, that the same parent-forms formerly inhabited both areas: and we almost invariably find that wherever many closely allied species inhabit two areas, some identical species are still common to both. Wherever many closely allied yet distinct species occur, doubtful forms and varieties belonging to the same groups likewise occur. It is a rule of high generality that the inhabitants of each area are related to the inhabitants of the nearest source whence immigrants might have been derived. We see this in the striking relation of nearly all the plants and animals of the Galapagos archipelago, of Juan Fernandoz, and of other American islands, to the plants and animals of the neighbouring American mainland; and of those of the Cape de Verde archipelago, and of the other African islands to the African mainland.'[1]

But in many cases Darwin himself admits that the expectations which follow naturally from his hypotheses are not borne out in experience and his arguments are designed to explain away these formidable, if (as he alleges) only apparent, exceptions. For instance, varieties can be intercrossed and the hybrids are usually more vigorous than their parents, but specific forms can seldom be crossed and, when they are, the offspring are generally sterile. If specifically different individuals had descended from common parents this should, surely, not be the case, and if it is the case, it

[1] *The Origin of Species* (London, 1929), Ch. XV, pp. 398–9.

should be evidence against the theory of common descent. Again, if different species have evolved from a common progenitor by the accumulation of slight variations, there should be fossil evidence of the whole range of intermediate forms, and even existing species should shade off into one another by small gradations. But neither was there in Darwin's day a complete graduated series of fossil remains (and even today it is far from complete) nor is it the case that existing species commonly present a gradual scale of differences. More frequently species are sharply distinct and no intervening forms are to be found.

Darwin's arguments to meet many of these objections are hardly inductive. The absence of intermediate forms among living species he explains by contending that, as the varieties which survive are those which have selective advantage, those from which they have varied would be at a disadvantage in the struggle for existence, and so would rapidly become extinct. This process cannot, in the nature of the case, be observed, because it is alleged to have happened in the distant past and the present evidence for it is the *absence* of the allegedly extinct species. It is hardly a valid inductive argument (according to the commonly accepted logical doctrine) that because AB is never observed AB must have existed in the past. Nor is Darwin's argument hypothetico-deductive in the approved form. Certainly he deduces the extinction of intermediate species from his hypothesis, but the actual occurrence of such extinction cannot be confirmed by observation and no attempt is made to confirm it by such means. The observed absence of certain specific forms—or the non-observance of them—is not, as has been said, evidence for their former existence. And further, the objection which Darwin is trying to meet—that if species were descended from a common source there would be intermediate forms—is also a deduction from the hypothesis, which, however, is disconfirmed by observation.

If the current doctrines were correct we should have to say that Darwin's theory was falsified by the empirical evidence and that it should be rejected as a scientific theory; and if Sir Karl Popper's view were correct, that one contrary instance is a sufficient refutation, that should have been the end of the matter for evolution as a scientific hypothesis. But this is so far from being the case that Darwin's arguments launched the theory upon a long, distinguished and triumphal career of scientific acclamation.

Much the same is true of Darwin's reasoning against other very formidable objections which he faced. The gaps in the geological record, he asserted, were due to the fact that sedimentary rocks were laid down chiefly in periods of subsidence when there was much extinction of species, whereas periods of elevation when little sedimentation occurred, were those in which much modification of species probably took place. These might be conclusions 'deduced' from the observable geological phenomena (along with other assumptions derived from Darwin's own biological theories) but they can hardly be claimed as valid inductive inferences, for no cases have or could have been observed, and the hypothetico-deductive procedure breaks down in the same way as before.

Of course, Darwin's arguments are not of the same character in every instance, but perhaps sufficient has been said to show that many of them would be difficult to accommodate under rubrics commonly approved by contemporary philosophers of science. Without labouring this point unduly, let us examine his procedures in our own search of an alternative theory. The hypothesis of evolution has now become so widely accepted that most biologists would claim that it was established fact. What sort of reasoning, we must ask, produces this sort of scientific conviction? It should by now be obvious that it is not induction of the traditional empiricist variety nor a hypothetico-deductive procedure of the kind hitherto set forth.

Perhaps the most noticeable feature of Darwin's arguments is that very few of them can be supported by direct observation. He does not, for the most part, contend that the processes he describes do or have occurred, so much as that they could have occurred, and that his hypotheses are not impossible. His defence of them is on the ground, mainly, that they provide a better explanation of the known facts than any other hypothesis. First let us notice the speculative character of the argument.

Analogical argument from the effects of domestic selection to those of natural selection proves, if anything, only that species could have arisen in this way, not that they did. The reasoning supporting the idea of natural selection itself is, as we have seen, largely deductive. So is much of the argument for its corollaries; for instance, that rarity tends to extinction (except where the species are few and isolated), that competition is most acute between closely allied species, so that, in the course of time, divergence tends to increase and intermediate forms to be eliminated. So he builds up his theory. But when he comes to defend it against objections he has to account for the observed existence of unlikely forms and for the peculiarly adaptive instincts of numerous species. In every case his evidence shows that the phenomena to be explained *could* have evolved as a result of natural selection, not that they did.

That some terrestrial animals could have become aquatic (and *vice versa*) by slow degrees, accumulating small favourable modifications of structure and habit, is supported by pointing to cases of birds and animals closely related to others with modified habitats, or ways of living, or both. Structures adapted to one way of life are shown to be possessed by varieties pursuing a different way of life. Again, wherever possible, Darwin points to intermediate forms. The evolution of organs of extreme complication and adaptive perfection (like the mammalian eye) is shown to be possible by pointing to a range of forms from pigment-spot to refracting lense. The persistence, on the other hand of unimportant seeming organs (like the giraffe's tail) is explained by seeking possible advantages which they might have given in the past or might still afford. The occurrence of varied, including sterile, varieties is given as evidence of the possibility in hymenopterous insects of neuter forms occurring which might then, because of gradually modifying habits, prove advantageous to the society. Analogous instances are sought (and found) of sterile varieties being produced by fertile forms the production of which can be (or in domestication has been) increased by selecting the fertile parents which most tend to produce them. The distribution of freshwater species over widely disparate regions is explained by the possibility of relatively obscure means of transportation (adherence to the feet and beaks of birds, wind, possibility of survival in salt water, etc., as evidenced by rare cases). The distribution of alpine species over widely separated ranges in the northern hemisphere is explained by the widespread arctic conditions obtaining in previous glacial periods and their gradual recession (leaving 'islands' of arctic species on the colder mountain tops). Other unexpected distributions are explained by an alleged alternation of glacial periods in the north and south.

All these hypotheses are supported by some empirical evidence, but none of it is direct observation, and the main hypothesis of evolution is not proved by it but is merely rendered feasible. Because seeds and spawn are sometimes carried over long distances adhering to the feet of waders, freshwater species could have been distributed in this way; because some seeds are found to withstand the effects of sea water for many days and to remain fertile, they could have reached off-shore islands adhering to driftwood; because there is a species of bee (Melipona) that makes cylindrical and spherical cells, the hexagonal cell of the hive bee could have evolved gradually by natural selection of slowly modifying instinct. There is of course much more relevant and interesting detail to these examples, but it does not amount to conclusive evidence in any one case, though it is evidence that (to use Bertrand Russell's phrase in a different context) 'points towards' the hypothesis to be proved.

So Darwin is able to claim with plausibility that evolution by natural selection is possible, and that the difficulties presented by the more complex phenomena are none of them wholly insuperable, because a process is conceivable in each case by which natural selection might have produced them. The overriding claim is that evolution is a better explanation of the relations between and occurrence of biological species than is special creation. If the latter had occurred, why should there be no Batrachians and terrestrial mammals on oceanic islands? Why should the species of plants and birds on such islands always be related most closely to those on the mainland from which they could most easily have migrated? Why, again, are species on islands separated by shallow channels more closely allied than those on islands separated by deep channels? Why should specially created species inhabiting widely separated mountain heights be alike? The stubborn opponent might simply reply, why not? But the scientist feels that a hypothesis like special creation, just because it is amenable to different and even seemingly contradictory examples is too arbitrary to serve as an explanation at all—and this gives us a clue to the nature of scientific explanation.

To explain is not simply to deduce from a higher level hypothesis. If it were, special creation would be a sufficient explanation of the forms of occurrence, distribution and sequence of species. But it would explain equally well any other forms, occurrence, distribution and sequence, and would therefore be a most unsatisfactory and inadequate explanation. In short, it could not explain why things are as they are *and not otherwise*. An adequate explanation is one that would exclude contradictory alternatives. The phenomena must not only follow from the hypothesis, but must follow as other possibilities do not. The explanatory theory must relate the phenomena in a systematic fashion so that they are mutually necessary, so that (as Copernicus insisted) if some facts were other than they are all the others would have to be different. When we can see how phenomena fit into such a system we find them intelligible and accept the theory setting out the system as an adequate explanation. And when this is the case we can appreciate the sense in which the hypothesis is deducible from the phenomena as well as the phenomena from the hypothesis.

Darwin's 'theory' to a marked extent satisfies these requirements and this is the secret of its success. He ranges over a vast area of facts, taxonomic, anatomical, embryological, ecological, paleontological, geological, geographical, which at first sight have little bearing on one another, and his theory provides a principle of organization which relates them systematically and enables us to understand the occurrence of some in the light of their relation to the others. So climate and temperature-range determining living-conditions are related to living form, physiological functioning and habit; the interdependence of species—flowering plants upon insects for fertilization, insects and herbivores upon plants for food, carnivores upon other species for prey, and thousands of other detailed and complex interdependences—is related to the struggle for existence, which is again shown to bear upon the adaptive character of specific forms (protective colouring, mutual simulation, developed organs of defence, etc.). Geological changes are related to geographical distribution and that again to known and possible means of transportation. The gradual modification of species is related both to the contemporaneous affinities of living forms and the successive series revealed in fossil remains. These last are related to the movements of the earth's crust and the way in which sedimentary rocks have been deposited. The process of heredity is related to morphology and the homology of organs, as well as to embryogenesis and the stages in the life history of the organism at which acquired modifications develop; and all this is reflected in taxonomy and the classification of living forms in 'groups under groups' according to a 'natural system'.[1] The evidence from each class of facts dovetails into that of every other—but only on the assumption that species have evolved by descent, with gradual modification and diversification, from a common progenitor. It is the way in which Darwin succeeds in

[1] Cf. *The Origin of Species*, Ch. XIV, pp. 351 ff.

relating all these varied classes of fact so that they bear upon and illuminate one another which gives his theory its overwhelming persuasive force and its explanatory value.

As remarked above no one of his arguments taken alone is conclusive,[1] and many prove only the possibility of his thesis rather than its truth. But the enormous variety of his evidence and the fact that it points towards the truth of his hypothesis from so many different angles, makes the thesis difficult to reject without denying a multiplicity of established facts. His procedure is similar to that of a map-reader fixing the position of a feature in the landscape by taking bearings upon it from a number of different directions. The convergence of many lines from a wide variety of compass points re-enforces the accuracy of the result. That Darwin himself realized this to be the typical form of scientific reasoning is evident from the following passage:

'It can hardly be supposed that a false theory would explain, in so satisfactory a manner as does the theory of natural selection, the several large classes of facts above specified. It has recently been objected that this is an unsafe method of arguing, but it is a method used in judging of the common events of life, and has often been used by the greatest natural philosophers. The undulatory theory of light has thus been arrived at; and the belief of the revolution of the earth on its own axis was until lately supported by hardly any direct evidence.'[2]

In short, scientific procedure is not generalization from directly observed occurrences, but is the construction of a system of interrelated facts and ideas built up by a progressive organization of experience, by reference to which explanatory hypotheses are deduced, facts are rendered intelligible and events may be predicted.

One point is to be repeated and stressed in concluding our examination of Darwin's exposition of his theory. The vast range of facts that have been but partially reviewed above would not, to the casual and uninstructed observer, reveal any systematic order. Fossil remains have no obvious bearing on the homology of

[1] The argument from the fact that more progeny are born than can possibly survive to the struggle for existence may be considered conclusive, but it is not sufficient to establish the theory of evolution by natural selection.

[2] *The Origin of Species*, p. 401.

organs; the periodic occurrence of glacial periods have no apparent relevance to the affinity of alpine with Himalayan plant life, and the alternation of ice ages north and south of the equator does not immediately link itself in the imagination with the modern occurrence of temperate species on equatorial highlands. These innumerable facts from different spheres of scientific interest, and the vast variety of observations which are made, not only by scientists but by all and sundry the world over, do not automatically fall into an orderly structure of mutual relationships. But Darwin is able to fit them together in systematic fashion by means of his evolutionary hypothesis. The main function performed by the theory is that of organization. It structures what would otherwise be a congeries of bewildering and chaotic factual appearances, giving us an intelligible order of reciprocally conditioning states of affairs.

vi. THE DISCOVERY OF THE POSITRON

Any contemporary example of scientific discovery of necessity presupposes a vast structure of background knowledge, and the point on which stress has hitherto been laid, that the discoverer never begins with a *tabula rasa*, is here too obvious to labour. C. D. Anderson's discovery of the positron was built on to and out of enormous monuments of scientific discovery to which J. J. Thomson, Lord Rutherford, Wilson, Planck and de Broglie, Schrödinger, Bohr, Millikan and Dirac are only some of the contributors. Even were I competent to do so, space would not permit my attempting to review the course of development in atomic physics associated with these names. The work of any one of them would serve to provide us with examples of scientific method from which we could derive an epistemological theory, but the reasoning which supported Anderson's discovery is especially interesting in its clarity, its concise elegance, and as an instance of elimination of hypotheses.

The actual observation which proved crucial in this instance was no more than a curved streak on the photograph of a Wilson chamber divided by a 6 mm. lead plate. We may set aside the fact that before anybody looking at such an object could derive any scientific conclusions from it, he must know at least that it is a photograph and that the dark band across its middle represents the thickness of the lead plate. But even an observer who knew this would make nothing of it unless he understood the purpose and the working of the Wilson chamber, and that by itself would avail him little without some understanding of all the complex theory of atomic physics which led up to the research on cosmic rays that Millikan and Anderson undertook. Before any significant observation could begin it had to be informed by, and read in the light of, all this complicated theory, and much more besides. The apparatus (designed by Millikan and Anderson) consisted of a vertical Wilson chamber in a magnetic field. The lead plate was placed in the chamber to serve as an obstruction to fast particles and to reveal their power of penetration. The streak in the photograph was the track of a particle marked by water droplets which resulted from the ionizing effect of the passage of the particle through supersaturated air. The track curved upward to the left, intersecting the lead plate, the curvature increasing in the upper half of the chamber. There was no more to observe and all the rest was 'deduction from phenomena'.[1]

There were four possible interpretations of the phenomenon in the light of what had hitherto been ascertained:

(i) it might be the track of a negative particle moving downward through the plate;
(ii) it might be the track of a proton moving upward;
(iii) it might be a combination of two tracks, one moving upward and one downward, of particles released by the incidence of a proton upon an atom of lead; or,
(iv) it might be the track of a particle of very small mass (of the order of the negative electron) but with positive charge, moving upward through the plate.

The energy of the particle can be calculated from the curvature of the track, if the mass is known, and the first hypothesis is immediately ruled out because, if the particle is negative and has the mass of an electron, the curvature would imply that it entered the lead plate with an energy of 20 million electron volts, and left it with 60 m.e.v., which is a physical impossibility.

The second hypothesis would require a proton moving above the plate with energy of 300,000 electron volts, but a proton with that energy has a range of only 5 mm.,[2] whereas the track of the

[1] Cf. C. D. Anderson, 'The Positive Electron', *Physical Review*, Vol. 43 (March, 1933), pp. 491–4, and his report in *Science*, LXXXVI (1932).

[2] As determined by Rutherford, Chadwick and Ellis in earlier measurements.

particle in the photograph extends above the plate for approximately 5 cm.

The third hypothesis implies that the upper portion of the track is that of a positive particle of mass similar to that of an electron (whatever the lower portion might be). This partially coincides with the fourth hypothesis, which agrees also with the calculated energies for an electron, and is possible because a particle of energy 60 m.e.v. would be slowed down by passage through a 6 mm. lead plate with a loss of energy of approximately 38 m.e.v. This had previously been established by Anderson in measurements of the energy loss of electrons on passing through a metal diaphragm.[1] Accordingly the track can only be that of a positive particle of similar mass to the electron, and is forthwith named by Anderson a positive electron (or 'positron' for short).

Corroboration came from the results of observations made by Blackett and Occhialini of particle 'showers', in which tracks were seen both with positive and with negative curvatures (i.e. both to the right and to the left, in a magnetic field), as well as some which were not deflected. In many of these the extent of ionization, as measured by the density of condensation along the track, were what would be expected of a particle with mass of the order of an electron, easily distinguishable from the heavier tracks associated with protons. A sufficiently large number of electron tracks with positive curvature were observed to confirm the existence of the positive electron.

The intimate connexion between observation and theory in this example of scientific discovery is incontestable. The relevant experimental data consist simply of the cloud photographs of the tracks in the Wilson chamber, and all that is visible on them are streaks and blurs. Only a trained physicist, or somebody with at least some inkling of the theoretical background of the experiments could have the smallest idea of what these streaks and blurs represent. The expert observer interprets them as the tracks of particles in a magnetic field. He knows of the latter from the experimenter's description of the apparatus used. The instruments and mechanisms have been deliberately arranged, with knowledge of the physical relationships involved, to bring about certain effects. It is this theoretical background knowledge that enables the experimenter to see the marks on the photograph as particle tracks and to recognize the particles as negatively or

[1] *Physical Review*, Vol. 43 (1933), p. 381.

positively charged, as electrons or as protons. Yet even the trained physicist could not recognize tracks with positive curvature as those of a positive electron before the course of reasoning outlined above had been followed through. Millikan actually did observe the track of a positron some time earlier, but he saw it as that of a proton—if a puzzling one, because it was thinner than it should have been. But Millikan suspected the ionization theory of error, not his identification of the particle.

All the data from which Anderson's reasoning proceeds is, therefore, what Hanson describes as 'theory-laden'.[1] They are far from being a collection of bare sense-particulars, or of propositions describing the occurrence of such particulars. The sense and meaning of any proposition in which any of the data can be stated, are saturated with theoretical implications and are moulded by the bearing upon them of structural principles ordering a ramifying body of factual evidence. Nevertheless, they are propositions describing the results of observation. There is a visible streak in the photograph, curving to the left, and it is visibly of smaller radius above the photographic representation of the lead barrier than below. The extent of the curvature, the length of the track and the amount of ionization can be measured, or computed from measurements made upon the photograph—processes involving further observations and affording factual data upon which the reasoning rests.

There are four possible explanations of the observed facts to be examined, and the reasoning, as we have said, proceeds by elimination of hypotheses. This has been included by logicians among the alternative methods of inductive inference. But the current logical theory assumes that the hypotheses to be considered have been derived from observation by enumerative induction, or have been arbitrarily proposed by the scientist. It assumes further that elimination is effected by inductive disconfirmation and that enumerative induction is again necessary to establish that the list of hypotheses is exhaustive. In the present instance none of these assumptions is true. The hypotheses are not arbitrary, nor are they derived by inductive generalization. It is true that the tracks of positively and negatively charged particles had been observed before and that these particles had been deflected appropriately in a magnetic field. But Anderson does not argue that because such observations had been made in the past this might just be another

[1] *Patterns of Discovery*, Ch. I.

instance. That charged particles are so deflected is not merely an inductive generalization. Electrodynamic theory requires that they should be deflected in definite ways and the magnetic field is deliberately introduced to identify them as positive or negative. That the track in Anderson's photograph might be of a negatively charged particle is, therefore, not derived from an inductive generalization, it is a possible interpretation of the observable fact by reference to theoretical considerations. That the particle must be negatively charged if it is travelling downwards, and cannot be negatively charged if it is travelling upwards, are deductive conclusions from universal theoretical majors and particular factual minors.

The possibility that the particle is a proton depends first on a decision as to its upward or downward movement, which is not possible from direct inspection of the photograph. But if the assumption is made that the particle is moving upward and is, therefore, positively charged, the hypothesis that it is a proton followed, not from any inductive generalization but from the contemporary theoretical belief that all positively charged particles were protons. One might argue that this was an inductive generalization (like 'All crows are black'), because no positively charged particles had hitherto been found other than protons. But this was itself largely the consequence of theoretical convictions. At least one positron track had been observed but theoretical limitations prevented its being seen as such. It was simply assumed that because electrical charges were of two kinds only, the same must be true of charged particles.

The third hypothesis, that there might be two particles, one positive and one negative could be derived from an inductive generalization based on past observations of the emission from metalic nuclei of several particles when struck by cosmic rays. In that case, the hypothesis should include the assumption that the positively charged particle is a proton, for no positively charged electron had ever before been identified. If this assumption is not made, the hypothesis would have to be excluded on inductive grounds; whereas, in the final outcome. it is precisely the inductively false element in the hypothesis which proves to be deductively necessary.

This inductively false hypothesis, that the particle is an electron with positive charge—inductively false, for all previously observed electrons had been negative—is the fourth and final hypothesis.

Obviously it has not been suggested by inductive considerations and is not, in the end, established by them.

That these four hypotheses are exhaustive is not quite the case, for it is possible to assume that the track is of a negatively charged proton. This cannot, however, be seriously entertained, not only because it is highly improbable on inductive grounds (the reason given by Anderson for ruling out yet another possibility that two negative tracks have accidentally coincided), but also because none of the characteristics of the observed particle except a possible *positive* charge points to its being a proton; and the theory of the day assumed that there were no negative protons. The exhaustiveness of our list of hypotheses, therefore, follows deductively from the premises that charged particles can only be either positive or negative, and that there are only two kinds, electrons and protons.[1]

The process of elimination is (with one exception) a series of disproofs, all of them deductive in kind. If the particle is negative it must be moving downwards. If it is moving downwards, the curvature of the track would indicate an increase in energy after passing the lead diaphragm. But penetration of such a barrier by a particle necessarily decreases its energy; so the hypothesis is disproved by *reductio ad absurdum*. Mutually contradictory propositions can be deduced from it.

If the particle is positively charged, it may be a proton. If so, its mass will be 1,000 times that of an electron. Therefore its energy, calculated from the curvature of the track above the lead plate will be 300,000 volts. A proton with that energy has a range of 5 mm., but the length of the observed track is 5 cm. This can be represented as hypothetico-deductive argument, though one can say also that from the curvature of the track one deduces an energy of 300,000 volts, while from its length one deduces another quite incompatible quantity; and so once again we should have *reductio ad absurdum*.

If there are two particles, the one moving upward must be positive, and it follows from the foregoing demonstration that it cannot be a proton. That is not a disproof of this hypothesis but of the previous one. If one of the particles presumed is positively charged but is not a proton, the fourth hypothesis is invoked.

So we have only one possibility left standing. But the scientist

[1] We now know, of course, that there are more; but any conceivable hypotheses about pions, muons or hyperons in the present connexion are entirely otiose.

is not content to affirm it solely for that reason. He proceeds, by a fresh marshalling of evidence to prove its truth. Past experiments and calculations have given values for the loss of energy by an electron penetrating a 6 mm. lead barrier, and these agree with the values calculated, in the present instance, if the particle is moving upwards and is therefore positive. Similarly, the thickness of the track indicates the correct amount of ionization for an electron. Thus of the four hypotheses, the implications of two involve contradictions; the third (that there are two tracks, one positive and one negative) coincides, so far as the positive particle is concerned, with the fourth, and that is confirmed by the deduced energy values and the amount of ionization.

The reasoning throughout has drawn equally upon theory and upon observation—both present and past. Conclusions from observation are in some respects deductive (as when energy values are deduced from the curvature of the track and the known or assumed mass of the particle), and in other respects inductive (as when the range of a proton is taken to be 5 mm. on the evidence of past cases,[1] and when values are determined for the loss of energy resulting from penetration of a metal barrier). But they establish incontrovertibly the conclusion that the observed track is that of a positively charged electron and not a proton. It is a necessary conclusion though it has been largely drawn from empirical data.

The cogency of the conclusion is the result of the convergence of the evidence, the calculated energy values, the quantity of ionization, the length, direction and degree of curvature of the track. These are all quite different aspects of the fact, but they fit together in such a way that their consequences are undeniable. Finally the conclusion drawn so inescapably from experiment confirms that derived deductively from pure theory. Dirac's relativistic wave equation of an electron in an electromagnetic field has two possible solutions, indicating both positive and negative kinetic energy. The latter is shown to be equivalent to a positively charged particle with the mass of an electron moving through the field. The theory is interpreted as establishing the principle of invariance under charge conjugation. So once again we have a system of interlocking evidence from presently observed fact, from past calculations and from theoretical demands. From

[1] This, however, also involves calculation and might be deduced from dynamical principles.

the established body of knowledge alternative hypotheses can be derived for the possible explanation of a present phenomenon. But their implications are such that some cancel out and a single conclusion is left standing that cannot be gainsaid. Its cogency is due to the convergence of the evidence, and to deny it would involve the demolition of a whole edifice of science constructed of close fitting, mutually supporting facts.

vii. FINDINGS

For the present, these six examples of scientific discovery will be enough, at any rate to enable us to draw some provisional conclusions. They can only be provisional for reasons which will presently appear, for the examples are, none of them, complete in themselves, but are extracted from a wider context which we shall presently have to examine.

(1) First, we may note the organizing function performed by theory. In each case the hypothesis, both that which is being disproved and that which is being established, functions as a pattern or organizing principle to reduce to order a host of otherwise confused observations. We have seen how successive astronomical theories performed this function, how Kepler used his own and Copernicus' theories to bring order into the confused accumulation of Tycho Brahe's observations; and we have seen how Darwin co-ordinated an immense range of facts with the help of the ideas of spontaneous variation in the progeny of a common ancestor and natural selection in the struggle for survival. Similarly Lavoisier's conception of chemical combination brought order into a mass of confused phenomena connected with calcination, respiration, combustion, as well as the solution of bases and metals in acids and the production of 'airs' by effervescence.

(2) Secondly, as organization is the function of theory, dissatisfaction with a theory arises from failure of organization—that is, conflict of evidence and contradiction in the implications of the doctrine. These are the stimuli to research. The 'facts' that refuse to fit and collide with the accepted theoretical requirements are those which we say 'cannot be explained', and to account for and incorporate which the old conceptual scheme has to be remodelled and surpassed.

As Copernicus rebelled against the use in the Ptolemaic system of different principles of explanation for different heavenly bodies, so Kepler rejected Copernicus' inconsistent vaccillations between heliocentrism and remnants of geocentrism. He then later wrestled with the conflicts arising in his own construction until he had reconstructed a harmonious network of facts and theories. Harvey does likewise. He castigates the absurdities of his immediate predecessors and recombines the very facts they had discovered, along with others of his own finding, in a pattern which elucidates them all. Darwin substitutes for an equivocal theory (special creation) which could explain equally all appearances, as well as their possible contraries, a theory which explains contrasts coherently by means of a single principle (e.g. the similarity of species on islands separated by shallow channels and the dissimilarity of species on islands separated by deep water).

(3) The consequences of this effort to resolve the contradictions is the construction of a system of facts—or evidence—interlocking and mutually corroborative, which compels assent to the conclusions that it requires and contributes to the building of a coherent conceptual scheme. So that the characteristic form of scientific proof is by corroboration (i.e. interlocking) of evidence, interplay of facts and consilience of hypotheses.

(4) What makes this possible is that between fact and hypothesis, observation and theory, no sharp line can be drawn. Pressures for adjustment and reconstruction come from both at once. They are not imposed on the existing theory from without by empirical discoveries which are wholly extraneous (however accidental, as is sometimes the case, the occasion of discovery may be). We have seen how 'fact' and 'theory' have gone hand in hand. Kepler's 'facts' were not so much observed as derived by the application to his data of the conceptual scheme with which he was working, and which his whole enterprise contributed to confirm. It was the general theory that made the facts (the calculated positions of the planet) what they were. Harvey's experiments with the tourniquet are as much dictated by the idea of circulation as *vice versa*. This is not to say that the theory predetermined the results of the experiment—certainly not—but (i) what facts are to be sought (what questions asked) depend upon the general direction of thinking and thus upon the theory under investigation, and (ii) more important, the nature and significance of the findings depend upon prior knowledge. How would one know that the coldness and pallor of the hand indicated a privation of blood from the artery, and that its swollen and lurid appearance

was caused by a stoppage of the veins, unless one understood the details of the anatomy? It is the knowledge of theory (*some* theory) that informs the observer of what he is observing and how it is relevant to his investigation. On the other hand, we have also seen that incomplete or faulty theory can blind a scientist to facts with which he is actually confronted, because what is actually appearing in the phenomena has to be theoretically interpreted before the 'fact' can be appreciated. Hence a positively charged particle can only be seen as a proton until a new structure of facts has emerged, either from pure theoretical reasoning or from systematic observation which makes a reinterpretation necessary.

(5) It is this inherent interpretation illuminating every observation (as it were) from within—its theory-laden character—that makes deduction possible from phenomena. It is because the facts as observed are systematically complex and (another aspect of the same truth) systematically related, that we can draw cogent conclusions from organized bodies of evidence. The structure and mutual disposition of the valves of the heart and the blood vessels is not independent of their function and consequent interaction. So the direction of the blood-flow can be deduced from them. The angle of refraction is not independent of the colour of the light. So the experimenter can deduce from the position of the prism and the shape of the image that different coloured rays are deflected through different angles.

What we seek is the peculiar logic of this 'deduction from phenomena', which is not the deduction of traditional and contemporary formal logic, nor yet inductive generalization from particular observations, but a conflation of both in a single logical process of deductive cogency drawn from an accumulation and convergence of empirical evidence. It is a combination of induction with deduction which transforms both. It is a logic by which inference can be made to new information by an *a priori* movement producing compelling conclusions by systematizing empirical facts. By virtue of its efficiency in integrating the observations its results are synthetic, yet, its deductive movement gives it necessary and universal force.

What sort of logic is this? Clearly a logic of construction, but how does that operate? Undoubtedly the ancient laws of thought must continue to hold—A must be A, and not both B and not-B. But these are no more than laws of tautology; they do not explain why, if A and B stand to each other in a certain relation, C must be thus and so. What principles govern transitions of that kind? It cannot be the logic of a formal deductive system in which variables are replaceable by simple particulars in external relation; for if it were the relation of A to B and C would throw no light on D or its relation to the rest of the assemblage. That this should occur the elements must themselves be complex. Their peculiar nature and the way they relate to one another must be interdependent so that there is a guiding link between one part of the *gestalt* and every other.

The elements of the scientific system, however, are the observed phenomena and it is from them that the 'deduction' proceeds. Part of the solution of our problem should, therefore, be found in the nature of observation itself. The old empiricists alleged that observations could be analysed into sense-data and that these were simple particulars externally related. From that it followed naturally that generalization by simple enumeration could, in the last resort, be the only method of discovering regularities, and that deductive reasoning could only be analytic. But if perception is not reducible to the apprehension of simple data, these logical theories will prove inadequate to explain scientific inference. Our problem therefore must be attacked, from one side at least, by seeking an adequate theory of perception as the form of scientific observation.

Before we do this, however, we must attack the problem from another angle. A wider and more comprehensive view of scientific advance has yet to be taken, for it has already been confessed that none of the examples we have considered is complete in itself.

(6) In every case we have found the scientist drawing upon a copious tradition of already accepted theory, even when he criticizes it adversely and though he is in the process of revolutionizing it. His hypotheses do not spring from his head like Athena fully armed from the head of Zeus, but are modifications or adjustments of former theories. Kepler's conceptions of the planetary orbit are first taken over from Copernicus and Ptolemy, and later only distortions or generalizations of theirs. The circular orbit is traceable to Pythagoras and beyond, and was common to Ptolemy and Copernicus; the ovoid is simply a circle with a bulge on one side; the ellipse is a generalized figure of which the circle is a special case. So we have here first an acceptance of a prior concept and then a progressive development or modification of it, under pressure of the demands from observation and calculation.

Harvey draws upon the Galenic tradition, even when he rejects it—for he never rejects it *in toto*, and his advances rest on an enormous mass of long discovered detail of anatomy and physiology. But, for our purpose, what is more important (and will be made evident in the next chapter), he had at his disposal the work of Vesalius, Fabricius, Cesalpino and Colombo, all of whom had made significant new discoveries leading up to his own work. It can hardly be doubted that if he had not studied at Padua he would less easily have conceived the idea of the circulation. He himself says that what first gave him the idea was the formation of the valves in the veins and these had already been discovered and some theory of their function, though the wrong one, had been put forward by Fabricius.

Newton, in his work on light, is in much the same case, if, in the examples we have discussed, less obviously. He too had distinguished predecessors in this field like Kepler and Descartes, and his debt to them is apparent, if not explicitly mentioned. But we have looked at the *Opticks* to find evidence of a special aspect of his method—his deduction from phenomena—and it is to the wider canvas of mechanics that we must go for a better view of the way in which he develops and advances upon prior discoveries.

Lavoisier not only confesses his own debt to his predecessors and reviews their findings but the very doctrine which he overthrew is one which in his earlier writings he constantly invokes—if only hypothetically:

'Most metallic calxes are reduced . . . only through immediate contact with a carbonaceous material, or some substance which contains what is called phlogiston', he says in the Easter Memoir.[1] And again 'As common air is changed into fixed air when combined with charcoal, it should seem natural to conclude that fixed air is merely a combination of common air and phlogiston.'[2]

Darwin takes advantage of the whole body of lore concerning the breeding of domestic animals, and calls to witness the discoveries of numbers of his contemporaries. His central argument comes from Malthus, and he found contributory evidence in the discoveries of the geologists and geographers, the observations of a host of naturalists and the reports of travellers. Even the notion of evolution itself had been introduced much earlier, by his own

[1] *Oeuvres*, II, p. 123. [2] *Ibid.*, first draft, 1774.

grandfather, Erasmus (among others) and had been discussed for a quarter of a century at least before Charles Darwin wrote.

Our example from modern particle physics begins *in medias res* and the fact that the scientists' concepts are no arbitrary invention is too obvious to require argument.

We may conclude, then, that hypotheses grow out of prior theories, and in specially puzzling situations, like that of Kepler or Millikan, the new idea dawns as the pattern of the facts begins to emerge: *'O me ridiculum! perinde quasi libratio in diametro non possit esse via ad ellipsin.'*[1]

(7) The important clue is this patterned nature of the facts. Even when relatively chaotic, in that confusion which demands rectification, the chaos is never total. There is always the prior theory. And in the half-dozen cases we have examined, what we have found is construction and reconstruction of systems. The process of discovery presents itself as a series of growing, or developing, structures—sometimes alternative and partly conflicting, sometimes mutually complementary. In the first case, the theory grows by successive rejections, yet each proposal is still, in part, a repetition of the one last found wanting. Thus Kepler proceeds from circle to ovoid and from ovoid to ellipse. In other cases the phases are more compatible, and what develops is the scope of the construction. So Harvey goes from fitting together the functions of the valves of vesicles of the heart, to the lesser circulation, and from that to the relation of the arteries and veins, from that to the estimation of the quantities of blood flowing through these courses in a given time, and so to the total picture of the vascular circulation. Darwin moves from the modification of domestic varieties to the struggle for existence and natural selection of wild varieties, from this to the geological record and the ramification of species from common progenitors, and thence to geographical distribution. All these systems finally combine in the general theory of evolution of the species.

The evidence so far gathered strongly suggests that the scientist moves by producing a series of progressively more adequate theoretical concepts. But each of these examples, as we observed earlier, is torn from its historical context. Each is a phase (often the culminating phase) of a longer development, involving a number of scientists, which is simply being continued in the thought and experimentation of the individual. It is this aspect of

[1] See p. 135 above.

scientific advance that we must consider next, for the logic of discovery is that running through the whole process. Our strategy will thus be a pincer movement advancing on the one hand from a theory of scientific progress, and on the other from a theory of perception. We shall find that the two movements coalesce and that the development from direct perception to scientific understanding is in some measure reflected in progression from more naïve to more sophisticated scientific theory. Meanwhile let us sum up the findings of this interim report on progress. What our case histories have shown is as follows:

(1) Scientific theories serve an organizing function, reducing a confused multiplicity of facts to order and unity.

(2) The stimulus to further advance is the conflicts and contradictions (in Aristotelian language the *ἀπορίαι*) in the hitherto accepted theory.

(3) The method of scientific proof is the marshalling of evidence into corroborative systems.

(4) Theoretical conceptions permeate the entire process of thinking, and facts always involve interpretation, so that no sharp distinction can be drawn between theory and observation.

(5) This interdependence of observation and theory underlies 'deduction from phenomena', and gives the lie to any rigid separation of inductive from deductive inference.

(6) Hypotheses are never conceived *ab initio*, arbitrarily, or independently of any logical process of discovery. They arise out of past theories, which they develop, and are suggested by new patterns of order among the facts.

(7) The entire process of establishing new theories is one of successive phases in which a series of structures emerges, sometimes mutually superseding and sometimes complementing one another in a wider system. (It will transpire later than these two kinds of sequence are not mutually incompatible and that there is in each some element of the other.)

CHAPTER VII

SCIENTIFIC ADVANCE

The popular conception, outlined in Chapter I, is that science advances by accumulation of new empirical discoveries requiring, in each major instance, a radical revision of the accepted description of the relevant aspect (or the whole) of the known world. The old theory is then abandoned as false and a new one conforming to the empirical evidence is substituted for it. This notion of scientific progress (I maintain) is wrong, and it is now time to examine the historical facts in order to substantiate that judgment. It is one which has already been made by T. S. Kuhn and persuasively defended in his book *The Structure of Scientific Revolutions*. With most of the theses and arguments of that work we shall find reason to agree, though on some points, which are by no means negligible we shall have to differ with its conclusions.

i. CONCEPTUAL SYSTEMS

That scientists do not work in an intellectual vacuum or produce theories from the recesses of their minds *ab initio*, for later empirical testing, we have already seen and might well have expected. But neither do they extract their theories simply by empirical generalization from laborious collections of detail. No science is possible, no research can be conducted and no advance can be made except by reference to, and subject to the requirements of, some conceptual scheme. It is the very nature of science to render experienced facts intelligible, and that is nothing other than to display them in a coherent and systematic order governed by definite interpretative, or organizing, concepts. The principles of this order or system constitute the conceptual scheme. Without it there is and can be no science, for theory cannot be derived or collected from any assemblage of uninterpreted facts. That theories are not derived in this way we have already found evidence. That they could not be, will become more apparent as we proceed; because, we shall repeatedly be forced to conclude, there are no uninterpreted facts, all intelligible experience being some kind of interpretation of some body of primitive sentience. The impossibility of raw facts Kuhn recognizes, in his own fashion, when he insists that there is no neutral observation language, no language independent of concepts provided by the interpretative system or world view—what he calls the 'paradigm' of science—established and operative at the time; and when he says that what the scientist finds in his laboratory investigations are not 'given' facts but 'concrete indices to the content of more elementary perceptions',[1] Kuhn is also aware that 'something like a paradigm is requisite to perception itself'[2]—a matter about which we shall have more to say below.

A scientific 'paradigm', according to Kuhn's first account of it, is laid down by the text-books of the period, based upon the exemplary discoveries and achievements of some outstanding investigators immediately prior to their compilation. The text-books inculcate the experimental methods and procedures of these men along with their theories. It is, of course, true that scientific text-books presuppose and embody an accepted conceptual scheme, and also, as Kuhn says, much of the attitude of scientists and historians to the fundamental ideas of science is the result of their education from, and their deference to, text-books. But the conceptual scheme is not derived from the text-books (Kuhn sometimes appears to suggest this, though he clearly does not really believe it). The text-books are derivative from the scheme. The scheme itself is the product of creative thinking by scientists, and, as Kuhn rightly observes, it involves a conception of the basic entities in the universe and their elementary ways of behaving and interacting, which it specifies systematically in accordance with a set of fundamental laws or interpretative ideas. A conceptual scheme of this kind is what Collingwood called a constellation of 'absolute presuppositions' and in past generations has been recognized as a metaphysical theory.[3] That controversial description need not here be pressed, and whether or not it is to be accepted can be decided later. What is more immediately apparent and less disputable is that all science proceeds under the dominance of some conceptual scheme such as has been described,

[1] Cf. *op. cit.* (*International Encyclopedia of Unified Science*, Chicago, 1962), Ch. I, pp. 125 f., and Ch. XII, p. 145.

[2] *Ibid.*, p. 112.

[3] Collingwood maintained that the absolute presuppositions of science were what metaphysics was about; that the task of the metaphysician was to discover them by logical analysis of scientific reasoning, and to trace their transformations in the course of history. Cf. *An Essay on Metaphysics* (Oxford, 1940).

and that the progress of science is marked by the periodic reformulation of its fundamental system of basic concepts which constitutes, for Kuhn, a scientific revolution. Such revolutions have unquestionably occurred, and the kind of progress which takes place between them is what Kuhn calls 'normal science',[1] to be distinguished from abnormal or revolutionary science as 'puzzle-solving' in contrast to the introduction of major transformations of 'paradigm'. He does admit that between major revolutions, minor ones in special areas are constantly occurring, the repercussions of which (we may add), though possibly latent at first, may prove to be widespread. Kuhn's distinction is not wrong, but like so many distinctions in epistemology must not be made absolute, for we may find that the two kinds of science differ as much in degree as in kind.

Examples of conceptual schemes dominating the practice of the sciences are easy to identify in history. From the third century B.C. to the end of the Middle Ages, science was conducted under the aegis of one of the most comprehensive and spectacular systems in the history of thought—that laid down by Aristotle. It was also one of the most successful; which accounts for its singular longevity. For a conceptual scheme is an explanatory system, and its function is to explain as wide a variety of facts as possible in terms of one or a few systematically related principles. We have already seen that 'explanation' consists in the successful incorporation of the 'facts' into the interpretative structure—their successful organization—and Aristotle and his followers succeeded in doing this, with apparent consistency, for an astonishing range of human experience.

Aristotle's theories together formed a body of doctrine which most writers would describe as metaphysical. But few would deny that at the same time they included theories rightly described as scientific, and that they inspired a vast amount of scientific and quasi-scientific research as well as some very fruitful scientific speculations. Aristotle certainly defined the fundamental contents of the universe, set out the systematic relationships in which they were arranged, and laid down the principles of interpretation (e.g. matter and form, potentiality and actualization, the four causes) according to which all experience was to be rendered intelligible. It was the paradigm of all conceptual schemes; and for ten centuries it worked. The reasons for its eventual failure, we

[1] *Op. cit.*, Ch. IV.

shall consider presently. First let us notice other examples of conceptual systems.

Aristotelianism, with its off-shoot, Ptolemaic astronomy, gave way, in the seventeenth century, to Copernicanism and the classical mechanism of Newtonian physics. This too was a system of fundamental concepts, not in any restricted field, but embracing the entire world, and it found expression in an acknowledged metaphysical theory.[1] Its characteristic doctrines were, first, an atomic or particulate conception of matter, the particles of which move under the action of forces dependent upon their mass and the distance between them. Secondly, it assumed the possibility of explanation of all qualitative characteristics in terms of quantitative relations between the atomic constituents of matter; and, thirdly, it drew a sharp division between matter and mind, leaning towards the former as ontologically prior. The philosophical outcome was either a dualistic metaphysics or, when the implications of the priority of quantitatively measurable characters were pressed to an extreme, a severely monistic materialism.

Within this over-arching conceptual scheme minor systems developed, appropriate to physiology, as introduced by Harvey, and chemistry, as conceived by Lavoisier; just as minor systems had evolved within Aristotelianism appropriate to astronomy, medicine and alchemy. Darwin's innovation and the rise of the conception of evolution may, from one point of view, be regarded as a new ramification of the prevailing system; from another it may be seen as the beginning of a new and revolutionary outlook, depending on the relative emphasis placed upon its mechanistic tendencies or its implications of teleology.[2]

Classical mechanism held sway, in a period of unprecedented scientific development, for two hundred years; and even today it dies very hard. It has been superseded by a conceptual scheme which is difficult to describe succinctly because it is relatively new and probably still in the process of formation. The physics of relativity and of quanta determine its form, along with a concept of organism ascendant in biology. Its typical philosophical expression is Whitehead's 'philosophy of organism', which

[1] Cf. R. G. Collingwood, *The Idea of Nature* (Oxford, 1945), Pt II, Sect. 8, pp. 103–5; E. A. Burtt, *Metaphysical Foundations of Modern Science* (Doubleday, 1955), *passim*; and E. E. Harris, *Nature Mind and Modern Science* (London, 1954), Ch. VI, pp. 118 ff.

[2] For discussion of this see my *Foundations of Metaphysics in Science*, Ch. XIII; cf. R. G. Collingwood, *Idea of Nature*, Pt III, pp. 133–6.

combines the influence of all three of these scientific trends. These are examples of major conceptual schemes, and science has never existed, and could not be pursued in its recognized form as systematic investigation of nature, except under the control of one or other such scheme. Nor could theoretical science, at any time, proceed appropriately or successfully under a conceptual system which had previously been superseded (though certain technical pursuits may do; as surveying may make use of geocentric astronomy, and mechanical engineering of Newtonian dynamics).

A scientific revolution is the major advance from one comprehensive conceptual scheme to another, but progress is not confined to such major revolutions for it goes on under any established scheme in the form of what Kuhn calls 'paradigm articulation' as well as in the form of minor revolutions, both of which build up towards the crises that lead to major transformations of outlook. We must now pay some attention to this process.

ii. RECOGNITION OF OBSERVED DATA

A comprehensive conceptual system is a world view. But it is more significant for our purpose to recognize it as a system of principles organizing the experience of the world. A belief in a chaotic world may, of course, always seem feasible, but we should hesitate to call it a world-view in the proper sense, and it would neither be science nor make science possible. For science is and may be defined as systematic thinking[1] about the world—a systematizing activity by which the world becomes intelligible and so intelligently manipulable. This became apparent in the last chapter, in our discussion of case histories, where we found the theory functioning as the organizer of the manifold of observed facts. But we found also that what the observed facts were was very largely determined by the interpretative system, and though there was a constant interplay between observation and theory neither was wholly prior, but they were co-ordinately dependent each upon the other. The theory not only sets the facts in order but, by systematizing, gives them intrinsic recognizable form.

The effect on perception of this interdependence is evident in three ways. First, perceived phenomena may be simply suppressed and overlooked for lack of a conceptual scheme in which

[1] The nature of system will be discussed in Pt III. For the present it must be understood in its commonly accepted sense, to be more carefully examined later.

to place them. Sunspots had been seen at least since the time of Charlemagne, and Stillman Drake[1] mentions a reference to them in Virgil. The Chinese (we are told by Joseph Needham[2]) had regularly recorded their appearance centuries earlier than Galileo. But it was only after Copernicus had opened the way to the belief in the mutability of the heavens that any scientific notice was taken in the west of these phenomena. Much the same must have been true of new stars, observations of which were often recorded in China, and it is hardly credible that Tycho Brahe's was the first in Europe. But he was the first in whose conceptual scheme they stood out as significant, and henceforth they were observed quite frequently.

Secondly, perceived phenomena are seen differently in different conceptual schemes. Comets before Copernicus were seen as meterological (sublunar) 'exhalations' in the air. After Copernicus they were seen as heavenly bodies orbiting beyond the planets. To the Peripatetics, as Kuhn points out, a swinging stone was a falling body constrained so that it fell slowly by stages; to Galileo it was a pendulum gathering in its fall inertial force equal to that exerted upon it during its downward course. Before Lavoisier, gases were all just air impregnated or contaminated with earthy vapours or smoke. After Lavoisier, they were recognized as separable chemical substances. This subservience of observation to preconception is ubiquitous in science. Under the influence of medieval bestiaries and travellers' tales, Dürer pictured a rhinoceros covered with scales and the most elaborately decorated armour. James Bruce, in 1790, referring to the inaccuracies of Dürer's woodcut and to improved representations by Buffon and other contemporaries, himself published, as a picture taken from life, an engraving of a rhinoceros which differs from the real animal (in the direction of Dürer's misrepresentation) as much as it does from Dürer's woodcut.[3] The early observers of electrostatic forces under the spell of Newtonian mechanics saw repulsion of like charges as a rebound, for attraction was all that was recognized.[4] To Millikan the first positron track was that of a proton, and Anderson and Nedermeyer, once Yukawa had postulated the existence of the meson, found

[1] *The Discoveries and Opinions of Galileo* (New York, 1957), p. 82.

[2] *Science and Civilisation in China* (Cambridge, 1959). Vol. III, pp. 423–9, 434–6.

[3] Cf. Ernst Gombrich, *Art and Illusion* (London, 1962), pp. 71 f.

[4] See Kuhn, *op. cit.*, p. 116.

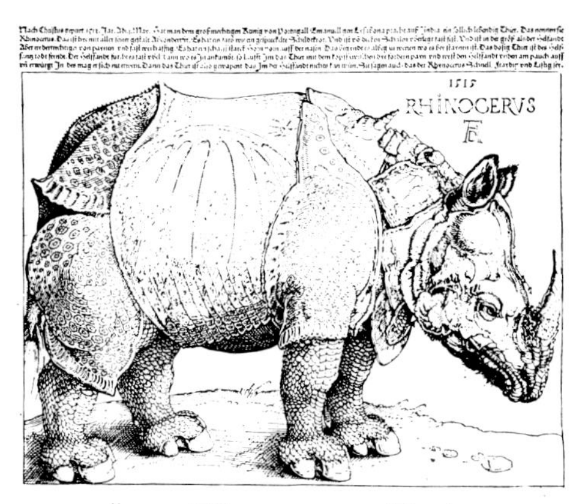

DÜRER: *Rhinoceros* 1515. Woodcut

HEATH: *Rhinoceros of Africa.* 1789. Engraving

African rhinoceros

examples, which, in fact, were *not* Yukawa particles. Is it impossible that similar tracks had been seen before, but had simply been left unnoticed as unidentifiable? After Sir William Herschel had discovered Uranus and its orbit had been successfully computed, it was found that the planet had been actually sighted twenty times before, and once, at least, when its movement among the fixed stars might have been noticed; but it was never seen as a planet and its movement was simply suppressed as a fact.[1]

The third way in which observation is affected by theory is in the complete failure to perceive what no concepts lead one to expect. Two quotations will serve to illustrate this fact. Charles Darwin, in his autobiography, writes of his tour in Wales with the eminent Cambridge geologist, Sedgwick:

'On this tour I had a striking instance how easy it is to overlook phenomena, however conspicuous, before they have been observed by anyone. We spent many hours in Cwm Idwal, examining all the rocks with extreme care, as Sedgwick was anxious to find fossils in them; but neither of us saw a trace of the wonderful glacial phenomena all round us; we did not notice the plainly scored rocks, the perched boulders, the lateral and terminal moraines. These phenomena are so conspicuous that . . . a house burnt down by fire did not tell its story more plainly than did this valley.'[2]

Mrs Arber bears witness in a different connexion to the same sort of experience:

'Every biologist must be able to confirm from his own experience that perception depends on preparedness of the mind, as well as on actual visual impressions. As a trivial instance, the writer may recall having been acquainted with Queen-Anne's-Lace (*Anthriscus sylvestris* Hofm.) for half a century, without noticing that the pattern of its growth is such that the main axis almost invariably terminates in a reduced inflorescence, which, in association with the grouping of lateral shoots below it, gives the plant a highly distinctive facies. When that visual fact had at last

[1] See Morton Grosser, *The Discovery of Uranus* (Harvard, 1962). Herschel himself first identified the object as a comet.

[2] *Autobiography*, ed. Nora Barlow (London, 1958), p. 70.

succeeded in forcing its way into the mind, any plant that came under observation was found to show this salient feature so strikingly as to leave the observer bewildered and humilitated at having been totally blind to it year after year.'[1]

In each case the failure of perception is due to a conceptual deficiency and not to perceptual inadequacy. Darwin and Sedgwick at the time were not familiar with the relevant geological hypothesis; and Mrs Arber at that stage had not developed her 'natural philosophy of plant form'.

iii. ARTICULATION AND PROLIFERATION OF SCHEMATA

The virtue of any system is its consistency and coherence, both within the set (or 'constellation') of principles, and in the coalescence of theory and observation. Further, as its function is to systematize and render intelligible as wide an area of experience as possible, these virtues must prevail over as extensive a range as can be reached by experience. The system must be comprehensive as well as coherent. But, in its first conception, scope and detail are lacking and the course of subsequent development has, for the most part, the purpose of supplying them.

The enunciation of a theoretical system, like the Aristotelian, the Newtonian or the Einsteinian, is in the first instance largely (if not wholly) schematic and abstract, and the period of Kuhnian 'normal science' is mainly one of application and the development of implications inherent in such application. It is, as it were, the period of absorption by the scheme of empirical experience, and we tend to call it (not inappropriately) a period of discovery of facts, in consequence. But it is all the time a process of development of implications, and their explication constantly leads to anomalies or conflicts within the system demanding further developments and modifications. The attempt to explain planetary movement in terms of the Aristotelian system revealed discrepancies requiring the introduction of epicycles. Nor was the Newtonian system exempt from such difficulties. The determination of the moon's motion according to Newton's theories, as well as that of several planets, revealed discrepancies which required

[1] Agnes Arber, *The Mind and the Eye* (Cambridge, 1954–64), p. 117.

refinement of its mathematical expression.[1] Similarly, by treating congruence as non-transferable (a possibility Einstein had overlooked), Weyl developed a geometry which so extended the theory of relativity as to be able to incorporate Maxwell's equations as symbolizing characteristics of the metrical field.[2] These modifications were not specially revolutionary and might well be regarded as examples of 'paradigm articulation'. For no conceptual scheme is perfect; none is completely self-consistent or comprehensive and the period of 'normal science' is one in which scientists seek to make the theoretical structure more complete, more widely applicable and (as Kuhn has seen) more coherent.[3]

As this process continues, however, difficulties arise which are not soluble by simple extension or adjustment of the principles of the system and contradictions appear that prove intractable. These are seeds of future crisis and the sources of eventual revolutionary change. For example the attempt to explain the flight of an arrow on Aristotelian principles led to the contradiction that it is both maintained in motion by 'antiperistasis', or air pressure from behind (since Nature abhors the vacuum), and brought to rest by air resistance from in front. Similar conflict emerged when thinkers tried to explain, by Aristotelian principles, the acceleration of falling bodies. Their speed increased, said the Peripatetics, as they moved nearer to the earth, through joy at the approach to their natural place. But this could not account for the fact that, of two bodies at equal distances from the earth, the one which had started its fall from a higher point moved faster. Incompatible conclusions were deducible from the theory in conjunction with the same facts.

But this kind of situation may also arise in consequence of

[1] Cf. W. Whewell, *History of the Inductive Sciences* (London, 1847), Vol. II, pp. 220–5.

[2] Cf. Sir E. Whittaker, *From Euclid to Eddington* (Cambridge, 1949), pp. 127–31; and A. d'Abro, *The Evolution of Scientific Thought from Newton to Einstein* (New York, 1950), Ch. XXXVI.

[3] See Kuhn, *op. cit.*, Ch. III, pp. 32–3: 'Therefore, from the Bernoullis, d'Alembert, and Lagrange in the eighteenth century to Hamilton, Jacobi, and Herz in the nineteenth, many of Europe's most brilliant mathematical physicists repeatedly endeavoured to reformulate Newton's theory in an equivalent but logically and aesthetically more satisfying form. They wished, that is, to exhibit the explicit and implicit lessons of the *Principia* in a logically more coherent version, one that would be less equivocal in its applications to the newly elaborated problems of mechanics.' Cf. René Dugas, *Histoire de la Mécanique* (Neuchatel, 1950), Bks IV–V.

minor revolutions within the over-riding conceptual structure. Long before the collapse of the Newtonian system, Dalton's discovery of chemical combination in proportionate weights led to the determination of differing atomic weights among chemical elements, which resulted in a conception of the atom in conflict with the one prevailing at the time in physics. The corpuscular theory of matter required the atom to be the fundamental particle and so the unit of mass and could not consistently accommodate the notion of atoms differing in weight according to their qualitative differences. This contradiction persisted to trouble scientists until contemporary theories of elementary particles and quantum physics were able to resolve the difficulty, only after a major revolution in scientific concepts.

The development of electrical theory continued for a long time within the Newtonian system, without disturbing it unduly. Mechanical models of electrical phenomena were satisfactorily forthcoming. Coulomb forces were found to obey Newtonian inverse square laws, and all proceeded normally until Faraday and Maxwell, without themselves realizing the impact their innovations would have, introduced the idea of the field which came to be viewed as a disturbance in the aether. This notion proved in time to be the thin end of the wedge that split the entire system by making a mechanistic account of the aether continually more and more difficult as the implications of field phenomena came progressively to light.[1]

iv. THE ORIGINS OF CHANGE

It is the emergence of contradiction within the system that both stimulates research and provides the drive towards change. The system progressively becomes a more thorough amalgamation of fact and theory, observation and interpretation, as it develops; and within it intelligibility coincides with integration. When conflicts are revealed and interpretations provided by the dominant concepts collide, intelligibility fails and new efforts to reorganize are called for.

The introduction into the Aristotelian astronomy of epicycles

[1] Cf. A. Einstein, *Relativity, the Special and the General Theory* (London, 1954), pp. 146 f.; A. Einstein and L. Infeld, *The Evolution of Physics* (New York, 1954), Ch. III; and E. E. Harris, *The Foundations of Metaphysics in Science*, pp. 52–3.

virtually contradicted the belief that orbits had to be circular, for the combination of deferents and epicycles, especially as the latter multiplied in number, produced orbits which were geometrically far from circular. Moreover, the alleged natural movement of the heavenly bodies became, in consequence, a 'violent' movement imposed on them by the complicated rolling of the epicycles upon the deferents. Small wonder that Copernicus complained of its conceptual incoherency.

The contradictions in Galenic physiology that moved Harvey to a new constructive effort have already been detailed, and the same was true of chemistry prior to Lavoisier. The confusions and conflicts that proliferated in phlogistic chemistry are most convincingly stated in Lavoisier's own criticism.[1] Stahl had held that phlogiston had positive weight and consisted of fire mixed with earthy substance in constant proportions. Some of his followers believed it to be a mixture in varying proportions and still others (e.g. Baume and Maquer) to be pure fire, or 'the pure matter of light'. But none of these variations of the concept enabled their authors to account for the augmentation of weight by combustion and calcination. Some tried to explain it (as had earlier been suggested by Boyle) by the entrance of pure fire into the burning material (often on the assumption that the particles of fire penetrated the containing vessel). But, as Lavoisier points out, that would require the element of fire not only to have weight (whereas, since Aristotle, it had traditionally been regarded as weightless) but even to be heavier than itself in combination with earth, for the ingression of pure fire had to do more than counterbalance the loss of phlogiston. The presence of phlogiston was the supposed source of colour in the calx, yet it was by loss of phlogiston from the metal that the calx was formed. Causticity was explained sometimes by lack of phlogiston, sometimes by its over-abundance. Here we have an egregious parallel to the situation of which Copernicus complained in astronomy, the appeal to contradictory principles to explain the same facts.

Lavoisier sums up his polemic as follows:

'. . . Chemists have made of phlogiston a vague principle which is never rigorously defined, and which, in consequence, is adapted to every explanation into which it is introduced. At times it is heavy, at others it is not; sometimes it is free fire, sometimes fire

[1] *Réflexions sur le Phlogistique, Oeuvres*, Vol. II, pp. 623–55.

combined with the earthy element; sometimes it passes through the pores of vessels, sometimes they are impenetrable to it. It explains at one and the same time causticity and non-causticity, transparency and opacity, colour and the absence of colour. It is a veritable Proteus who changes in form at every instant.'[1]

The same kind of complaint was made at the turn of the last century about the conception of aether, which focused the difficulties and ultimately caused the collapse of classical mechanism. But mechanism, in spite of its apparent clarity, simplicity and facility of explanation, was full of contradictions from the start. Its basis is the notion of a corpuscular reality—the old atomic concept of Leucippus and Democritus, of atoms moving in a void. The only source of explicable movement among them then is impact. Action at a distance is unintelligible. But Newton's theory of gravitation introduced precisely this unintelligible source of motion without replacing the other, which is nevertheless incompatible with it. Newton himself and many of his later followers throughout the reign of the classical mechanics, were sorely troubled by this conflict of ideas and cast about fruitlessly for a mechanical explanation of gravitation.[2] At the end of the *Principia*, Newton hints at a possible resolution of the difficulty by the introduction of the idea of a 'subtle spirit', suggesting a kind of aether, a conception which he further discusses at some length in *Quaestio* 28 of Book III in the *Opticks*; and later the phenomena of electricity and magnetism made irresistible demands for the postulation of an aether. Yet none of these ideas were or could be made self-consistent.

Even the idea of motion by impact was unintelligible on closer analysis. How does one body convey its own motion to another by impact? Presumably as a result of their elasticity. But elasticity is explained by the compression into a lesser space of the elementary particles of which the body is composed. The fundamental atomic corpuscles were not so conceived. They had to be solid through

[1] *Op. cit.*, p. 640.

[2] Cf. Newton's letter to Bentley, February 25, 1692–3: 'That gravity should be innate, inherent and essential to matter, so that one body may act upon another at a distance through a vacuum without the mediation of anything else through which their action or force may be conveyed from one to another is to me so great an absurdity that I believe no man who has in philosophical matters a competent faculty of thinking can ever fall into it.' *The Correspondence of Sir Isaac Newton*, ed. H. L. Turnbull (Cambridge, 1961), Vol. III, p. 254.

and through. How then could they be elastic? Yet if they were not, how could impact transmit motion, unless the two bodies remained continuously in contact? It is this unintelligibility of the transference of motion by impact on which Hume insists in his analysis of the idea of cause and effect.[1]

Thus the very foundations of mechanism were crumbly and incoherent. It demanded action only by impact, which was unintelligible, and also action at a distance, which was more so. Then aether was introduced to bridge the gap, but brought with it only more confusion. The original atoms moved in a vacuum, but the void was now to be filled with a new matter which was not corpuscular. Yet the plenum was not to impede the movement of particles. So, for mechanics, aether had to be completely permeable by matter without disturbance of its parts. But for electrodynamics it was to be the medium of electromagnetic waves, and, as such, it required a high degree of rigidity, an assumption colliding head on with that of the utter fluidity needed to allow bodies to pass through unimpeded. Further, as gross bodies traversed it without resistance, there could be no mechanical interaction between material atoms and ethereal substance; but light, which was assumed to be wave motion in the aether, was found to move at different velocities through air, through glass and through water. A mechanical explanation of these phenomena implied that there must be mechanical interaction between material particles and the ether. There were contradictions at every turn; but for a long time they remained below the surface and the progress of science was not unduly checked by them. Only when their accumulation became an intolerable obstacle did the shock administered to its fabric by the Michelson-Morley experiment demolish the entire structure.

V. THE TRANSITION PROCESS

Such difficulties and conflicts must not blind us to the fact that these conceptual schemes were systematic to a remarkable degree and did have spectular explanatory power within certain limits. We shall presently see that, even after the revolutions in which they were superseded, they were not wholly abandoned, and that the transition was never as sudden or total as it later appeared to

[1] *Treatise of Human Nature*, Bk I, Pt III, ii and vi. See also Emile Meyerson, *Identity and Reality*, Ch. II.

have been, but occurred in stages and was almost as much evolutionary as revolutionary.

(*a*) Even the change from Ptolemy to Copernicus was less cataclysmic than it is usually deemed to be. Quite apart from its anticipation by Aristarchus of Samos, whom Copernicus quotes,[1] Occam had already disputed Aristotle's view of the difference between the sublumary and the celestial spheres in both substance and physical laws. Jean Buridan had questioned the doctrine that the earth was at rest in the centre of the universe and Nicole Oresme had maintained, in 1377, that it rotated on its axis in the immobile sphere of the heavens. By the end of the sixteenth century men's mind had been prepared for change by the revival of interest in the works of the Platonic Academy with their flavour of Pythagoreanism, by the reaction which had erupted against Aristotle, and the growing preference for the atomism of Epicurus and Lucretius.

More significant still was the innovation in Aristotelianism itself, made by the Parisian scholars of the thirteenth century, of the theory of impetus—the doctrine that motion once imparted to a body was retained and died out slowly (as a heated body cooled by degrees). This overcame the problems both of flying missiles and of falling bodies, for the former could then be said to persist in motion by the impetus imparted by the original impulse, and to come to rest in consequence of air resistance, while the latter would accelerate, in proportion to the time or the distance fallen,[2] by the impetus continually added during the falling movement. It was no great step from this position to imagine a body unimpeded by air pressure continuing in motion indefinitely by its own impetus, uniformly in a straight line (or, when appropriate, in a circle).[3]

Both Descartes and Galileo learned this theory as students and its influence is clearly apparent in Galileo's reflections on the behaviour of pendulums and his argument that 'a plummet hanging from a cord, which, removed from the perpendicular

[1] Archimedes' report of Aristarchus' theory had been resuscitated at the Renaissance, so that the possibility of the alternative to Ptolemy and Aristotle was already there.

[2] As proposed by Albert of Saxony. See *A General History of the Sciences*, ed. René Taton (London, 1963), Vol. I, Pt. III, Ch. 7, p. 509.

[3] Jean Buridan did, in fact, assert, against Aristotle, that the revolution of the heavens was due to impetus impressed upon them originally by God, after which they revolved perpetually without resistance. See Taton, *op. cit.*, p. 509 f. Compare Descartes's theories in *Le Monde*.

(its state of rest) and then set free, falls toward the perpendicular and goes the same distance beyond it . . .' by what has, earlier in the same passage, been called 'impetus'.[1]

The Copernican revolution itself was accomplished in stages. First came Copernicus' heliocentric proposal and the attempt to construct the corresponding model of the heavens; then Tycho Brahe's partial apostacy, with his contribution of more profuse and more accurate observations; then Kepler's laws of planetary motion, derived almost as much as discovered, in his attempts to combine Copernican ideas, Brahe's observations and his own conviction of the mathematical harmony of the universe. Galileo concurrently laboured to produce a new dynamic, and related the times and distances of motion of bodies falling, both vertically and along inclined planes. From these investigations the law of inertia emerged as a limiting case. Newton's consummating contribution was to combine all these partial and piecemeal advances into a single, systematic mathematically correlated system which subordinated them all logically to a minimal set of explanatory principles.

Thus one system, the Aristotelian, in the course of its development, by revealing within its own structure contradictions (what Aristotle, himself, would have called 'aporiai'), is transformed, by successive efforts at more coherent thinking, into a new system, so different in aspect, so much more comprehensive in detail and so much more elegant mathematically that the transition stages are forgotten and the whole process presents itself as a revolutionary change—which indeed it is, though there has been no clean break in the progression.

This is the pattern or paradigm of all scientific revolutions. The earlier conceptual scheme, as it is articulated, reveals its defects. In the ensuing endeavours to repair it, marked, as Kuhn observes, by a proliferation of hypotheses, all of them, in some degree, related to, and attempted developments of, the old one,[2] partial and somewhat disconnected discoveries are made, until finally some thinker of exceptional ability sees the new *gestalt* which unites the work of his distinguished predecessors in a new and

[1] *Dialogue Concerning the two Chief World Systems*, trans. Stillman Drake pp. 22 f.

[2] We observed, as does Kuhn himself, the persistent Aristotelian and Ptolemaic elements in Copernicus—his adherence to the circle, his persistence in the idea of natural motion, his surrender to epicycles, and the like. See above, Ch. V, 1(a).

more satisfactory structure. This is the true significance of Newton's confession that, if he had been able to see farther than others, it is because he had stood upon the shoulders of giants.

(*b*) Harvey, for all his originality, insight and experimentation, likewise builds upon the work of earlier innovators. He has the genius to see the pattern into which the parts of the jig-saw fit. Many of them he provides himself, but the main stages of discovery were contributed by others. Vesalius initiated the practice of independent dissection and investigation unshackled by slavish adherence to Galen. He first commented on the toughness and solidity of the septum of the heart and raised doubts about its permeability. Colombo discovered the lesser circulation. Fabricius drew attention to the valves in the veins (which, by Harvey's own confession, first gave him the idea of circulation). None of these men could see the import of their own discoveries, but they provided the pieces of the jig-saw. It was not only the Copernicans who were, in Koestler's phrase, 'sleep walkers'. The fore-runners of every scientific revolution grope in the dark and catch dim and disconnected glimpses of a fragmented picture which springs to light only when someone sees the ordering principles that unite the whole. Harvey achieved this unifying insight and a new era in physiology and medicine began.

(*c*) The story is repeated a century later in chemistry. The shafts of light that illuminate the disjointed fragments are each shed by a different contributor. Lavoisier himself lists them in the first section of his *Opuscules*. Von Helmont and Boyle had long since observed the existence of a special sort of 'air', which indeed had been noted even by Paracelsus and called 'Spiritus Sylvestris'. Boyle called it 'artificial air'. By 1674 John Mayow had discovered that combustion and respiration both had similar effects upon the atmosphere,[1] and Joseph Hales, much later, in numerous ingenious experiments, found that Boyle's 'artificial air' was associated with burning, with fermentation and other processes. He saw that it was somehow contained or 'fixed' in solid substances and called it 'fixed air'. He evolved a method of collecting it, and other 'vapours', over water, which was later improved by Joseph Black, using mercury (as the gas was soluble). Black showed that 'fixed air' existed as a separate variety of air, though he was not properly aware of its independence of other gases. The same was true of

[1] See *Memoirs of Dr Joseph Priestley* (London, 1806), Vol. I, p. 236.

Cavendish who discovered still more 'factitious airs', and even of Priestley who carried pneumatic chemistry much further.

None of these workers were clearly aware of what they had discovered. They interpreted their findings in the light of the existing theoretical principles, which were a sort of late modification of Aristotelianism. Phlogiston, derived from and closely related to Aristotle's[1] element of fire, was the concept which shaped their thinking, along with the belief that air was another element; so that all the various gases they discovered were seen as air, tinctured or tainted with other matter. The differences were interpreted as depending mainly upon the amount of phlogiston the air contained, accounting for its ability or inability to burn or support combustion.

By 1770 a welter of somewhat confused information was available, all of it germane to what Lavoisier subsequently achieved. Cavendish had isolated hydrogen and tentatively identified it with phlogiston, because it burned. Yet if it was phlogiston, how could it in combustion emit itself, phlogiston? Another 'air' emitted from charcoal (our CO) also burned. This was known, and the relation between the two remained uncertain. Priestley had collected oxygen from mercuric oxide, but he thought (because it supported combustion) that it was 'dephlogisticated air'. Yet nitrous oxide which he had also isolated behaved similarly in some ways, though not in others, so that at one stage (somewhat inconsistently) Priestley thought that his new gas (oxygen) might be 'phlogisticated nitrous air'. Black's fixed air would not support combustion or respiration and so was termed 'phlogisticated', but nitrogen had been known (not, of course, as such) since the time of Boyle and Hooke and was also regarded as 'phlogisticated air'. That substances increase their weight when burnt was noticed by Jean Rey as early as 1630 and the idea that metals took something from the air during calcination had already been suggested before Lavoisier. Guyton de Morveau had conducted a series of systematic experiments to establish finally the phenomenon of augmentation and his findings were reported to the French Academy in 1772 by Maquet, in Lavoisier's presence.[2]

All the factors were available for the construction of the new theory, as they had been for Kepler seeking the orbit of Mars, as

[1] Or, more correctly, Empedocles' element.

[2] See Guerlac, *op. cit.*, p. 136.

they had been for Newton evolving the laws of mechanics, or for Harvey searching for the function of the heart. In Kepler's words, there lacked 'only the architect who would put all this to use according to his own design. For . . . the multitude of the phenomena and the fact that the truth is deeply hidden in the particular details are obstacles to progress.'

(*d*) The same is true of Darwin's relation to his predecessors. In this case the prior conceptual structure was a mixture of Aristotelian tradition and religious belief, crystallized in the ordered framework of the Linnaean taxonomy. The mainstays were the fixity of species and the precise teleological adaptation of animal and plant form to the way of life and conditions of existence of the creature. The articulation of this system involved the collection of specimens, their minute description and comparison, and resultant classification. But in the course of this process fossils had to be accommodated, allied varieties had to be distinguished from separate species, and apparently purposeless affinities between the inhabitants of off-shore islands to those of the mainland, as well as differences between those of neighbouring islands in a group, had to be explained. The attempt to do the first by postulating a series of antedeluvian catastrophies raised awkward questions. What was the source of each new fauna, a special creation, or, as Cuvier alleged, the migration of survivors from distant regions. If the former, how is the evident (if not altogether unbroken) continuity and gradual change in fossil forms to be explained; if the latter, how the differences between extinct and living species? Again, if exact adaptation by an intelligent creator was the rule, how does one account for the intermediate and less accurate examples, such as terrestrial geese with webbed feet, Melipona which makes spherical cells in its honeycomb, the fully formed, yet apparently functionless toe bones of the pig, and vestigial organs of all kinds?

Answers to most of these questions had been suggested by Darwin's predecessors and all the main ideas of his theory had been put forward by one or other of them before the appearance of *Origin of Species*. The conception of a continuity of forms goes back to Aristotle and had been reaffirmed by Leibniz, and even Linneus admitted the possibility of modification as a result of hybridization, crossing and degeneration. Buffon, in his middle period, rejected the cruder view of teleology and taught the mutability of species under the influence of environmental change.

He suggested that affinities of type indicated common origin, drew attention to the effects of artificial selection in domestication and anticipated Malthus by postulating a struggle for existence in which the stronger survived and the weaker were eliminated. He also alleged that the direct action of the environment produced new structures which were transmitted by heredity.

Darwin's grandfather, Erasmus, followed Buffon in his assertion of a struggle for existence, but he denied that adaptation was the result of creation and maintained that it was acquired. He believed consistently in evolution, making man no exception, alleging that life was spontaneously generated as 'specks of animated earth', that it developed first in the seas and evolved by stages to its modern forms. He put forward the view of accidental variation (for instance the chance pull of a muscle opposing thumb to forefinger) which, providing some advantage, is further developed by use and perpetuated by heredity. He gives as evidence of possible evolution of species the type of development observable in the tadpole, the effects of artificial selection in domestication, and of interbreeding, and the influence of climate on varieties of the same species in different regions.

A thorough and systematic theory of evolution was put forward by Lamarck[1] who believed that environment acted directly upon plants to produce changes and upon animals through their nervous systems to create new wants and excite new efforts to satisfy them. Changes of structure, he thought, are induced by habits resulting from these efforts and are passed on by heredity. In later developments of his theory he set out a system of classification based on evolution and the divergence of genera and species from a common root. Lamarck was also one of the pioneers of geology. He opposed the theory of catastrophic transformations and supported that of slow gradual change. With environmental change, however, he believed immutability of the species to be impossible and he denied special creation.

The doctrine of inheritance of acquired characters was rejected by Lamarck's contemporary and colleague Geoffroy St Hilaire, who still adhered, however, to the view that the sole source of specific modification was environment. He thought the state of the atmosphere had an effect on respiration which was far-reaching in its consequences for the functioning of the organism and more beneficial in some cases than in others. This led to a theory of

[1] Jean Baptiste Pierre Antoine de Monet, Chevalier de Lamarck.

natural selection by which the favoured individuals survived and the less favoured died out. He also anticipated the discovery (even later than Darwin) of accidental mutations, suggesting sudden changes by 'saltations' which could be the result of chance causes. He sought, and found in embryology and the study of sports and monsters (teratology), evidence that present forms had descended from others now extinct, and alleged a universal unity of plan of composition among animals.

The great opponent of many of these views, especially that of gradual change and transformism was Georges Cuvier who believed in fixed species and cataclysmic geological changes at distant epochs in the past. Nevertheless, as a founder of palaeontology and a believer in organic holism he contributed much to Darwinism. He pointed out homologous organs in related species and taught that the members of each major group were constructed on the same ground plan, stressing similarity of function as determining structure (in opposition to St Hilaire's structuralism). His theory of catastrophic changes carried with it belief in the extinction of fossil forms, many of which he discovered and described with great accuracy. He also maintained the view of extensive migration, and though he refused to admit modification of the species, this theory naturally draws attention to the effects of changed environment upon the migrating species, and it is clear that Darwin's argument, that a slightly modified form might fill a niche in a new environment better than older types and so be selected, owes something to Cuvier's ideas.

Goethe[1] and Oken also worked out theories of evolution, based largely on a comparative study of skeletons and a somewhat exaggerated belief in the importance and modifiability of the vertebrae. Though T. H. Huxley demolished this theory, he himself proposed an evolutionary view of the development of the skull and the vertebrae from earlier forms, which proved more satisfactory. Meanwhile von Baer worked upon and developed the science of comparative embryology and ontogenesis which Lamarck and Geoffroy St Hilaire had already begun.

Add to all this the momentous work in geology contributed by Charles Lyell who evolved the theory of the sedimentation and folding of the strata, the effects of glaciation and the occurrence of the ice ages, and we see that Darwin found all his material ready to hand. By the early decades of the nineteenth century

[1] Goethe believed in the descent of man from lower species.

evidence for evolution was forthcoming from botany, zoology, palaeontology, embryology, and geology, but none by itself was sufficient to establish the theory. It was the genius of Charles Darwin that wove them all into a single pattern, that saw all together and recognized their convergence. Yet, like Newton, if he saw farther than others it was because he stood on the shoulders of giants.

The struggle for existence was taught by Malthus, who was anticipated both by Buffon and Erasmus Darwin, both of whom also put forward the idea of selection and pointed to the evidence of domesticated varieties. The modifying effect of environmental changes was recognized by nearly all of these thinkers, even Cuvier, who did not believe in transformism. The survival of the fittest was adumbrated by Geoffroy St Hilaire, and also (in agreement with Cuvier) the extinction of earlier forms. Cuvier stressed besides the importance of migration, and, again, with Lyell and other geologists, provided the evidence of the rocks and the fossil record, while von Baer offered that of embryonic development. To all this Darwin made his own direct contribution from his own multifarious observations. Even so, the French accused him of eclecticism. In that there is no fault if the eclectic welds the borrowings from many diverse sources into an overwhelming body of corroborative evidence for a new revolutionary theory of more powerful explanatory value than those it supplants. This, as we have seen, was precisely what Darwin achieved.

(*f*) Einstein's relation to his predecessors is analogous to that of Newton's to his, of Harvey's, of Lavoisier's and of Darwin's. Stepping-stones to the theory of relativity were being laid down from the early decades of the nineteenth century. The conception of the field in the theories of Faraday and Maxwell marks the beginning of the departure from strict mechanism, the breakdown of which was effected by a steady erosion from that point on. The successive discoveries of phenomena, like the Doppler and Zeeman effects, the aberration of starlight, the Fresnel-Fizeau convection coefficients, all of which had to be accommodated in one self-consistent theory; and the continually increasing difficulty of reconciling the phenomena of electromagnetism with classical mechanics, called into being a series of theories of the aether accumulating difficulties and contradictions which we have already noticed. But each new venture contributed something which was later incorporated into the Einsteinian synthesis. The

Fitz-Gerald contraction, the Lorentz transformations and Poincaré's Principle of Relativity each found a place in the new system. Minkowski's four-dimensional continuum, and the geometries of Gauss and Riemann provided the mathematical framework. Einstein's creative contribution was, once more, architectonic. It was a reorganization of concepts and a reinterpretation of categories. The ideas of space, time, simultaneity, velocity and measurement all required reformation giving rise to a new conceptual scheme. Yet all the new ideas retained some continuity with the old in a way which we shall presently see to be important.

vi. INNOVATION AND CONSERVATISM

By now it is sufficiently apparent that new hypotheses are not derived by generalization, of a simple enumerative kind, from particular observations, if only because without the theoretical background the observations themselves are lacking—quite apart from the invalidity of the logical process. It should be equally obvious that new hypotheses are not independent postulates derived either from mysterious psychological depths or from *a priori* metaphysical speculations, unrelated to the experienced facts until these have been deduced from them (in conjunction with initial conditions) and then checked by experiment. They are the fruits of an organizing activity on the part of the discoverer, combining and integrating the partial developments of an earlier conceptual system that his predecessors have made in their search for consistency and comprehensiveness. So Newton, familiar with the general idea of inertia due to Descartes and Galileo, could assume that a body free of all impressed forces ($\Sigma F = 0$) would move uniformly in a straight line. Aware that the planets did not move uniformly in a straight line he could conclude that $\Sigma F \neq 0$ and that some force bent their orbits into a curved figure. He was in possession of Kepler's law: the square of the mean period of revolution is proportional to the cube of the mean distance ($T^2 \propto r^3$), and of Huygen's law of centrifugal and centripetal force, $F \propto r/T^2$. Combining these it follows that $F \propto r/r^3$, or $F \propto 1/r^2$. Add to this his own postulate that Force = mass times acceleration ($F = ma$) and the law of gravitation emerges: $F = G (m_1 m_2)/r^2$. This gave him an organizing principle by which to link the phenomena of the tides, of falling bodies and projectiles, of hydrodynamics and of planetary motion into a single system.[1]

Mutatis mutandis, the same sort of thing was done by Harvey with the discoveries of Vesalius, Fabricius and Colombo; by Lavoisier with those of Hales, Black, Cavendish, Priestley and Guyton de Morveau; by Darwin with the ideas of his grandfather, of Buffon, Lamarck, Cuvier, Geoffroy St Hilaire, Malthus and Lyell, and by Einstein with the contributions of Maxwell, Mach, Lorentz, Fitzgerald and Poincaré, as well as the mathematics of Minkowski, Gauss and Riemann. In smaller compass Kepler performs a similar task in his re-structuring of Tycho's data. Adams and Leverrier calculated the whereabouts of Neptune from the measurements and observations made by numerous other astronomers of the perturbations of the orbit of Uranus, and Anderson constructed his body of evidence for the existence of the positron out of the findings of Rutherford, Chadwick, Millikan and others, as well as his own. The new theories were all 'abductions', in the sense that they were drawn from and directly related to observed phenomena, but they were also constructive syntheses dictated by the pattern of organization of these facts in the framework of the conceptual schemes with which the scientists were working. This framework is no fixed or static structure; it is continually in process of development, either by articulation of detail or by combination and integration of developments in separate areas often involving radical reformulation of concepts.

Change is not initiated primarily by the unheralded discovery of new empirical data. More usually unfamiliar facts, when they are present, are overlooked, misperceived or misinterpreted, unless the theoretical framework demands them, or, by its inherent discrepancies, raises questions that draw attention to them. It is these internal defects that underline anomalies and raise the questions, answers to which reveal new facts and demand theoretical reconstructions. The scientists who achieve such

[1] Cf. N. R. Hanson, 'An Anatomy of Discovery', *The Journal of Philosophy*, Vol. LXIV, No. 11; June, 1967; pp. 341 f. and 347: 'Newton's Law . . . does not have to do with any *object* at all. Rather this Law delineates the structure of phenomena by providing a conceptual framework for parameters. Newton's L . . . is a form-giving structure which enables us to link descriptions of planetary perturbations, eclipses, tides, missiles and falling apples in a way that discloses how all such happenings interconnect conceptually; they are all thereby rendered intelligible to that extent.'

advances are those with the most synoptic vision, with the best sense of the interrelatedness of things, and with the greatest capacity for abstracting the essential principles of order from the complexities of the mutually connected particulars.

Both these accomplishments are necessary—both the ability to see things together and to abstract general principles—because the latter by itself give only a bare skeleton, whereas the scientific system is a totality, or conceptual whole, of observed phenomena ordered and interpreted by means of a system of abstract laws. The laws take the form, for the most part, of mathematical formulae—functions of co-variant quantities—which give form and significance to the observed facts and permit inferences from some elements in the structure to others. This is possible because the system is polyphasic;[1] that is, because the laws are functions of covariants, as has been said. Such functions are unitary principles which display themselves, when substitutions are made for the variables, in a wide variety of different exemplifications, and so are multi-faceted. In many of the examples we have studied this is obviously the case. Kepler's theory is stated in three mathematical laws of planetary motion, Newton's in still more general laws, Einstein's in others yet more complicated. Though laws in the biological sciences lend themselves less readily to mathematical formulation, the function of Harvey's principle of circulation and, yet more obviously, of Darwin's natural selection is the same as that of the mathematical equations. They are general rules that permit of multiple exemplification. They all serve as what Gilbert Ryle calls 'inference tickets', and are the unifying principles that reduce the manifold of particular facts to order.

But the scientific order is neither an accumulation of 'facts' nor an unchanging pattern of interpretative conceptual principles. The system develops in stages, each of which is a phase in a two-fold process of greater articulation and wider and more thorough synthesis. Sometimes it takes the form of further specification and more detailed application of already established theoretical principles; sometimes it is brought about by the combination into a new pattern of small innovations or departures from the older established scheme. And throughout the process the source of theoretical invention and the motive force of scientific advance is the inherent drive of the scientific intellect towards

[1] Cf. *The Foundations of Metaphysics*, pp. 82 ff. and *passim*.

conceptual unification and all-embracing explanatory efficacy. This trend is repeatedly revealed in history. It has just been illustrated by the example of Newton's unification of the sciences of celestial and terrestrial mechanics. In like manner, Maxwell united the fields of optics and electrodynamics, and Einstein and Weyl fused the laws of mechanics and of electromagnetics into a single system. The quantum theory has bridged the gap between physics and chemistry and advances in organic chemistry and microbiology are fast merging the fields of the physical and the biological sciences. But this unifying trend is only the over-all manifestation of the synthesizing activity of scientific thinking in the process of discovery—an activity essentially of organization which does not merely amalgamate collected empirical material but strives to relate it in its minutest detail within a systematic conceptual structure.

It is this urge to conceptual unity that underlies both the construction by the great thinkers of new schemata and the resistance which their contemporaries almost invariably show to accepting their innovations. When an established system, in the course of its articulation, reveals its inadequacies; when scientists are prompted to make *ad hoc* postulations and suggest complicated additions to theory in order to remedy these inadequacies, and when their efforts result only in more confusion and contradiction, the man who succeeds in finding a new and more coherent synthesis seeks to put it in the place of the old. But the old system would never have been adopted if it had not, in its day, been effective in explaining a wide range of experience; and comprehensive explanatory power is so important that no system with any plausible pretentions to it can ever be lightly abandoned. It follows that the main body of scientific opinion is always conservative and new theories often meet with stubborn opposition.

Resistance to new discoveries is easily understandable and is even scientifically proper, not only because it ensures salutary caution and excludes theories that lack secure foundations in evidence, but also because any major innovation which threatens an established conceptual system threatens the entire basis of scientific investigation. We have already seen that a conceptual background is indispensable to scientific endeavour and the existing scheme can never be abandoned unless and until a more adequate one has been provided. For this reason scientific theories are never falsified simply by the discovery of anomalous phenomena.

They are only superseded when a substantial body of evidence validates an alternative and more viable theory.

The most notorious cases of obstructive conservatism, the resistance of the Peripatetics to Galileo and the Copernicans, and the stubbornness of Priestley in his refusal to give up phlogiston, were by no means instances of sheer obscurantism and pigheadedness. The Aristotelian-Ptolemaic doctrine was one of the most comprehensive and consistent explanatory systems ever constructed. In terms of a single principle, the relation of matter and form, with its associated and implied ideas (potentiality and actualization, the four causes, dispositionalism and the 'logos' or ratio of combination) Aristotle explained the genesis of the physical elements from the basic opposites, chemical combinations, the phenomena of change and movement, generation and corruption, the motion of the heavens, the relation of body and soul, the interrelation in living forms, of nutrition, sensation and locomotion, reason and intelligence, and the principles of moral action. Aquinas, by incorporating this system into the teachings of the Christian Church, gave rational support by means of its scientific principles to religious belief. By his one innovation, Copernicus threatened the whole of this impressive structure with destruction. If the earth were not the centre of the universe, the whole theory of motion and natural place was undermined and this was conceptually connected with the movement of the spheres, the ultimate origin of which was God, the unmoved mover. If the earth were in motion, all this had to be radically revised. But even worse, Copernicus' theory entailed the removal of the sphere of the fixed stars to an incalculable distance from the sun (in order to explain the absence of observable parallax), and for the Aristotelians it was the movement of this sphere (caused directly by God) through its contact with the sublunary layers of the elements, that produced motion and change among terrestrial entities. This movement was the cause of mixture among the elements, the 'logos' or form of which gave their special character to the ὁμοιομερῆ (chemical compounds). These again became the proximate matter of soul, the higher forms of which were sensitive and intelligent. The entire gamut was of a piece—a continuous series of forms in which change at any one level threatened to disorganize the whole scale. By introducing a new cosmological schema Copernicus undermined the Aristotelian theory of motion, and the entire complex of scientific, ethical and religious belief was thus put in jeopardy. Small wonder, then, that seventeenth-century scholars and clerics should regard Galileo, in his advocacy of heliocentrism and his endeavours to replace Aristotelian dynamics with a new-fangled theory of falling bodies, as a dangerous revolutionary. Their resistance was on the side of organized science and devout well-ordered society.

Priestley's devotion to the phlogiston theory was similarly motivated, though the theory itself was of less far-reaching significance. Nevertheless, it had introduced light and order into a welter of confusion and obscurity handed down by mediaeval alchemy to the scientists of the seventeenth century. Doubtless even alchemy had some scientific contribution to offer, but a glance at the writings of Paracelsus easily reveals the mixture of superstition and naïve empiricism that constituted the sum of chemical knowledge in his day. He tells us:

'By the mediation of Vulcan, or fire, any metal can be generated from Mercury. At the same time, Mercury is imperfect as a metal; it is semi-generated and wanting in coagulation, which is the end of all metals. Up to the half-way point of their generation all metals are Mercury. Gold, for example, is Mercury; but it loses the Mercurial nature by coagulation, and although the properties of Mercury are present in it, they are dead, for their vitality is destroyed by coagulation. . . .'[1]

And there is much more of the same sort. How much more enlightening than this is a theory that explains the similarity in the properties of metals by their common incorporation of phlogiston, and of calxes by their loss of it on heating. The phlogiston theory gave, for the first time, a general conception of chemical combination and analysis and the possibility of something like a chemical equation.

$$\text{Calx} + \text{Phlogiston (from Charcoal)} \rightarrow \text{Metal}$$
$$\text{Metal (heated)} \rightarrow \text{Phlogiston} + \text{Calx.}$$

It also explained how air supported combustion by absorbing the emitted phlogiston, and ceased to do so when it became saturated or 'phlogisticated'. Here was a unifying principle that could

[1] From *Coelum Philosophorum*, in *Hermetic and Alchemical Writings of Paracelsus*, trans. by A. E. Waite (London, 1894).

produce order out of the chaos of alchemical 'canons' and 'principles', with their 'waters, oils, limes, sulphurs, salt-peters, alums, vitriols, chrysocollae, copper-greens, atraments, auri-pigments, fel-vitri, ceruse, red earth, thucia, wax, lutum sapientiae, pounded glass, verdigris, soot, crocus of Mars, soap . . .' and so on *ad lib.*, listed by Paracelsus.[1] Despite all the difficulties later revealed by the work of the pneumatic chemists, one can readily sympathize with Priestley's reluctance to give up a doctrine which had in fact proved to be the mainstay of the first truly scientific movement in chemistry.

Moreover, not even the major revolutions bring about total transmutation of concepts. Aristotle's notion that movement occurs only when some force (pull or push) is operating is corrected by Newton to the rule that acceleration occurs only as a result of the impress of force. The medieval conception of impetus survives in Newtonian physics as the classical notion of momentum (a term used by Galileo loosely to mean something between the two). The circular motion attributed by the Ancients to the heavens was no sheer error. It is and remains the rotation of the Earth on its axis.

The idea of combustion as a process of chemical change in which substances are combined, gaseous effluvia given off and in which air plays an essential part, was valid both before and after phlogiston had taken its place among scientific myths. What changed was the conception of the chemical elements taking part in the process. That this change made all the difference in the world is not to be denied, for a new configuration always does alter the entire aspect of its components, even though some of them may also have been present in the old; but the discoveries of the pneumatic chemists led to the new conception by successive steps and they were made, for the most part with the help of the idea of phlogiston, without which much of the experimental work might not have been attempted.

Guerlac rightly contends that to view the chemical revolution simply as tantamount to the overthrow of the Becher-Stahl phlogiston theory of combustion says at once too little and too much. 'It exaggerates the break with the past; it neglects the accumulated body of old and recent factual knowledge that was absorbed unaltered by the new chemistry. . . .'[2]

Similarly, Darwinian biology retained the taxonomy of Linnaeus

[1] *Op. cit.* [2] H. Guerlac, *op. cit.*, p. xvii.

though the relation between genus and species had been transformed (and we should not forget that the whole system of classification was an inheritance from Aristotle and the ancients). Not even contemporary physics, with all its revolutionary departures from absolute space and time, solid indestructible atoms and action at a distance, has cut itself wholly adrift from classical mechanics. Newton's laws appear as special limiting values of Einsteinian formulae; the Principle of Least Action of Maupertuis and the Principle of Least Time of Fermat are combined by de Broglie in his wave theory of matter and Schrödinger's wave equation is based upon the Hamiltonian.

In fact, the progress of science, though marked by periodic revolutions, is nevertheless a continuum. There are no clean breaks, even when there are radical transformations. And, throughout the process, the constant characters of scientific thinking are its systematic and architechtonic form, its persistent aim at theoretic unification and its unrelenting battle against inconsistency and conceptual conflict.

vii. SYSTEM AND DEVELOPMENT

In the last chapter we found that the creative scientist drew upon already existing knowledge and built upon it both by destructive criticism and by development of its implications. In this chapter we have gained more insight into this process by the recognition that scientific research proceeds within the structure of an existing conceptual scheme which it develops both by detailed articulation and by modification and reform demanded as the result of this articulation.

New ideas are suggested to the discoverer by his reflection upon and manipulation of the existing structure, the inconsistencies of which (revealed in the course of its articulation) had stimulated such reflection. So Kepler manipulated the circular orbit and, stimulated by its discrepancy of 8′ of arc, converted it into a circle with a bump on one side (an ovoid), and that again finally into an ellipse. The pneumatic chemists progressively modified the conception of air from that of a uniform element to an element that could be tainted or tinctured, then they distinguished phlogisticated from dephlogistigated air, fixed air from inflammable air, and ultimately Lavoisier came to see it as a mixture of distinct gases.

In every case we found the theory functioning as a principle of organization reducing a manifold of observed phenomena to order, and the method of establishing a new hypothesis, we saw, was a process of construction correlating the evidence in a coherent system. In this process we found an intimate interdependence between observation and theory, which has now been further illuminated by the realization that observation is always conditioned and directed by the operative conceptual scheme. A satisfactory theory of this feature of scientific method will require a closer study of the nature of observation and the perceptual assurance on which it rests. The construction of a new theory by organizing diverse evidence is, in effect, a reorganization of the phenomena in a more systematic and consistent manner and is a part of the process of articulating a wider system and a phase in a larger process of development of the conceptual scheme as a whole. Properly to understand the logic of discovery and verification in the more specific instances, therefore, it will be necessary to grasp the nature and principles of this larger process in which the lesser is a phase. Scientific revolutions are transitions from one conceptual system to another and the large-scale progress of science presents itself as a series of intermittent revolutions which is nevertheless a continuous process. The intervals between revolutions are periods, as it were, of gestation in which one conceptual scheme, by internal specification and external expansion gives birth to the next. Each new advance is a new synthesis, and each major step in the process is a structural whole connected to its neighbours by minor structures developed during the intervening period. For example, the Aristotelian system is transformed into the Newtonian by way of the theories of impetus, heliocentrism, Kepler's laws of planetary motion, Gilbert's theory of magnetism, Galileo's laws of falling bodies and Descartes's theories of vortices and of inertia. The classical physics becomes the physics of relativity and quanta by way of Maxwell's equations, Lorentz's electromagnetic theory of matter, on the one hand; and J. J. Thomson's theory of electrons, Planck's theory of quantal radiation, the Rutherford-Bohr theory of the atom, de Broglie's and Schrödinger's wave theory of matter, on the other. Thus the process regarded serially is a succession of developing theoretical reorganizations, and seen in cross-section is a system of theoretical principles ordering a multitude of phenomena.

Each theoretical system is a polyphasic unity—a unitary

principle or function exhibited in multiple applications and actualizations—and the whole movement, so far as it displays uniformly the characters we have ascribed to it and is an expression of a continuous nisus toward comprehensive and unified theoretical systematization is itself polyphasic. For each theory exemplifies the same general character of an organized whole and develops by a similar dialectical process of emergent aporiai and their solution, as does the over-all process itself. The logic of this dialectical process is what we need to discover if we are to achieve an adequate epistemological theory of scientific explanation and scientific discovery. Our attempt to expound such a theory will begin with the investigation of perception as the source of empirical knowledge and will then consider the logical process by which this is elevated to the level of science. Finally, we shall have to assess the claim of science to be knowledge of the real world and examine the notions of objectivity as attributable to science and of reality as imputed to the world it studies.

PART III

EPISTEMOLOGICAL

CHAPTER VIII

PERCEPTION

i. THE EPISTEMOLOGICAL CRUX

The importance of observation in science is so great and its relation to theory so intimate, so complex and so variously interpreted by philosophers, psychologists and historians, that no clear insight into the matter is possible without a full study of perception. I cannot do complete justice to the subject in one chapter but shall try to give an account sufficient to serve as some guide for the construction of a satisfactory theory of science. Accounts of perception have been offered by physiologists, psychologists and philosophers and I have already discussed some of the physiological and psychological theories in *The Foundations of Metaphysics in Science*.[1] In some measure, what follows here will be a continuation of that discussion and will draw upon its results. But in the last resort the only theory of perception that will be useful and adequate for our purpose will have to be an epistemological theory, so attention will be given below to the views of philosophers. No epistemological theory is likely to be sound, however, which neglects either physiological or psychological facts (so far as they are known) and no interdict is to be placed on contributions which may be forthcoming from those sources.

In everything that follows the essential epistemological questions, to which answers must be sought, should be kept in mind; and about perception what we chiefly need to discover is how sense-experience can give us knowledge of the material world. As it is commonly held (whether rightly or no) that knowledge, to be knowledge, must be true, a further question naturally arises. How do we distinguish, among sensuous experiences which appear to reveal facts about the external world, whether or not they are veridical? These are the central epistemological questions and any theory which evades either of them fails of its purpose. We must note, also, that the second question casts no doubt upon the fact that we do make the distinction between veridical and non-veridical perception. It asks only for the criteria by which we can

[1] In future references abbreviated to *Foundations*.

do this unerringly. It is not answered, therefore, by the insistence that unless we sometimes, or generally, perceived veridically we should never know that we suffered illusion (and thus could base no arguments upon its occurrence and nature). That contention is true enough, for if we could not distinguish true from false, we could identify nothing as illusory. The important point, however, is not that what we perceive is frequently true of the world, but is to identify the reliable criteria of veridicality.

ii. COMMON VIEWS OF PERCEPTION

The all too common view of science, as rooted in and derivative from some form of pure and theoretically neutral sense-observation, rests on certain preconceptions about the character of perceptual knowledge itself as *fons et origo* of all knowledge whatsoever. The kind of account which would be given might go somewhat as follows. Knowledge, especially the sort to which the natural scientist aspires, is knowledge of the world around us, and our only access to this is through our senses. It is by sensation and by no other means that we have direct contact with things in our environment. Sensations of touch (i.e. of direct contact) are primary, smell and taste are modifications or specializations of this, and sight and sound which apprehend distant objects are faculties specially adapted to receive and record impulses transmitted through space (electromagnetic and atmospheric vibrations) which reveal the position and qualities of distant objects. This is a somewhat up-to-date and sophisticated way of describing the situation which is held to give rise to knowledge of the world, but the view in its main outlines has been current since ancient times, though earlier writers would probably have spoken of 'species' or 'idola' in place of electromagnetic vibrations.

Sensation is thought of as a set of ways of transmitting information. The external world is a vast concourse of qualified objects of which the character, structure, movements and changes constitute a mass of information. This, or at least some of it, has to be conveyed to our minds if we are to know it, and the primary means of conveyance are the physical influences impinging upon the sense organs from the objects which the information is about. The sense-organs are, of course, only receptors, and the information is transmitted through the nerves to the brain, where it is recorded as perception, stored in memory and made available as knowledge. This, roughly, is the sort of account we find in Locke.

But as our minds are active on their own account, they 'operate upon' the information coming in through the senses. They reflect upon it, separate and distinguish the impressions (made by certain sensory processes) that represent special qualities, and recombine them in various ways. They 'associate' some with others and on appropriate occasions revive sensations earlier experienced but no longer actually being received from without. Accordingly, apart from directly received sensation, our knowledge is largely factitious. We construct for ourselves a representation of the outside world out of the materials that we receive through the senses, and this representation may or may not accurately correspond to the external facts. Scientific theories are part of the constructed representation and their accuracy, therefore, like that of any factitious knowledge, can be checked only by reference to the original source of information, the deliverances of sense. *Sine experientia scientia non potest.*

In their raw and unprocessed form, as they are originally received, the deliverances of sense are simple, particular qualities—Locke's 'simple ideas of sensation' or the 'sense-data' of modern theories; and these are the original building blocks out of which all knowledge, and science as the most faithful representation of the outer world especially, is constructed.

Not all of this account of the matter is utterly false and all of it is plausible and seductive, so that in more or less elaborate versions it has dominated the minds of many philosophers throughout history and is the sort of story which, so far as he is interested in it, appeals to the average unphilosophical man. But it contains a number of assumptions which on scrutiny prove to be untenable. Two of these, closely connected together, are (i) that sensation is a process of conveying, in some sort of coded form, information already existing complete in the outside world, into the brain where it is reconstructed as knowledge; and (ii) that the deliverances of the separate sensitive faculties of vision, hearing, touch, smell and taste are the sole and original sources of perceptual knowledge. A third assumption which proves to be untenable is that these deliverances are particular sense-data of which we are explicitly aware and can isolate by analysis from more complex perceptual experience.

The first two of these assumptions I have done something to

discredit in *Foundations*.[1] There I have argued that physiological theories, treating sense processes as the transmission of coded information, lead to self-refuting epistemological theories of perception, which are not themselves entailed by the physiological evidence, but to which physiologists are sometimes led by the tacit adoption of epistemological presuppositions. Philosophers, on the other hand, tend to be misled, by a superficial acquaintance with certain physiological facts, into too ready an acceptance of theories of the sense-data variety. In that book also I have argued (Chapter XVI) that the special senses are differentiated out of a more primitive form of sentience by a process of organization, so that what they produce is something very different from a train of bare and simple particulars. We shall have occasion later to revert to this last point, but I have no wish to repeat here what I have written in the earlier book, and beg leave of the reader simply to make use where necessary of the conclusions reached in that discussion. The notion of sense-data, however, is so intimately involved in recent and contemporary philosophical discussions of perception that some space may usefully be given to an examination of the arguments pro and con.

iii. SENSE-DATA

To philosophers in search of a theory of knowledge a natural and promising way to begin seems to be to analyse the more complex objects of our awareness into their simplest elements, and those who look for criteria of validity and truth are apt to try to distinguish those elements in our experience that are least liable to error and so least susceptible of doubt. As errors are usually the result of faulty composition, these two lines of inquiry tend to converge and the simplest elements are identified as the most indubitable. In dealing with perception, if we begin with the sensuous presentation to consciousness of the complex objects of everyday experience, the first method of procedure leads us to analyse out of these the simple qualities of colour, sound, scent and the like, as the elementary units; and the second quest leads to the recognition that while we are sometimes mistaken in identifying the objects we perceive, we can hardly be mistaken in our immediate awareness of any one of these simple qualities. It is always possible to misjudge the shape, size, distance, colour or identity of any

[1] Chs XVI, XVII and XIX.

object seen, felt or heard, but it is not possible to doubt that, while one is actually aware of a patch of colour, or a sound, or a feeling of pressure, that that particular sensation is what one experiences. Numerous philosophers have, therefore, identified these simple qualities as the primary elements of perceptual knowledge and have called them by various names, 'simple ideas of sense', 'sense-impressions', 'sensa', 'sense-data', 'sensibilia'. They then proceed to explain how, from these, our developed perception is built up and how they and the constructions we make with them are related to the physical things of the world.

Perhaps the most persuasive consideration in leading thinkers to identify the fundamental constituents of perceptual knowledge as sense-data has been their reflection upon our experience of perceptual errors[1]—doubtless in their search for the indubitable factors, or what Russell called 'hard data'. Among such experiences some of the most impressive are visual (and other) illusions and delusive appearances, because objects are presented in them as directly as in any perceptual experience, yet they diverge, in specific ways, from the physical objects actually there. Consequently, something is, or seems to be, perceived where nothing (or something quite different) actually exists. Yet while we are thus misled as to the actual situation, we are nevertheless (in Locke's phrase) invincibly aware of the immediately sensed qualities involved. What we are perceiving in such cases is obviously not the physical object, though we certainly do perceive something, Moreover, this something is qualitatively like what we perceive when the physical object present is correctly represented —that is why we are misled. Therefore whatever it is that we perceive when we are victims of illusion and delusion must be of the same general character as what we perceive when we are not mistaken; yet it cannot in both cases be the actual qualities of presented physical entities. It must therefore be some other sort of entity which we may call a sense-datum because it is given in sense.

Sense-datum theorists usually go on to point out that a large proportion of the perceptual data we receive is in some sense or in some measure illusory. Visual objects are nearly all distorted perspectivally. Their apparent sizes and shapes vary with their position, though, of course, their actual sizes and shapes do not vary in the same way, if at all. Their apparent colours change with

[1] Cf. *Foundations*, pp. 349 and 370.

their illumination. Sounds differ in intensity according to their distance from the percipient and change pitch if the sounding object is in rapid motion away from or towards him. In numerous ways, on reflection, we are compelled to admit some discrepancy between what we directly sense and what we know or later discover the object to be like. So these philosophers come to hold that what we directly sense can never be the physical object itself, that we never perceive anything except sense-data. Then either the existence of physical objects is denied altogether and they are held to be constructions out of sense-data, or else their existence is supposed to be inferred from the way in which sense-data occur. Neither alternative turns out to be tenable, and a reaction against theories of sense-data is hardly surprising.

That physical objects are constructions out of sense-data either is or leads to phenomenalism, a doctrine which has been ably refuted by several contemporary writers. The most telling arguments against it have been admirably presented by Professor R. J. Hirst and I need not repeat them here.[1] The view that we infer from sense-data to external objects has been widely and variously criticized. Its main defect is its dependence on the causal theory of perception which breaks down inevitably in a system which makes the causes, to which inference is supposed to be made, unknowable *ex hypothesi*. For in that case the relevant causal laws could never be discovered.[2] The fundamental flaw in the theory is that sense-data are incurably subjective. Some writers have argued vigorously for unsensed *sensibilia* but none have succeeded in avoiding either confusion or self-contradiction or both.[3] Consequently sense-data must be conceived, if one is to be consistent, as *sensed* data. This is so if only because the basic assumption of the theory is that the data of sense are the ultimate source of empirical knowledge. They cannot then reveal the existence of what is not included in their assemblage—that is, what is not sensed. But all physical objects have been excluded by the theory from this assemblage, they are all inaccessible things-in-themselves and no knowledge can therefore ever consistently be claimed, on this type of theory, about the external world, about relations, whether of resemblance, causation, or belonging, between sense-data and physical things.

[1] See *The Problems of Perception* (London, 1959), Ch. IV. I have said something in criticism of the theory, as it appears in Berkeley, Price and Ayer, in *Nature, Mind and Modern Science*, Chs VII, XV and XVI.

[2] See *Nature, Mind and Modern Science*, Ch. VI and *Foundations*, Ch. XIX.

[3] Cf. *Nature, Mind and Modern Science*, Ch. VII.

iv. CRITIQUE AND MERITS OF SENSE-DATA THEORIES

The entire sense-datum position with all its variants has in recent years come under severe and widespread criticism, by philosophers most of whom seem to have overlooked the decisive criticisms by which it had already been disproved in the early decades of this century (if not before). In fact, the credit for disposing of it belongs as much to Kant as to any later philosopher. The arguments of most of its modern opponents are, moreover, unsound even when their conclusions are commendable, and it will help us to appreciate their errors, if we note briefly, before attending to their criticism, the lasting merits of the theory. Sense-datum theorists are primarily concerned about the distinctions between the veridical and the illusory, between the indubitable and the corrigible, and these are matters essential to any epistemological treatment of perception. When they go on to seek the relation between sense-data and physical objects their underlying aim is to identify that relation which enables us to tell when our perceptual experience truly represents the character of the external object. Whatever may be the defects of such theories, therefore, they cannot rightly be accused of epistemological irrelevance or convicted of evading the central problems.

Further, what cannot be denied is, that unless something appears to the percipient, nothing is perceived.[1] It is also well established, by physiologists and psychologists, as well as by common experience, that what appears is seldom identical with, and usually very different from, what we take to be actually the case. (For example, what appears to a person on the ground looking at the pyrimid of Cheops can never be more than two triangular shaped sides but he takes the object to be a solid four-sided pyramid.) When we perceive, whatever we perceive, our perceptual awareness always goes beyond what is directly presented to us. This is a fact of fundamental and far-reaching importance, to which the sense-datum theory quite rightly draws attention, even though the explanation of perception that it offers is incorrect.

A number of modern philosophers have become, somewhat belatedly, aware that something is amiss with the conception of sense-data and it has been brought under cross-fire from several angles. But the volleys from most (if not all) of these attackers have

[1] Cf. *Foundations*, p. 349.

misfired or have failed to hit their mark because the critics have, in the main, missed the essential truth about perception that the sense-datum theory has recognized. This essential truth is that what could ever be directly given or passively received falls far short of what we normally perceive, so that perception inevitably involves some activity over and above the mere acceptance of unprocessed data. By emphasizing the given element, however, the sense-datum theory, in its turn, distracts its proponents from the very feature of perception that best supports their theory and gives them pretext for making their basic (but unwarranted) distinction between 'hard' and 'soft' data. No satisfactory theory can be given which overlooks the essential element of interpretation that every act of perception involves, and the presence of which in every such act is unquestionably revealed by all the available evidence. It will be my object to establish this fact in this chapter, and to show that the activity concerned entirely absorbs everything analogous to data, which, so far from being 'hard', are merely the vague rudiments of what is subsequently articulated as a cognizable object.

V. ACHIEVEMENT

The view made popular among analysts by Ryle is that verbs of perceiving are 'achievement' words and, as Ryle's own assertions make clear, acts of perception are the accomplishment of some objective which requires training, skill and sometimes a degree of effort—an effort that may fail or go wrong. But however effortless it might be at other times—and, for the most part, usually is—it is the outcome of a process or activity, by means of which, presumably, some piece of information has been attained. The clear implication, therefore, of calling verbs of perceiving achievement words is that perception is not and cannot be the immediate and intuitive apprehension of a given, atomic, simple datum. On the contrary, it is now more common for philosophers to maintain that perception is an intentional process, and that the use of verbs of perceiving implies the actual presence of a physical object (or state of affairs) of which the perceiver becomes aware in perception. No such awareness could be the result of an immediate apprehension of a simple datum, for no physical thing is simple, still less is any actual state of affairs. The least of these is a complex object which could not be presented *in toto* in any one perceptual act, or to any one of the senses, or even to all of them together.

An ordinary solid object has sides not all of which can be seen at once, unless it happens to be transparent. It has an inside which is usually (e.g. when it is opaque) not accessible to any of the senses. It has causal properties and is liable to causal influences of various kinds (compare Descartes' discussion of the wax in the second Meditation). When we say that we perceive it, we mean that we perceive it as a physical object having all these properties. But certainly they are not all immediately apparent to us. Yet what we perceive is not just one or two plane surfaces, or a patch of colour, or a complex of tactual qualities, or a sound, or any group of these in conjunction. It is a physical thing. The perception, therefore, could not possibly be a simple immediate receptive awareness, but must be an achievement—the end result of an activity which makes us aware, in a single act of comprehension, of a large number of qualities and properties, aspects and characteristics of the complex object. Moreover, we comprehend them not simply as a conglomeration but as systematically interrelated in precise ways. What is true of a physical thing is even more forcibly true of a situation or state of affairs, which includes numbers of such physical things as well as temporal changes and successive events.

That physical things and situations are indeed the objects of perception is not merely the common belief of 'the plain man', as reflected in ordinary language, it is also the considered view of philosophers and scientists. Hence it follows of necessity that perception must involve some complex mental process of which it is the achieved result. The question is whether it is a process (more or less conscious) of construction out of elementary sense-data or one of some other kind. The former is ruled out if no evidence can be found—as none can—of any apprehension of data of the required sort. We must therefore explore the alternative.

A sense-datum was held to be a sort of replica or reflection of some character in the physical world reproduced in the brain (or mind). Quite apart from the evidence available from physiology that no such reflection occurs in the brain and the numerous other considerations which make any such view untenable, no mere replication of external qualities, even though complex in structure, would by itself amount to a perception. The reason for this is that perception is the acquisition of knowledge about an object and no copy, photograph, or replica of an object, though it may serve as a means to the acquisition of knowledge, can by itself constitute knowledge. That involves at least recognition of the replica *as* a representation of the object—an achievement always quite impossible, *ex hypothesi*, on any sense-datum theory (for nothing is recognizable as a replica unless the original is also available for inspection). The proper understanding of the nature of perception depends on the clear realization of this fact, that perception is a form of knowledge—and, of course, the acquisition of knowledge is an achievement. That is why verbs of perception are achievement words.

D. M. Armstrong recognizes this in his book *Perception and the Physical World*. He asserts that an act of perception is one of acquiring knowledge, or belief, about particular facts in the physical world by means of the senses. But, as we have already seen, no knowledge of any particular fact is a simple matter. Whatever fact is perceived, is perceived as 'the fact that' something is the case—to use an example dear to Armstrong, 'that the cat is on this mat'. It is not possible to know or believe this unless one already knows what sort of things cats and mats are and how they are, and can be, related to one another and to other things. It also involves prior knowledge of spatial and temporal relations and the awareness of a spatio-temporal scheme into which both cat and mat can be fitted. In fact, even so simple a perceptual situation as that in which I see the cat on the mat, presupposes a great mass of background knowledge without which the act of perception could not take place. Further, it is an achievement for which (it is assumed in the example) I use my senses. It therefore presupposes also the physiological and psychological activities that result in sensory awareness and enable me to combine it with this mass of background knowledge in order to acquire the belief about the cat's situation. To understand the nature of perception, therefore, we must take into consideration all of these contributory factors, the physiological, the psychological and the epistemological.

vi. DISCREPANCY BETWEEN 'DATA' AND PERCEPT

Both historically and theoretically the notion of sense-data was prompted by physiological considerations.[1] But when we appeal to

[1] Cf. Hobbes, *Leviathan*, Ch. I, and Locke, *Essay*, Introduction, 2, and Bk II, Ch. VIII, 13.

modern physiology we find no grounds for postulating anything of the sort.[1] What we do find is a wide discrepancy between the character of the stimuli and of the neural processes on one side and that of the perceptual object on the other. Equally there is no similarity between the physiological processes and the external object as described either by common sense or by the natural sciences. Locke was partially aware of this when he remarked on the dissimilarity between the feeling of pain and the division of the flesh by a sword. We do not, in fact, perceive stimuli, nor do we perceive neural processes and neither is in the least like the objects that we do perceive. We shall find psychologists most emphatic (for the experimental evidence is copious) that we do not 'see' retinal images nor perceive objects in the way in which they are projected upon the retina. Still less are we aware of anything similar to the corresponding neural processes in the brain.[2]

Further, in vision, as also in the other senses, the stimuli are almost always ambiguous in their relation to the perceived objects. An elliptical patch of colour may relate to a flat elliptical object, to a flat circular object tilted, or to a solid object partly shaded. A high pitched sound might indicate the cry of a child, or of an animal, the call of a bird, the creak of a door, or the sound of a machine. These ambiguities are ubiquitous and the knowledge we acquire through them must be accounted for otherwise than in terms of resemblance.

Two factors are instrumental in bridging the discrepancy between stimulus and percept. One is the activity of organizing diverse responses into a coherent system and the other is the mediation of past experience. How these two factors function we shall discuss further below. They are not mutually independent for the first is the process by which the mediating experience is

[1] Cf. *Foundations*, pp. 318–22.

[2] Cf. R. Borger and A. E. M. Seaborn, *The Psychology of Learning* (Penguin Books, Harmondsworth, 1966), p. 118: 'Human sense-organs register light stimuli of varying area, intensity, wave-length and duration, sounds of varying frequency and amplitude and so on. The incoming proximal stimuli are constantly changing, so that from second to second the person receives differing total physical stimuli from the environment. Despite this, perception is of a stable world of solid objects with few, if any, discontinuities.'

Cf. also E. H. Gombrich, *Art and Illusion*, p. 45: 'What we get on the retina, whether we are chickens or human beings, is a welter of dancing light points stimulating sensitive rods and cones that fire their message into the brain. What we see is a stable world. It takes an effort of the imagination and a fairly complex apparatus to realize the tremendous gulf that exists between the two.'

generated and the experience, as it grows, contributes to the process of organization. But that process is primal and can be traced well below the level of consciousness at which one can properly speak of 'experience'. There is a wealth of physiological evidence of processes intricately and minutely integrating unconscious responses to afferent impulses, and adjusting efferent reactions to produce co-ordinated behaviour in the organism, which, while serving an important function in its own right, also contributes indispensably to our conscious perception.

Impulses from the labyrinth of the inner ear, are co-ordinated in the brain with innumerable kinaesthetic stimulations from muscular tensions throughout the body and varied pressure on the joints due to the pull of gravity. The reactions to them (in man) are similarly co-ordinated in muscular responses enabling him to maintain an upright posture. This activity is almost wholly unconscious; but it results in sub-conscious and semi-conscious sensations which affect our awareness of the vertical and the horizontal in space.[1]

Similarly, unconscious adjustment of the eye muscles focusing the lenses and accommodating the eyes both to the intensity of the light and to the distance of the object—reflex responses to dazzle and the degree of sharpness of the retinal image—contribute to the acuity of vision. There are countless movements and adjustments of the eye essential to visual perception, some of them wholly unconscious and little known even to the physiologist, others almost on the verge of awareness, as when a bright patch or sudden movement on the periphery of the visual field causes reflex movements of the head and eyes to bring it into focus.[2]

The apparent weight of objects stimulates unconscious adjustments of muscular effort when we prepare to lift them. Here,

[1] Cf. M. D. Vernon, *The Psychology of Perception*, Ch. 8; and J. J. Gibson and O. H. Mowrer, 'Determinants of the Perceived Vertical and Horizontal', *J. Exper. Psych.*, 1948.

[2] Cf. also *Foundations*, pp. 373–4: 'The perceived position, for instance, of a red patch depends upon the relation to each other of its images on the two retinae, on the accommodation of the eyes, the reflex reactions of the eye muscles, the position of the head, and so forth. All these bodily facts are represented in the brain by neural impulses, some below the level of consciousness but others occasioning identifiable somatic sensations. "In other words, the spatial setting of the red patch is derived from my body; and the proof of this is that when this bodily machinery goes wrong I no longer see the red patch 'there'. I see two red patches, or the red patch goes round and round, or though I see it, I simply cannot find my way to it" ' (Russell Brain, *Mind, Perception and Science*, p. 32).

obviously, past experience mediates the subconscious reaction but the conscious act takes no account of it. And similar unconscious adjustments of the head and ears (more obvious in other mammals than in man) minister to the reception of auditory stimuli.

This unconscious, sub-conscious and partially conscious integration of impulse and reaction is part of the self-adjusting, self-maintaining activity of the organism that I have called auturgic. Its polyphasic character and its continuity with conscious processes were a main theme of discussion in the second and third parts of *Foundations*. It is a process typical and pervasive of life and one which at a determinate threshold of intensity and integral unity is felt. It is the continued articulation and organization of this feeling that constitutes consciousness, which emerges in the form of perception.

vii. SCHEMATA

There is a marked convergence of psychological theories on the view that perception is a process of organizing and structuring the elements of what is often termed the sense-field.[1] This is the mass of sentience which is selectively brought to consciousness by attention. 'The progress of learning', says J. J. Gibson, 'is from indefinite to definite, not from sensation to percept'.[2] What precedes consciousness proper is a vague mass of feeling confused as to modalities and without definite reference, and the auturgic process of the organism, in pursuit of self-maintenance, progressively articulates this sentience in accordance with schemata which seem to be partly innate and partly constructed in the course of the process. Among the most important of these schemata is the body schema by reference to which sensations are localized in the body of the percipient himself. This does not seem to be wholly innate but is constructed in the course of early experience.[3] Other schemata are elaborated in infancy which enable the child to recognize and locate spatial objects and gradually to build up the framework of a spatio-temporal world within which developed perception takes place.

[1] Cf. *Foundations*, Ch. XX; and F. Allport, *Theories of Perception and the Concept of Structure*.

[2] *Perception of the Visual World* (Boston, 1950), p. 222.

[3] Cf. Russell Brain, 'The Concept of Schema in Neurology and Psychiatry' in *Perspectives in Neuropsychiatry* (ed. D. Richter, London, 1950), and other references in *Foundations*, Ch. XVII, p. 338.

Sir Frederick Bartlett in his experiments on perception presented subjects momentarily with groups of objects ranging from simple arrangements of straight lines, through more complicated patterns to drawings of familiar objects. He found that even with the simplest material, as he put it, 'interpretation often runs beyond presentation'. Some subjects reproduced a simple reversed N (N) as the correctly printed letter complete with *serifs*, and an incomplete square was reproduced whole. More complex designs, though non-representational, were given names spontaneously and automatically by the percipients. 'The presented visual pattern', says Bartlett, 'seemed at once to "fit into" or "match" some preformed scheme or setting'.[1]

In an extremely interesting discussion, E. H. Gombrich maintains that the art of representation proceeds by the adoption of schemata which are tested against visual effects and are progressively corrected to attain greater verisimilitude.[2] He quotes F. C. Ayer who writes, in a doctoral dissertation on the psychology of drawing:

'The trained drawer acquires a mass of schemata by which he can produce the schema of an animal, a flower or a house quickly upon paper. This serves as a support for the representation of his memory images and he gradually modifies the schema until it corresponds with that which he would express. Many drawers who are deficient in schemata and can draw well from another drawing cannot draw from the object.'[3]

This implies that the perception of the object equally requires a schema, for if that is lacking it cannot be copied (whereas another drawing already embodies a schema).

For our immediate purpose the processes of testing and modification are of special importance. Both the psychologists and the art historian bear witness to it. Gombrich explains at length how artists (impressionists and charicaturists especially) explore our perceptual habits for the clues and schematic correlations which enable us to perceive objects, facial expressions, and the like, and

[1] *Remembering, A Study in Experimental and Social Psychology* (Cambridge, 1967), pp. 19–20. Vernon reports similar results with children. See *op. cit.*, p. 88.

[2] Cf. *op. cit.* (London, 1962), Ch. V, p. 126: 'Without a schema which can be moulded and modified, no artist could imitate reality.'

[3] *Loc. cit.*

experiment with methods of reproducing them in paint so as to create the same (though in large measure an illusory) effect. J. J. Gibson[1] and others speak of cues which enable us to recognize objects and which in large measure account for the phenomena of constancy; and yet other psychologists, like Postman and Bruner,[2] maintain that perception is a constant process of testing, confirming and falsifying and revising 'hypotheses'. The hypothesis and the schema mentioned above are much the same thing, and the process of testing and modification is that of learning by experience. We must examine the epistemological implications of all this presently; for, clearly, if scientific procedure is one of evolving, testing and establishing hypotheses, the view that the touchstone by which they are tested is sense-perception must be significantly affected by the discovery that sense-perception is itself a process of forming and testing hypotheses.

viii. CONTEXT

To this apparent paradox we shall return, but first we must note a further development of the schematic and structural character of perception. That the percept is a product of spontaneous organization and is always a configuration with certain assignable characteristics such as simplicity, closure, balance and 'goodness' has been tirelessly demonstrated by Gestalt psychologists, and is consonant with a vast mass of evidence, as well as compatible with other theories.[3] But beyond this, it has been shown that what is perceived is in large measure determined by context, even though the elements on the periphery remain unnoticed by the percipient. Parts of a surface in relief may be made to appear either recessive or protruding according to the angle of incidence of the light, but if the percipient looks at the surface through a peep-hole, which prevents him from seeing the direction of the light, the appearance of shadows and relief is eliminated and the surface appears flat with light and dark patches of colour.[4] The percipient, however, need not consciously take note of the direction of the illumination. The effects are the same even though this feature of the situation remains unnoticed, as long as it is visible. If a blue disc with a

[1] See *The Perception of the Visible World* and cf. *Foundations*, pp. 407 ff.

[2] Cf. Blake and Ramsey (eds), *Perception—an Approach to Personality* (New York, 1957), and *Foundations*, pp. 411–12.

[3] Cf. *Foundations*, pp. 393–400.

[4] See Vernon, *op. cit.*, p. 130.

sector cut out is rotated in front of a yellow background, an observer will perceive it as a blue film in front of the yellow surface. But if he looks at it through a tube, or a narrow aperture in a screen, which excludes everything except the rotating disc, he cannot distinguish the blue film from the yellow background and sees only a single neutral grey surface.[1]

The perception of movement is very dependent upon the relation of the moving object to its stationary background. If the background is uniform and homogeneous the movement appears slower than when the background is variegated.[2] Piaget found that children under 8 or 9, were influenced by their impression of the total situation in perceiving the movement of objects within it so that they judged the ascending car of a funicular railway to have moved further than the descending car, because its movement gave the impression of greater effort. Likewise two objects arriving together at the same point were judged to have moved at the same speed regardless of the difference of the times or distances they had travelled.[3] 'Whether or not actual movement is perceived, and what type of movement is perceived', says Vernon, 'depend on the relation of this change to the general surroundings, or to other changes; and also on certain expectations on the part of the observers.'[4]

In Bartlett's experiments, not only were complex diagrammatic figures seen as wholes and readily recognized even when inverted or otherwise changed in position, but when drawings of familiar objects were presented, appropriate details were supplied by the subjects despite actual inability to see them. In reaction to the picture of a closed gate surmounted by a notice on which the actual writing could never be distinguished in the time of exposure, 80 per cent of the observers immediately guessed that the notice said 'Trespassers will be prosecuted'. One said 'I seemed to see it vividly. It is foolish, I know, because I can't see the writing, but I seemed to see "By Order" written underneath.' The words 'By Order' did actually appear in the drawing along with the rest of the words correctly 'guessed'. Questions may be raised whether this result was due to suggested imagery supplied by the subject, or just plain inference from the closed gate to the probable content of the notice. But undoubtedly, whatever the source of the suggestion, it arose out of the context and affected the detail of the perception.

[1] Vernon, *Op. cit.*, p. 79. [2] *Op. cit.*, p. 144. [3] *Op cit.*, p. 150. [4] *Op. cit.*, pp. 155 f.

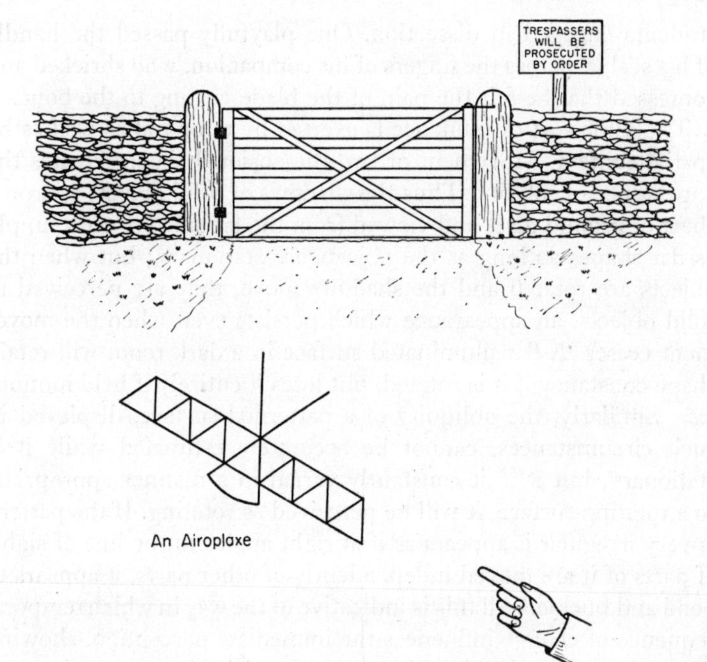

The same effect was observed when a line-drawing suggesting an aeroplane was shown on a card with a hand in the lower right corner pointing up at it. The drawing was labelled 'Airoplaxe', but while the majority of subjects instantly identified the drawing as an aeroplane none of them observed the mis-spelling; and one of them, who came from a city where air attacks had recently been expected (the experiment was carried out in war-time), saw the hand as an anti-aircraft gun.[1]

'Practically always . . .,' writes Bartlett, 'the subject first reacted to a picture as a unity, getting an immediate general impression of its significance, or compositions or distribution of light and shade, and only then examining the details analytically.'[2]

The sensation of the handle of an instrument drawn across the fingers is ordinarily perceived as such, and occasions no discomfort; but in a special context the perception may be very different. Blanshard[3] quotes from Lewes Gratiolet's story of two medical

[1] Bartlett, *op. cit.*, pp. 27 f. [2] *Op. cit.*, p. 28.
[3] *The Nature of Thought* (London, 1948), pp. 119 f.

students engaged in dissection. One playfully passed the handle of his scalpel across the fingers of his companion, who shrieked and confessed that he felt the pain of the blade cutting to the bone.

The same sort of influence is exerted by temporal as well as by spatial context. Movement of two-dimensional shapes creates the appearance of solidity. Thus the shadows of solid objects cast on a sheet of ground glass and viewed from the far side are seen simply as flat shapes so long as the objects are stationary; but when the objects are rotated and the shadows move, they are perceived as solid objects, an appearance which persists even when the movement ceases. A flat illuminated surface in a dark room will retain shape constancy if it is rotated, but loses it entirely if held motionless. Similarly, the obliquity of a patterned surface, displayed in such circumstances, cannot be accurately estimated while it is stationary, but it if it constantly varied in a manner appropriate to a rotating surface, it will be perceived as rotating. If the pattern is very irregular it appears as if at right angles to the line of sight. If parts of it are moved independently of other parts, it appears to bend and buckle.[1] All this is indicative of the way in which temporal sequence of change influences the immediate perception, showing that the perceptual activity is constantly one of integrating impressions whether successive or co-temporaneously spread out in space.

In art this effect of context is unmistakeable and indispensable. We react in general to gradients of light and shade and not to absolute intensities, and when we interpret pictures this is the invariable rule. The glint of light on the water, in Turner's 'Dutch Fishing-boats', and the sunlight on the distant cliffs are made luminous by their relation to the gloom of the threatening storm cloud. 'It is particularly the impression of light, as we know,' Gombrich tells us, 'that rests exclusively on gradients and not, as one might expect, on the objective brightness of the colours.'[2]

In the centre of Constable's incomparable landscape *Wivenhoe Park*, there is an irregular shaped dark smudge of paint surrounded by an equally irregular white dab. Seen in isolation it represents nothing distinct; but seen in the picture as a whole it is instantly recognized as a cow lying down by the water's edge under the trees on the far side of the lake. Gombrich draws attention to the way in which Altdorfer reduces the shapes of angles, in the background of his painting 'The Virgin Amidst Angels', to 'a series of luminous dots which we surely would not read without knowing

[1] Vernon, *op. cit.*, pp. 148–9. [2] *Op. cit.*, p. 48.

their context'. Another example, akin to many visual illusions, is the Frazer spiral, in which a series of concentric circles are made to look like a receding helix by means of a background of light and dark interlacing curved stripes, which are tapered in breadth to create the effect of recession.[1]

One might object that examples have been given only of artificial situations, contrived in a laboratory, or by artistic subtlety, and of exceptional cases which cannot be treated as typical. But this objection has little force. There are indeed limits to the reliance one may legitimately place on psychological experiments and they are unavoidably artificial. But they have the merit of separating the particular aspect of the phenomenon under investigation from inessential accompaniments and permitting observation of what takes place under controlled conditions. Moreover, in most of the experiments quoted the psychologists have taken care to select as subjects persons least liable to be disconcerted by the unnatural circumstances, and allowance has been made for divergence from the normal. The examples from art are especially persuasive, because the skill of the artist consists precisely in his success in simulating the conditions of normal perception. The allegedly exceptional cases may be, in significant ways, special, but they are less exceptional than they seem. Who has not had the experience of seeing a noxious spider walk across the floor only to discover, on closer investigation, a feather blown by the draught. How often does one start at the sound of a terrified child's cry in the night, only to discover two tom cats fighting on the tiles. The difference, in each case, between the error and its correction, is a difference of context in which the object is perceived.

A more serious objection might be that, in every case, we can distinguish between the misperception and what is objectively true. The psychologist knows that there is a colour-wheel revolving in front of a contrasting background, and he knows (for he has arranged) the contrasted colours. When features are overlooked in a situation they can usually be discovered by closer scrutiny or by resorting to another sense. When blanks are automatically filled, we can always find out subsequently and somehow that they are blanks after all. What is seen as a whole can afterwards be analysed in detail.

All this is true, but what does it signify beyond the fact that

[1] See Gombrich, *op. cit.*, pp. 180–5, reproductions 181, 182 and 184.

perception has stages in which we come progressively nearer to the truth, a process shortly to be examined? It does not prove that we have any access, *other* than perception, to the external facts, by which we correct perceptual errors. The psychologist's perception is not privileged—though he may be privy to conditions unknown to his subject. What he knows to be the case he also has to perceive. And when we correct our own initial misperception, it is only by finding new perceptual evidence in the acquisition of which we are subject to the same hazards and conditions as we were in making our initial error. Indeed our major task must be to bring this corrective process clearly to light and to lay bare the criteria by which we identify some perceptions as mistakes.

What we can so far assert with confidence is that, in all cases, the perceived object is an organic whole. Its character depends on its structure and the organization of its context, and it changes concomitantly with changes in its surroundings. What occupies the focus of attention is modified by alteration of its spatial background, and the present percept differs according to what came before it and what is expected to follow.

ix. INNATE AND ACQUIRED SCHEMATA

We have next to consider the question whether and to what extent the schemata, or hypotheses, which guide the activity of perceiving are inherited or to what extent they are learned; although, whatever the outcome, the principle remains the same, so long as we admit the auturgic character of evolutionary process.[1] For if the schematic structure of experience is innate and is imposed by the activity of the mind, it must be because it has developed in the course of evolution; and if it is acquired it is equally a process of adaptation in the organism's effort to maintain itself in the environment. There are good reasons to believe that schemata are not wholly either innate or acquired for while the general tendency to organize into structures, which is typical of living activity, is clearly inherent, the form which the structures take changes in the course of experience, and the changes (though partly the result of maturation) are certainly in very large measure the product of learning and experience. Let us look at the way in which certain perceptual skills are generated in an attempt to clarify this matter.

[1] Cf. *Foundations*, p. 391, on 'nativism' and 'Empiricism'.

A new-born child will start and blink its eyes if suddenly exposed to a bright light, but at first the movements of the two eyes are not regularly co-ordinated and the eyes do not follow moving objects in the child's vicinity. Within six or eight weeks co-ordination takes place and the child is able to follow moving objects with his eyes, but this is as likely as not an effect of maturation. How much can we conclude from these facts to the actual perceptual experience of the child (if any)? The ability to follow an object with the eyes does seem to be evidence for an ability to distinguish the moving object from its background and this, taken along with evidence from a range of psychological experiments on both children and adults, would suggest that the figure-ground relationship is very early experienced. The capacity to perceive this distinction may well be innate[1] and its achievement may depend on no more than adequate development of muscular movements and neural co-ordination. But, beyond this, it does not follow immediately that very young infants perceive objects in depth or have any direct experience of spatial relations.

But, we shall find, the perception of visual depth and the awareness of spatial relations generally do depend on schemata which may well be inherited and may be brought to bear spontaneously. Infants learn to recognize objects and their spatial relations as much by movement and practical activity as by visual impression (though, of course, in the normal sighted child, the two go hand in hand). We learn to perceive at the same time as we learn to perform bodily actions, a process which again is continuous with the subconscious behavioural expression of primitive sentience.[2] In this process we tend to recognize, place and correlate sensory elements in our experience in accordance with formulae (or schemata) which we apparently inherit and which we test and modify in the course of experience.

The very young infant's sensory experiences are, as Vernon asserts, 'all things that happen to him'. That they take the form of perception of independent objects is almost certainly not the

[1] Post-operative cataract patients have been found immediately capable of this discrimination. Cf. R. Borger and A. E. M. Seaborn, *op. cit.*, p. 117. The reference is to M. von Senden, *Space and Sight* (London, 1960).

[2] Cf. *Foundations*, pp. 325 ff. and pp. 336 ff.; M. D. Vernon, *op. cit.*, pp. 18 f.; A. Gessell and F. L. Ilg, *The Feeding Behaviour of Infants* (New York, 1937), *Infant and Child in the Culture of Today* (New York, 1942); Piaget, *La Naissance de l'Intelligence chez L'enfant* (Paris, 1948), etc.

case. What William James described as 'blooming, buzzing confusion', Vernon translates as 'a random set of lights, noises, touches, tastes and so on'. At regular intervals groups of these are associated together—the warm soft touch of the mother's breast, the feeling of innervation or impulse and vague kinaesthetic sensations consequent upon grasping and sucking, the pleasant taste of milk, the felt relief of hunger, frequently occur together. But we must not fall into the error of thinking that these are felt as separate sensations. 'The young child', Vernon later admits, 'tends to perceive situations as a whole.' Here she is already referring to more developed perception than that of the week-old infant, and we may be confident that the sensory experiences of early infancy are indiscriminate masses of feeling rather than conglomerations of separately felt sensations. In fact the primitive sentience of the infant is most probably felt as a single complex totality within which, though there are differences of quality, there is no discrimination, but all variety is blended into one fluid, changing sense-content with a dominant tone of pleasure or pain. Even much later the different sense modalities are sometimes difficult to separate in direct experience—notoriously taste from smell, almost always pleasure, pain and emotional tone from exteroceptive sensations, and even occasionally visual and auditory experiences.[1]

Like most psychologists, Vernon assumes that the regular association of certain similar sense-experiences produces habits of association enabling the child to identify and distinguish particular material objects. Locke too alleged that we call by one name the complex idea of sensation which is formed when simple ones go constantly together. But the formation of such complex ideas and the acquisition of skill in identifying objects is not so straightforward a matter. Noises and visual appearances often go together though they have little connexion in reality (take, for example, the twinkling of the stars and the sound of the crickets mentioned in the last footnote). Other sensed qualities may be separated though they actually belong to the same object, as when we can smell the scent of wistaria though the flowers may be out of sight. It is in this process of identifying and distinguishing, associating and grouping that schemata come into play—primarily spatial schemata, of which one of the most important is the body image.

[1] At the age of three, the author identified the chirp of crickets at night with the twinkling of the stars.

It is clear that no recognition of bodies as external is possible unless a distinction can be made between sensations that originate from our own bodies and those resulting from influences that impinge upon us from without. Susanne Langer contends for a fundamental distinction between what is 'felt as impact' and what is 'felt as autogenic action'.[1] (Possibly she has in mind Whitehead's doctrine of 'vector feelings',[2] and we are reminded also of Locke's declaration that we are invincibly conscious of the actual entrance into our minds of simple ideas of sensation.) But is this difference ever directly, or originally, felt? We criticize Locke for his dogmatic assertion running counter to common experience, for we have no invincible consciousness of the actual entrance of ideas such as he alleges, and, apart from physical thrusts strong enough to upset our balance or shake our stability, we never feel incoming stimuli as impact. There is plenty of evidence to the contrary. Until we have learned to schematize our experience and to separate the internal from the external, we are very apt to confuse what comes through the extroceptors with what originates centrally and introceptively. Do we immediately and naturally distinguish between the sudden bang and the startled pang of fear that it excites? Is there a felt distinction between the tactual sensation of a blow in the stomach and the nausea that accompanies it? Or, to revert to Berkeley's familiar instance, do we separate in sense the warmth of the fire from the pleasure it gives us, or the pain of a burn from the heat that caused it? Do we feel as impact the relaxing warmth of a hot bath or the soothing harmony of the second movement of Beethoven's Sonata, Op. 111? If these examples are thought too complicated by centrally excited feelings, consider simple visual experiences of coloured patches which occur sometimes due to light impinging upon the eye, sometimes as after images, and sometimes with equal vivacity are stimulated entirely centrally and are 'seen' with closed eyes. The direct experience is much the same in each case and not infrequently these differently stimulated experiences are phenomenally indistinguishable. If in adult life we distinguish such experiences only with difficulty and after reflection, how much less likely is it that similar discrimination is possible for the young infant?

Yet such discrimination is essential for the perception of objects as external and every percipient comes to make it. We can

[1] See *Mind, An Essay on Human Feeling* (Baltimore, 1967), p. 23.
[2] *Process and Reality* (Cambridge, 1929), p. 121.

but assume that it is an achievement of the living activity which somehow orders and correlates the sensory elements. The probability is that this activity is at once both perceptual and practical. The child reacts reflexly to certain stimuli, and senses simultaneously both the reactive movement and the afferent influence. It grasps at bright objects and turns its head and eyes to them if they appear on the periphery of the visual field. These and many other instinctive movements are inherited and they are all felt. The child's attention, attracted first to one and then to another compact mass of feeling, selects and isolates elements within a complicatedly felt situation and he spontaneously imposes upon them and fits them into schemata by means of which they can be ordered both in time and in space. Among these the body schema, in and upon which sensations can be located, must be among the first; though it cannot be altogether prior to the formation of others, for the general schema of solid body must be concurrently formed. Yet, until such localization of sensation is effected the distinction of inner and outer, what belongs to oneself and what one receives from the surround, cannot be made. The priority of this distinction to the recognition of objects as spatially external to oneself is a logical necessity, not just a psychological fact.

Merleau-Ponty, in an argument which recalls Kant's transcendental deduction, shows that the body image cannot be derived as secondary.[1] It cannot be the product simply of collecting images of the bodily sensations or summing experiences of localization, because its existence as a single whole is the condition of any or all of these. Awareness of the several parts of the body and successful localization depend on reference of sensory data to the body image felt as a whole. The body image is (in Merleau-Ponty's words) 'a total awareness of my posture in the intersensory world'. The spatiality of the body, he asserts, must work downwards from the whole to the parts. We cannot become aware of the latter first and build up the former out of separately apprehended parts, for without the whole there can be no localization of sensation and no distinction of parts, as parts of my body.

Thus body image is in a real sense '*a priori*' in spatial experience, which is not to say that it is an immediate awareness at birth (though in many animals it may be), or that other schemata, necessary to the perception of material bodies as such are not

[1] Cf. *The Phenomenology of Perception* (London, 1962), Ch. 3, pp. 98–103.

concomitant with it. In fact, they must be, if we are to become aware of our own body as related in space to others.

The formation of schemata of solid bodies, in one of which our immediate sensations are located, is doubtless facilitated by the inherent tendency to distinguish figure from ground. We do this not only visually, but also auditorily and, to a lesser extent, tactually. The bird's note stands out against the background of rural noises, the voice of a speaker against the background of extraneous sounds and the light-switch for which we feel in the dark stands out tactually against a variety of pressures and temperature sensations. This figure-ground relationship is itself a schema (or *gestalt*), and it is undoubtedly one fundamental factor in perceptual discrimination of spatial position.

Depth perception is another factor. It is to some extent connected with subconscious sensations of eye movements and to some extent with the disparity in the positions on the two retinas of the corresponding images of projected objects. Somehow, mysteriously, the direct sensation of these processes is mentally transmuted into the perception of visual depth. We are not aware of focusing our eyes, nor of the disparity of the two images, but simply of apparent distance in the direct line of vision. One might expect disparity between retinal images to result in blurred or double vision, but it does not. Yet it is no geometrical index of the mutual distance relations between objects, for visual depth is perceived equally well (by a curious illusion) when two photographs of the same view taken from points slightly apart, are presented simultaneously to both eyes in a stereoscope; though the actual objects seen (the photographs) are in the same plane.[1] When we perceive in this way, what is happening? Are we translating muscular sensations into visual sensations? If so we have no awareness of the former or of the transition. Do we infer from one to the other?—How could we? For neither are we conscious of the alleged premises, nor could mere sensations serve as premises of an inference, if we were. Is it not rather that we spontaneously impose a spatial schema upon the sense content, which we then progressively adjust in experience by co-ordination with concurrent or subsequent sensations of the same objects in other modalities? Vernon states categorically that stereoscopic

[1] If the positions of the photographs are reversed, moreover, the stereoscopic distance relations of the apparent objects are not altered. Cf. D. M. Vernon, *op. cit.*, p. 133.

vision from fused disparate retinal images is unlearnt and develops (presumably by normal maturation processes) in the first six months of life. If so, and if it is accepted as evidence of the spontaneous imposition upon visual data of a spatial schema, we may conclude that such schemata are inherited mental equipment.

The process of association and correlation would not be enough by itself to generate schemata; for, as we have seen, the sense contents are not originally distinguished, and it is only distinct elements that can be associated or correlated. Schemata must be imposed as the media of distinction and correlation in one. To distinguish is at once to relate (and *vice versa*). Thus it is the articulation of primitive sentience, by the imposition upon it of schemata, which both creates distincta within it and makes relations possible. This imposition is surely the auturgic activity of dynamic organization that is inherent in living process. Moreover, it is a logically prior necessity for the perception of a coherent and orderly world of objects—the *leit-motif* of Kant's transcendental analytic.

Another consideration which must convince us of the priority of schemata is the account which psychologists give of the phenomena of size constancy. This, we are told, is due to our compensating for the reduced size of the retinal image by our direct awareness of the distance of the object. This direct awareness is conveyed to us, they say, by so-called distance cues, like the differences of colour, shading, clarity and relative movement which we learn to correlate with distance. But, we could not learn to correlate these cues with varying distances unless we knew the distance relations of objects by other means. Tactual and kinaesthetic stimuli would not serve alone because in themselves they contain no element of distance. When we walk (or crawl) we are aware of progress through space only by correlating our tactual and kinaesthetic experience with visual. A child reaching out for a coloured bauble is already aware of distance and must already be unconsciously applying a spatial schema. The only means at our disposal, therefore, for discovering distance relations, apart from the so-called 'distance cues' listed above, are visual depth (when the distances are relatively small) and apparent difference in size. But if we judge distance from the size of the image we cannot account for size *constancy* by the judgment of distance; and unless we know distances by some independent criterion, we cannot learn the significance of 'distance cues'. To perform any of these very necessary and important tasks we must have *originally* some spatial schema into which to fit our percepta.

Once the schemata are available the process of distinction, comparison and correlation can be carried out and we can (as we undoubtedly do) learn from experience how to adjust and modify our schemata in order to make our objects coherent and intelligible as an ordered world. We must conclude, therefore, that the mind, continuing the typical life process of self-organization and self-maintenance at unconscious levels, has an innate capacity to schematize the contents of its primitive sentience, and that in the course of its experience it progressively modifies its schemata when the objects with which they provide it fail to fit together. This is precisely what Gombrich finds to be the case in the development of the visual arts, and is supported by a great body of evidence set out by the psychologists of perception and learning. We shall proceed forthwith to scrutinize some of this evidence and to examine the process of development of perceptual skills.

X. INFLUENCE OF PAST EXPERIENCE

Spontaneous organization of primitive sentience ensures that the objects of perception are never bare sensa, but are always structured spatio-temporal percepta. That we perceive complex objects directly is, therefore, true; but that the material entities of the physical world are immediately given to our everyday observation is far from correct. For such observation, not only maturation, but also learning and in some cases special training is needed. The extreme difficulty of visual shape recognition experienced by congenitally blind cataract patients after their sight has been restored, and the length of time required to accomplish it, testify to the necessity of learning and practice for quite ordinary perceptual skills. This difficulty is greater than can be explained by post-operative shock, or the strain of having to rely on artificial lenses for focusing, or defect of retinal development (the last, in fact, is not found in cataract patients). Similar inability is displayed by chimpanzees and other animals that have been reared from birth in the dark, or in diffused light that admits no patterned stimuli, when they are later introduced to normal lighting conditions. Even when such subjects have acquired considerable tactual skill in discrimination, it does not transfer to sight.[1]

[1] Cf. Borger and Seaborn, *op. cit.*, pp. 110–13 and 120; M. von Senden,

Even more convincing proof of the contribution of learning to perception has been given by experiments on visual displacement, in which subjects have worn goggles that reversed, inverted or slanted the image. The immediate effect was complete disorientation and inability to co-ordinate bodily movement with visual impression. But after a few days' practice not only was co-ordination between hand and eye restored but in time some subjects experienced normal vision.[1] The results of these experiments make plain why we perceive the world visually in normal spatial relation despite the fact that the retinal image is inverted in relation to the rest of the body. So long as neuro-muscular organization is adequate, our perception of the world is indifferent to the relative position of the retinal image, and is the fruit of a mental activity, not the product of merely mechanical replication.

The psychologist defines learning as change of behaviour (more or less permanent) which is the result of experience; and behaviour is 'informed activity'.[2] It is informed by perception, and therefore learning is simply the influence of perception on behaviour in the course of experience.[3] But perception itself is an activity and no mere receptivity, thus no sharp separation is possible between perception and behaviour. Both involve the activity of the organism as a whole, and learning is the modification of this activity in the light of its stored and recorded past. The influence of learning on perceptual skill, and its indispensability, reinforce the judgment that perception is a constant activity of organizing, not directed only upon immediate sensory presentation, but temporal as well as spatial, assimilating present appearances to past information, and modifying past habits in the light of present discoveries.

That this is the case is revealed by a variety of experimental results and historical exemplifications. Perception is never instantaneous. Tachistoscopic exposure of objects to percipients proves that they need time to perceive them. They do so by stages:

op. cit.; A. H. Riesen, K. L. Chow, J. Semmes and H. W. Nissen, 'Chimpanzee Vision after Four Conditions of Light Deprivation', *American Psychologist*, Vol. 6, 1951.

[1] Cf. Borger and Seaborn, *op. cit.*, pp. 113 ff.; I. Kohler, 'Experiments with Goggles', *Scientific American*, Vol. 206 (1962), pp. 62–72; J. G. Taylor and J. Wolpe, 'Mind as a Function of Neural Organization' in *Theories of the Mind*, ed. J. Scher (New York, 1962), p. 232.

[2] Cf. *Foundations*, pp. 345 ff.

[3] Learning must not be confused with memorizing what has already been learnt for future use in new situations. A child 'learning' a poem by heart is an example of the latter.

first reporting an awareness of a vague 'something', then of 'something of a certain kind'; then a guess is hazarded that 'it might be so-and-so', then perhaps this is corrected—'No, it is really such-and-such'—until confident assertion can be made of what the perceived object is.[1]

The gradual process of recognition is obviously mediated by past experience and established knowledge. 'It might be so-and-so' poses a hypothesis. 'It looks like something of such and such a kind' is an effort to classify—that is, to place the object in some known category. If the subject is given some prior indication of what to expect, he perceives in accordance with his expectation. Fig. 1 shows how persons presented with a rough outline drawing and told what it was supposed to represent, later reproduced it much more nearly like the object named than that actually presented.[2]

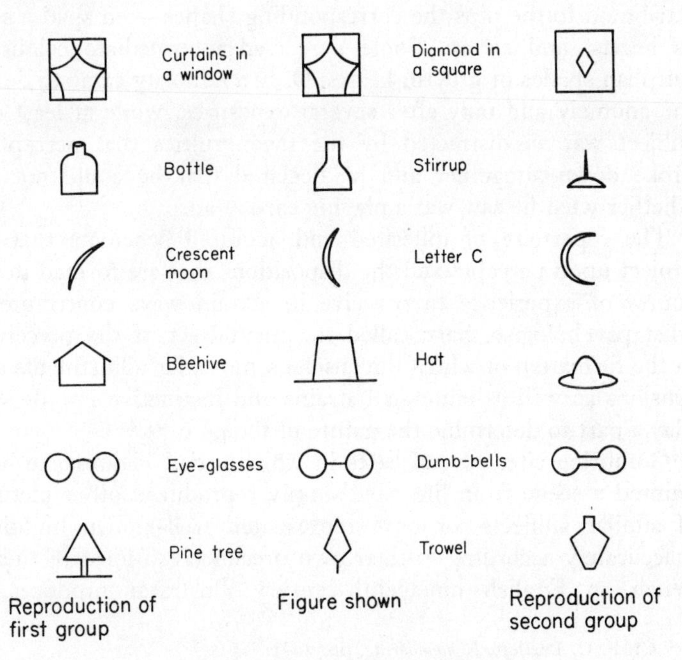

(From M. D. Vernon, *Psychology of Perception*, p. 37)

Other subjects shown highly simplified drawings tended to identify them with familiar objects, to name them themselves,

[1] Cf. Vernon, *op. cit.*, pp. 31 f. [2] *Op. cit.*, p. 37.

and then later, when required to reproduce the drawings from memory, to depict the objects with which they had identified them in a more detailed or conventional fashion.[1]

Experiments with familiar objects give the most convincing evidence of the influence of acquired knowledge upon present perception. If a subject is shown playing cards of different sizes in conditions which exclude indications of their distance, he will perceive a half-sized card as farther away than a full-sized card displayed at the same distance, but he will take the size of both to be normal.[2] Still more remarkable is the effect of displaying in a tachistoscope playing cards with the colours of the suits reversed. In these experiments, some subjects reported the denominations of the cards correctly but attributed to them their normal colours—red spades, for instances, being seen as black, black hearts as red. Others reported the colours correctly and attributed to the pips the corresponding shapes—red spades seen as hearts, and so on. Some perceived intermediate colours—purplish spades or greyish hearts. Only a minority came to detect the anomaly and only after several exposures, while at least one subject was so distracted by the incongruities that perception broke down altogether and he declared that he could not tell whether what he saw was a playing card at all.[3]

The repertoire of inherited and acquired schemata that we project upon percepta and the dispositions that are formed in the course of experience to perceive in certain ways contribute to what psychologists have called the mental set of the perceiver, in the formation of which unconscious muscular adjustments and tensions as well as emotional strains and instinctive impulses all play a part to determine the nature of the percept.[4]

Gombrich cites several cases in which artists claiming to have painted a scene from life have simply reproduced other pictures of similar subjects, or have represented well-known buildings inaccurately according to their own preconceived ideas of them.[5] When an English nineteenth-century illustrator produces an

[1] Cf. F. C. Bartlett, *Remembering*, pp. 20–1.

[2] Cf. A. H. Hastdorf, 'The Influence of Suggestion on the Relationship between Stimulus Size and Perceived Distance', *Journal of Psychology*, 29, 1950, p. 195.

[3] Cf. J. S. Bruner and L. Postman, 'On the Perception of Incongruity', *Journal of Personality*, 18, 1949, p. 206.

[4] See *Foundations*, pp. 400–4.

[5] Cf. Gombrich, *op. cit.*, Ch. II.

engraving of Chartres Cathedral accurate in most respects except that he gives it decorated Gothic windows instead of Romanesque, are we to think that he merely imagined this divergence in the architectural style, that he misremembered it, or simply that he misperceived?[1] Especially instructive are the pictures which Gombrich reproduces of Derwentwater, one by a Chinese artist, and another by an English engraver of a century earlier.[2] These pictures were certainly drawn from direct observation, but such different representations could hardly proceed from similar perceptual experiences, unadulterated by past experience, convention and habit.

Spontaneous imposition of schemata merges with the influence of past experience to produce normal perception. We have argued that distance perception by means of vision must to some extent depend on the imposition of a spatial schema, but, as the Ames demonstrations[3] have proved, visual stimuli are spatially ambiguous—and we noted earlier that this is generally true of all proximal stimuli. The spatial schema does not definitely determine the relations of the objects. For more complete determination we must depend on experience and the correlation of cues the significance of which can only be learned. The impressions gained from Ames' rotating trapezoid window and from his distorted room are those of normally shaped objects. They are supplied by expectation and habit. When we see rectangular objects obliquely they project a trapezoid image, but we habitually take them to be rectangular because the expectation that they will be is borne out by other experiences of them (tactual and visual). These other experiences are also cast in spatial moulds, but we adjust them one to another and adopt as normal those that most commonly combine without conflict. This habitual expectation is at work in our perception of Ames' rotating trapezoid, so that we see it as rectangular but oscillating. The same account applies to the distorted room, until we test its appearance by practical activities such as bouncing balls against the walls—in other words, by comparison with the results of applying our schemata to other modes of perceiving.

In much the same way constancy phenomena depend upon experience—that of the significance and relevance of distance cues.

[1] Gombrich, *op. cit.*, p. 63, reproductions 48 and 49.
[2] *Op. cit.*, p. 74, reproductions 63 and 64.
[3] Cf. *Foundations*, pp. 405 f. and the references given.

We must see objects spatially and in depth and also feel them in terms of a spatial schema before we can recognize their relative distances at all. But as soon as we do we can correlate and remember aspects of their appearance which indicate distance. Then distance cues may subconsciously modify our apprehension of their shapes and sizes to compensate for perspectival distortion and foreshortening. Perceptual constancy is not just given, it has to be learnt. Young children are found to report more accurately the projected shape of objects than do adults. 'Perhaps therefore,' writes Vernon, 'we must conclude that children gradually acquire schemes of the relationships of sizes of objects to their surroundings at varying distances.'[1]

The mutual adjustment of spatial schemata in the course of experience is undoubtedly no more than the continuation of the spontaneous process of organization already working at more elementary levels. The combination of impressions resulting from the imposition of schemata in different sense modalities and on successive occasions must be coherent. The various spatial readings must fit together; and part of the process of fitting them is that of learning to co-ordinate hand and eye, a process which, as the Ames experiments demonstrate, modifies the perception. For after seeing the specially constructed room as regular, we are first shocked by the disproportionate size of familiar objects introduced, and then discover the mutual disposition of surfaces by bouncing balls and similar practical tests, until the distorted shapes ultimately dawn upon visual apprehension. The most outstanding testimony to this process is the gradual restoration of normal visual orientation to subjects wearing inverting and reversing goggles. 'This occurred most quickly when they made voluntary movements', writes Vernon. 'Thus . . . additional non-visual information was utilized in learning a new way of perceiving'.[2]

That we develop perceptual skills by the mutual adjustments of schematized impositions to fit together without conflict, is also supported by the psychologist's account of distance perception. This, we are told, is achieved by the reading and combination of 'distance cues', but, Vernon remarks:

'Although none of these cues, other than binocular disparity and convergence [of which, be it noted, we are not usually aware],

[1] *Op. cit.*, p. 98. [2] *Op. cit.*, p. 164.

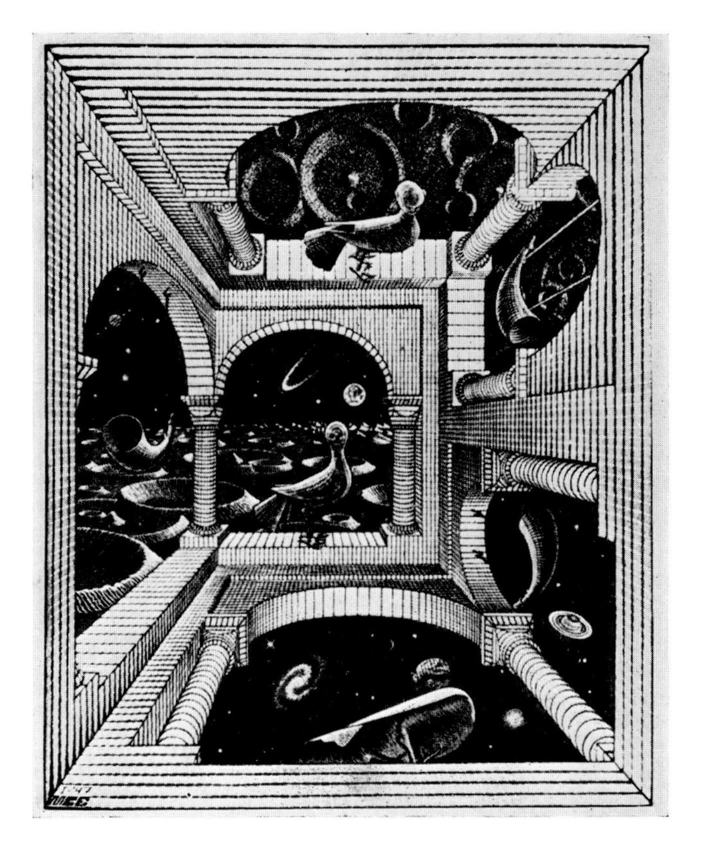

ESCHER: *Autre Monde*. 1947. Woodcut

PLATE 2

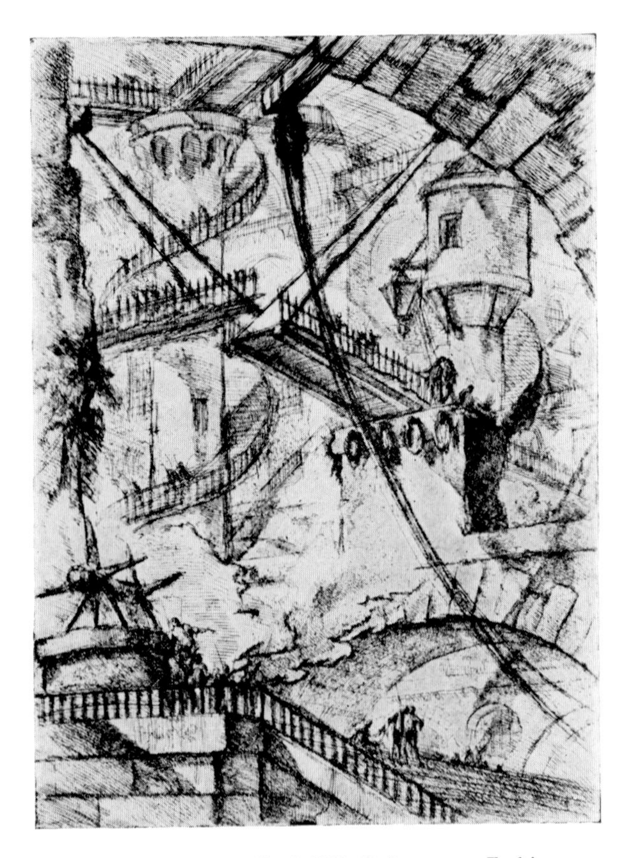

PIRANESI: '*Carceri*', pl. VII. Before 1750. Etching

gives complete or reliable evidence as to the distances of objects, in normal everyday life many of them combine together to produce *corroborative evidence* which enables us to make reasonably accurate judgments of the distances of near objects . . .' (my italics).[1]

Similar testimony, in many ways even more impressive and convincing, is offered by Gombrich. 'It is the guess of the beholder', he says, 'that tests the medley of forms and colours for coherent meaning, crystallizing it into shape when a consistent interpretation has been found.'[2] The artist by using the common conventions of perspective drawing can create an illusion of accuracy which is nevertheless paradoxical in terms of normal experience. The example that Gombrich offers is M. C. Escher's woodcut, *Autre Monde* (1947) in which the perspective looks impeccable, but the directions, 'up', 'down', 'right' and 'left' are indiscriminately combined. A similar exercise in coherent perception is offered by Piranesi's etching *Carceri* (*c.* 1750). The main point of these examples is that while the immediate impression is of recognizable objects their detailed spatial relations are ambiguous and puzzling, because they do not fit together.

Ambiguity from another source is that of size-distance relationship. It is notoriously impossible to judge whether an object is small and near, or large and distant, without corroborative information. Yet we never fail to see an object to some extent 'out there', even if it is an after-image. This projection of visual images is another of our innate schemata, but to become aware of the size and distance of objects more is required—what Gombrich refers to as 'contextual aids'. Pictures reproduce (for the most part) the ambiguities of perspective, so vividly illustrated by Ames, and leave the viewer to read into them depth, direction and distance; but the artist assists our perception by providing additional cues in light and shade, definition and vagueness. Even so, alternative readings are possible, as in the case of the photograph of a staircase which can appear in three different ways. But what narrows the possibilities is the limit of consistency. As Gombrich puts it: 'The more evidence of spatial situation is taken in, the less possible will

[1] *Op. cit.*, pp. 131 f.

[2] *Op. cit.*, p. 204. Note in passing the natural assumption both of Vernon and of Gombrich that perception is a matter of 'judgment' and 'interpretation', a topic to which we shall shortly return. See Gombrich, pp. 189, 191, 198, 219, 221 and *passim*.

it be to accept the alternative reading. The consistency test will be put to increasing strain.'[1]

We have identified schemata, in our earlier discussion, with the 'hypotheses' postulated by some psychologists as what we test and confirm, or disconfirm and modify, in perception.[2] What produces the confidently accepted percept is the consilience of hypotheses, the convergence of assumptions and the mutual corroboration of cues. What Gombrich calls 'the true marvel of the eye' is characteristic of all perception, that in it we interpret, with incredible rapidity and remarkable precision, 'the interaction of an infinite number of clues'. We do this even in apprehending the subject of a painting:

'We could not make sense of Constable's *Wivenhoe Park* without the well-proven assumption that grass is as a rule sufficiently uniform in colour for us to recognize the modification due to light and shade, that Lilliputians rarely populate the English landscape and that therefore the small mannikins are far away, and that even fences are generally built fairly even in height so that the tapering off must indicate increasing distance—all these interpretations are found to dovetail together and support one another so that a coherent picture emerges.'[3]

And what is true of interpreting pictures is even more true of interpreting nature. There is not only projection, depth perception, and comparative size of objects, light and shade, and definition, but also texture and movement (our own movement as well as that of the object). There is not only vision but also touch and hearing. And it is the corroboration and mutual 'dovetailing' (as Gombrich calls it) of all of these that is effected in a single flash of insight to give us perception of the presented scene.[4]

[1] *Op. cit.*, pp. 226. Cf. also p. 239 f. on cubism: 'Where it succeeds is in countering the transforming effects of an illusionist reading. It does so by the introduction of contrary clues which will resist all attempts to apply the test of consistency.'

[2] Cf. *Foundations*, pp. 412 ff.

[3] *Op. cit.*, pp. 230 f.

[4] Cf. also Gombrich, *op. cit.*, p. 234: 'These . . . cases . . . help to explain, I believe, why we still experience some kind of illusion when we see a picture on a wall or in a book—from a point, that is, where the perspective should go wrong. Here as always we first read the picture for consistency, and this consistency, the interaction of clues, is not wholly upset by our changing viewpoint. The painting may cease to be consistent with the world around it, but it remains closely knit within its own system of references.'

But this apparent immediacy of the product does not exclude a long and perhaps arduous process of learning by which the skill in integrating the interpreted clues was acquired. If Bruner and Postman are correct in their 'hypothesis theory' of perception, and if expectations and assumptions play so large a part in direct perception as they maintain, it is because we have been schooled by experience to expect certain patterns and concomitances. And even developed adult perception, these psychologists assure us, is a process of testing hypotheses and modifying them if they do not gibe with the information supplied. The source of this information, moreover, can only be sentience, as that is construed in already established schemata.

In all this there is, of course, no intention to suggest that, in directly perceiving familiar objects, we consciously or explicitly go through any noticeable process of ratiocination. The status of the implicit processes that are (and on the evidence, must be) involved will be discussed below. Nor is anything that has been said above in conflict with the contentions of the Gestalt school that structures are naturally and immediately perceived without 'unconscious inference'. (Again the question of the occurrence of such inference will be taken up anon.) The existence of innate schemata and the inherent tendency of the mind to schematize is sufficient to vouch for the truth of the gestaltist affirmation of configural perception *ab initio*. And the instinctive compulsion to 'simplicity' in perception that they allege, and which constrains us to see, for instance, rhomboid outlines as receding rectangles, may be traced to the same source. Because our minds are agencies of organization, they find an organized pattern 'simpler' than an incoherent jumble. But again what seems simpler to us may be merely what is more familiar. We more commonly experience receding rectangular surfaces and solids (buildings, courtyards, roadways and the like) than vertical rhomboids awkwardly balanced on one corner. If so, simplicity may itself be a product of learning.

xi. INTERPRETATION

If perception is so intimately dependent upon context (the spatially remote) and learning (the temporally past), how, we must ask, do these operate in present awareness? We have already seen that the two factors are not mutually independent, for what is learned are in large measure correlations and combinations arising out of structured contexts. It would seem, also, that our question is more easily answered in the case of the spatially remote than in the case of what is past and gone, because if, as has been maintained, we perceive wholes, the concentration of attention upon one portion need not be assumed to exclude the effect and influence of the rest. Gestalt psychologists argue vigorously to the contrary.

Yet in cases where a perceived relationship leads to misperception, as when the traveller in one train, standing next to another in a station, sees the adjacent train moving and has the impression that it is his own that is in motion, we may wonder what internal process causes the illusion to arise. If the train in which the perceiver is sitting moves out of the station he does not as a rule think that the station is moving and that he is not. What is the difference in the two cases? The psychologist tells us that in the first, because the passing train completely fills the field of vision framed by the window, the percipient tends to take it as stationary and to interpret the changing spatial relations that he sees as due to movement of his own train. In the second case, because he knows that buildings are normally static, he takes perceived changes in relation between his own train and the station as due to the movement of the train. But this explanation reduces the effect of context to that of past experience, and so poses afresh the question: How can past experience operate in present awareness?

A ready answer is that, as this occurs in practical activity, like swimming, riding a bicycle, playing tennis, or playing a musical instrument, where continued practice improves performance, so perception, which is a mental activity, is improved and elaborated by practice. What is operative here, it may be said, is habit-memory akin to that effective in the acquisition of any other skill. It seems to be an original characteristic of living processes that they preserve traces (of some mysterious kind) of their past which progressively modify the manner and organization of their functioning. This habit-memory is nothing consciously recalled. It is merely a disposition built up by the repetition of a certain conjunction of activities making their performance easier, more probable in similar circumstances in future, and more efficient.

In this account there is undeniable truth. Much, if not all, perception is dispositional. It is this feature of it that so impressed

Hume—that our judgments (e.g. of causation) are largely a matter of custom. This accounts also for the tendency to perceive in accordance with expectations which past experience (Hume would say constant past conjunctions) has built up in us. I should differ from Hume at least in wishing to substitute, for his constant conjunctions of particular impressions and ideas, the integration of schemata.

But what is a mental disposition and how does it operate in and upon consciousness? Habitual activity like riding a bicycle is (or becomes) unconscious and automatic and may be attributed to some, as yet unknown, modification of neural organization. No doubt the same is true in large measure of learning to perceive familiar objects. We recognize them at once and without effort. But whereas in learning to ride a bicycle my behaviour becomes more unconscious as my skill increases, until I can ride without using the handlebars and read a book as I go, perception is itself a conscious act and ceases to be such only by ceasing to be perception.[1]

When I misperceive the train as moving and rightly perceive the station as unmoving, it is because I am *conscious* of certain changes of spatial relation, which I see. Something further is needed to cause me to *interpret* these seen changes as the movement of the train rather than of the station; and it is hard to think of this as a neural organization built up by past experience (though there must be some such concomitant neural development), if it causes me to adopt one interpretation in the case of the passing train and precisely the opposite interpretation of a closely similar experience in the case of the station. It must not be forgotten that what I perceive is not just motion or change, but *that* the train is moving, or *that* it is not. I am acquiring by means of the senses knowledge about external objects; and knowledge can be expressed only in propositions. To perceive that p, is to judge. Our question thus becomes: How do past judgements affect present judgements? It may not be wrong to say that they engender dispositions to judge in certain ways, but this tells us little. If I have a disposition to judge that p, and somebody asks me why I am so disposed, I am apt to give a reason. In short, I am apt to state my opinion as the conclusion of an argument or inference.

[1] 'Unconscious perception' might in some special instances be a permissible phrase, but it would have to be used with caution and serious qualification, and still smacks strongly of *contradictio-terminorum*.

Accordingly, some philosophers who have clearly seen that perception always is, and never can be less than, judgement have asserted that it results from 'unconscious inference'. The inference is unconscious, because perception takes place with apparent immediacy and we are unaware of the process by which it is reached. That there certainly is some process can hardly, on the evidence, be denied; and nowadays most philosophers hold views of perception which imply inescapably that the percept is the result of an activity or skill, which may be done better or worse, and terminates in an achievement of some sort. A judgement issuing from grounds would answer to this description. It would be an achievement resulting from the exercise of a skill, and it might be accomplished with more or with less success. Further, the psychologists seem to support this idea, whether they do so inadvertently or deliberately. Not only is the doctrine of unconscious inference an old one in psychology, but, though it was for a time scornfully rejected (and in the form advocated by Helmholtz this fate was no doubt deserved), it has recently been revived with more plausibility by the 'probabilistic' and 'transactional functionalists'.[1] Moreover, as already observed, psychologists like Vernon and Bartlett speak freely of interpretation, inference and judgement in their descriptions of perceiving processes, and Gombrich, the art historian, does so without qualm or hesitation.

Apart from the famous expositions of the doctrine by Bradley and Bosanquet, one of the most cogent and competent advocacies of this theory of perception (as interpretation, unconscious inference and judgement) has been set out by Blanshard in *The Nature of Thought*,[2] and it has been restated with admirable clarity and convincing argument by C. A. Campbell in his essay, 'The Mind's Involvement in "Objects"', in J. Scher's volume, *Theories of the Mind*.[3] I shall not repeat their arguments at length but will mention certain salient points, before considering the formidable attacks that have been made upon this so-called 'idealistic' view of perception.[4]

[1] Cf. *Foundations*, pp. 392 and 405 ff.

[2] London, 1948.

[3] New York, 1962.

[4] This label applies vicariously to the theory under investigation. Its main proponents have called themselves, or have been called, 'Idealists', but there is nothing specially Berkeleyan or subjectivist about the theory; and it would seem odd to classify psychologists like Brunswik, Cantril, Bruner, Postman, Ames and many others who have held related views, as Idealists. Still less does it seem

Crucial to the decision of the issue, whether or not perception involves judgement or inference (or both), is the way in which we understand these terms. In common parlance they are loosely used, and philosophers have defined, or understood, them in different ways. For C. A. Campbell, who follows Bradley and Bosanquet, a judgement is an assertion or affirmation of a proposition implicitly claiming truth, and Blanshard would undoubtedly agree. The proposition is what is asserted, the judgement is the mental act of assertion. Campbell explains that 'judgement' is to be preferred to 'proposition' as the term to use, because a proposition, the object of judgement, taken alone is an abstraction; and if we try to divorce it from the act of affirming, we are apt to be led astray into raising pseudo-problems about its ontological status—questions that at one time gave Bertrand Russell so much needless trouble. Both Campbell and Blanshard regard the act of judgement as an intellectual act, but it plays, they believe, an indispensable part in perception, which also contains an unavoidable sensory element.

Other philosophers tend to consider nothing less than verbal, or some other suitable symbolic, statement essential to judgement, and modern logicians speak rather of 'sentences' than of 'judgements' or 'propositions'. As expressions of intellectual acts they must be set out explicitly as formulae of some kind. Inference, likewise, is often interpreted as requiring explicit steps, statable in something like syllogistic form or as the transformation of sentences in a deductive system.

If judgement and inference are taken always to require such explicit forms, then, of course, the allegation that perception is judgement, or that it involves implicit inference looks like nonsense. For perception does not take sentential form, and implicit inference (on these terms) becomes a blatant contradiction. But no such criticism is fair to Blanshard or Campbell. Their theory of judgement and inference is very different, and is that which Bradley and Bosanquet expounded with great subtlety. Judge-

appropriate to include under that name modern philosophers whose views on perception have implications fairly obviously pointing to a judgement theory—e.g. Ryle, Quinton, Armstrong, or Merleau-Ponty. Cf. A. M. Quinton, 'The Problem of Perception', *Mind*, LXIV (1955), pp. 28–51, and 'Perception and Thinking' in *The Proceedings of the Aristotelian Society*, Sup. Vol. XLII (1968); G. Ryle, 'Sensation', in *Contemporary British Philosophy*, third series; D. M. Armstrong, *Perception and the Physical World*; and M. Merleau-Ponty, *The Primacy of Perception* (Evanston, 1964).

ment, for them, is the self-development of an object in thought, and inference is its continuation; and thought, as Blanshard defines it, 'is that activity of mind which aims directly at truth'. Truth is knowledge of the real, and, if this comes to us at all through the senses, the sensuous object must convey some information. This may be to any degree implicit or unarticulated, and the business of thought is to explicate. But information can be conveyed in nothing short of a proposition; therefore the sensuous percept must be at least implicitly a judgement; and if the information it conveys depends upon grounds inference is already inchoate within it.

Unless something is taken to be something, there is no perception. To perceive *X* is, at the very least, to perceive it as *X*; and to perceive something as *X* is to perceive a state of affairs, the character of which can be expressed only as a proposition (*that a* is *X*). The act of perceiving, therefore, so the argument runs, is an implicit affirmation of that proposition. It is implicit judgement.

Moreover, perception is not always veridical, it may be true or false, so, Blanshard maintains, it must be judgement; for by definition a proposition is what is either true or false and the affirmation of a proposition is judgement. Perception involves meaning, and meaning involves reference—the reference of a presented object to a system of fact which gives it significance. Here we have all the makings of judgement and inference, and it is on these characteristics of perception that Blanshard's theory relies. We shall see later that its strength lies precisely in this insight.

But in nine cases out of ten we perceive that *a* is *X* through something else. I perceive that the train is moving by seeing changes in the relation of the window frame to the objects outside. Thus my implicit judgement that the train is going is mediated. I have 'unconsciously' inferred it from what I see. Similarly, when I see a speck in the sky and hear a hum, I implicitly infer that an aeroplane is passing overhead. (This is Blanshard's example, which nowadays might be improved by substituting vapour trails for the speck.)

It has been objected that perception, unless it is called in question and becomes critical scrutiny, is ordinarily unreflective, a mere acceptance or taking for granted which comes short of affirmation, and so cannot be called judgement. But here again Campbell and Blanshard reply that what is accepted is at the very least a belief and to accept a belief is at least *implicitly* to affirm something as fact. C. D. Broad objects that one need only say that to take something for granted is to behave *as if* one affirmed it; but then the judgement is implicit in the behaviour, for the behaviour is a response to a perceived situation and the perception, its mental counterpart, which gives it its special character, cannot be 'behaviour as if'. To say that one perceives as if one judges, on the other hand, would be the same as to say that one judges implicitly in perceiving.

The summary I have given is hardly an adequate statement of the theory but it is enough to indicate its main tenets. Let us turn now to the critique of its opponents. The strictures of Price and Broad from the point of view of the sense-datum theory weigh less today because that theory is in decline. What is most telling in this critique has been taken up again by another opponent of implicit inference who himself rejects sense-datum theories and accuses Blanshard of falling into the same trap as their exponents. This more recent critic is Professor R. H. Hirst. While Price argues that if perception is interpretation there must be something—some datum—to interpret, Blanshard distinguishes the speck in the sky from the aeroplane perceived as if the former were the datum (as Price would have alleged).[1] But, according to Professor Hirst, they are both wrong, for what is perceived is neither sense-datum nor speck, but an aeroplane. True, it looks like a speck from here, but it is no less an aeroplane for all that and we perceive it as such immediately and without any 'implicit inference'. Professor Hirst admits that we may in certain circumstances perceive a speck and judge that it is an aeroplane, but the perception is not the judgement nor the judgement the perception and in more usual instances we simply perceive the aeroplane directly. He insists that we do not distinguish datum from object when we perceive, and any such distinction is made, if ever, only for special (theoretical) purposes. Without it, he asserts, inference is impossible.

Before indulging in closer analysis we may remind ourselves that to say that we perceive an aeroplane but that it looks like a speck is to pose the very epistemological problems that prompted the sense-datum theory. What the object looks like is what we see. If we perceive an aeroplane and it looks like something else, what we perceive is different from what we see. Either we must say

[1] Cf. *Problems of Perception*, pp. 234 f.

that the perception is direct and that we judge that the object looks like a speck, or we must say that the appearance is immediate and that we judge it to be an aeroplane. The latter is the more plausible, for the appearance is surely the more direct of the two. When we make mistakes it is because we judge the 'immediate' appearance to be of the wrong object. So the problem of the relation between the appearance (or the 'look') and the object of which it is the appearance is just as pressing as any concerning the relation of sense-data to physical things. Part of the trouble is that phrases like 'appears to be', 'looks like', are loosely used, sometimes referring to the appearance and sometimes to that which appears. Great caution will be needed, therefore, to find a solution that is neither a sense-datum theory, nor an inference theory, and is nevertheless adequately explanatory. Professor Hirst's own theory has considerable merits, but itself implies something very near to what he criticizes and professes to reject. First let us examine his criticism.[1]

There are marked differences, Hirst contends, between perception and judgement that prevent our identifying them with any plausibility. (i) Perception involves no deliberation or consideration of evidence, judgement does. (ii) Judgement is possible in the absence of its object, perception is not. (iii) When judgement is false and the error is discovered the judgement is corrected; in the case of perceptual illusion even the knowledge of the facts does not eliminate the illusory effect.

(i) and (ii) obviously refer to explicit judgement and Blanshard denies that the judgement of perception is explicit. We have already seen that to regard judgement as necessarily explicit destroys Blanshard's position and he would be the first to agree. But it does not follow that judgement cannot be implicit. When I brace myself to lift a suitcase I neither weigh evidence nor deliberate but the muscular adjustment *implies* the judgement (or knowledge) that the suitcase is heavy. If it is empty, I was deceived in my perception of it, i.e. the implicit judgement was false. To this point we shall return. It is not enough to say, with Broad, that in such cases we simply behave in accordance with expectation and are surprised if our expectation is not fulfilled. For what is expectation but implicit judgement?—What I expect is always '*that* so-and-so'. Hirst's first objection, therefore, even so far as it is admissible, does not damage Blanshard's case.

[1] *Problems of Perception*, Ch. 8, Sects 3 and 4.

His second point is true enough but is even more irrelevant. That judgement is possible in the absence of its object (e.g. a judgement about Julius Caesar or the antipodes) does not entail that it is impossible in the presence of its object. Blanshard does not contend that all judgement is perception, only that perception is one sub-sub-class of judgements—those judgements made in the presence of their objects which are, moreover, implicit. So we must dismiss this argument as well.

Judgement may be corrected by discovery of the truth (though it must be admitted that this is not always the case with stubborn partisans), but perceptual illusion seems to persist even when we know the facts. In the Müller–Lyer illusion the lines still look unequal even though we know they are not. First, we must note that what is true of this kind of illusion (which may be in part due to unconscious physiological organization, as in the case of colour contrasts), is not true of all misperception. When I take a bush in the mist for a man (one of Hirst's own examples), and am then assured that it is only a bush, I no longer see it as a man. (Though I may still see in what respects its appearance resembles that of a man.) Secondly, once we see that a judgement is false, the reasons for affirming it have been removed, but the discovery of the equality of the lines in the Müller–Lyer illusion does not remove the effects of their context upon our perceiving. Now we have, not a correction of a false judgement, but a conflict between judgements, either both implicit, or one implicit and one explicit and based on different evidence. But, thirdly, even the Müller–Lyer illusion can be corrected with close attention and sophisticated viewing. We can, if we try hard enough, see the lines as equal.[1] A rectangle superimposed upon a set of converging straight lines looks distorted (and does so even if the lines and the rectangle are presented separately but in rapid succession). If we know that the shape is rectangular we still see it as distorted. But if we think of it as a vertical frame standing on a floor made of parallel boards, we can overcome the illusion. This is strong evidence of the dependence of perception on interpretation, turning Hirst's evidence against himself and in favour of Blanshard.

Hirst complains that there is no introspective evidence of unconscious or implicit judgement or inference in perception. In

[1] This is true of some, though not all, visual illusions; but when it is not, the above consideration, that the deceptive evidence is still present, always holds.

fact, he says, the suggestion that there is 'goes flatly against the introspective evidence' (p. 230). Blanshard might reply that we could hardly expect introspective evidence of the unconscious

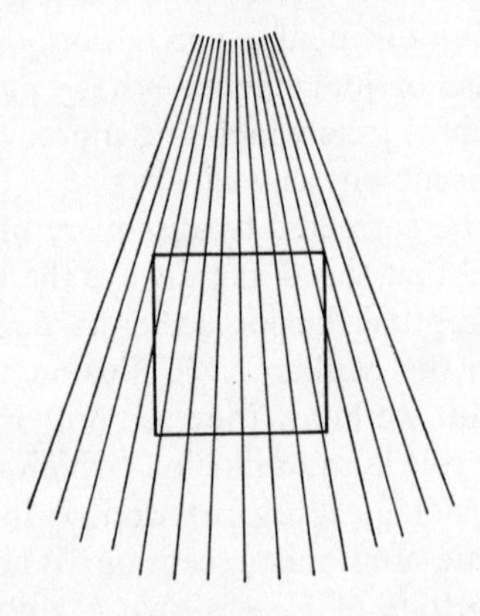

and the implicit, yet examination of the evidence we have might nevertheless require us to postulate its existence. What introspective evidence have pyschoanalysts of repressed complexes in their subjects? But this is not the best answer to Hirst's contention because that contention is false. Here is Bartlett's record of the reports by subjects in his experiments on perception:

'Order, or plan, of construction now at once became a more dominant factor. Symmetry, similarity, sameness, difference and progressiveness were all reacted to readily. Very interesting in this connexion was the use of the phrase "to have an impression", or alternatively, "to have a feeling". Each of these expressions was extremely common. All the structural relations illustrated were constantly described as "felt": "I had an impression that the figure was symmetrical"; "I had a feeling that the figure was growing more complex"; "I had an unconscious assumption (this subject seemed to mean precisely what others meant by 'impression' or 'feeling') that the figure was progressive". The plan of construction and of successive change, being thus "felt", was readily used as a basis of inference, and hence as a guide to observation: "I got an impression that the figure was symmetrical, though I did not notice the details. I built on that, looking for an addition or an omission, and then inferring others in other parts of the figure." Here we can see the "scheme" or "setting" coming in again, used now not to aid immediate identification, but as a basis of inference, and playing a dominant part in the reproductions effected.'[1]

In another experiment subjects were presented with a series of figures terminating in the drawing of a crown.

'Practically all the observers', Bartlett reports, 'said "It will be a crown", and began to infer the nature of the coming changes. At once a specific attitude, containing a large mixture of expectancy, was set up, and by it the perceptual act was directed and determined.'

To find reliable introspective evidence one may need the controlled conditions of the laboratory, but these reports leave little doubt as to its availability. Let us take just one more example, this time from Gombrich:

'The result is exactly the opposite of the experience I described as the sorting out of clues in Piranesi's *Carceri.* There we tried out various interpretations until we found one which fitted a possible world, however fantastic. [Note that he is speaking not of interpreting the picture in words but of *seeing* it as . . .]. It is a point of cubism, I believe, that we are constantly teased and tempted into doing this but that each hypothesis we assume will be knocked out by a contradiction elsewhere so that our interpretation can never come to rest and our "imitative faculty" will be kept busy as long as we join in the game.'[2]

Can one try out interpretations and assume hypotheses without some kind of judgement or inference? And all this is (apparently) involved and required simply to see and visually to appreciate pictures.

Hirst alleges that Blanshard confuses judgements arising out of perception, or judgements involved in describing what we perceive, with the perception itself. Campbell strenuously denies this. Indeed judgements are needed to describe, and others do arise out of perception (else there could be no science), but not all of these are the judgements alleged to be involved in the perception

[1] Bartlett, *Remembering*, p. 24.

[2] *Op. cit.*, p. 240. Cf. his earlier reference to 'illogicalities' in Escher's woodcut and in Piranesi's etching, pp. 205 f.

itself. If I see a man approaching and say to myself, 'That must be Smith', the judgement arises out of the perception and is not the perception itself. But the perception itself is (and surely cannot but be) the implicit judgement (i.e. the awareness) that some man is approaching. If I say 'How beautiful that tree looks in the sunlight', I am describing what I perceive, but what I perceive cannot then be less than that there is a tree before me lit up by the sun; and here there are two implicit judgements entailed by my explicit description.

Bartlett's subjects were not simply describing what they perceived, but how they perceived it; and when the judgement was a description, or expectancy, of what was, or was to be, perceived it was direct evidence of inference. When I exclaim 'There's the bus', the exclamation does indeed arise from a perception, but what could the perception itself have been other than an apprisal of the *fact that* the bus was approaching, which could hardly be expressed otherwise. If it is not expressed—as it need not be—does that change its logical character? When I see that it is the bus I have judged implicitly.

But Hirst strikes at the very root of Blanshard's argument. He denies that perception is, or can be, true or false. That he admits is the character of judgements and propositions, but perception is an activity like playing the piano or playing tennis. It can be done well or ill, rightly or wrongly, but it cannot be true or false, because truth and falsity depend upon the relation of propositions to the world they are about, which does not obtain between activities and the world in which they are performed. Now it would seem odd, and is probably not Hirst's intention, to deny that judgement and inference are activities, and still odder to assert that if they are they cannot be true or false. Some philosophers might do so. They might argue that only the propositions that are their objects can be true or false, and that to affirm a false proposition is just to affirm badly. No such argument, however, is likely to appeal to any philosopher who lays stress on ordinary usage.

But if Hirst's distinction between activities and propositions is valid, why do we speak of striking a false note on the piano? (i.e. false to the score). And is a 'wrong note' so different from a 'wrong answer' or a 'wrong conclusion'? Is the relation between the performance of a sonata and the score altogether different in kind from that between a proposition (or, better, a statement) and the fact about which it is made? In tennis do I never misjudge the speed or direction of the ball, and, if I do, is this misjudgement not integral to my bad play? Finally, as we said before, performatory skills are not, as a rule, skills in acquiring knowledge. Perception is precisely that. Knowledge may be largely dispositional, but not when it is being exercised. Perception is the actual exercise, in a present situation, of knowledge. And that is surely nothing other than judgement. But in this case it is judgement based on sensuous experiences, a fact that gives perception its distinctive character.

This entire theory of perception, however, is condemned by Price, Broad and Hirst as an over-intellectualization of perception. Blanshard replies with a sort of *tu quoque*, these philosophers under-intellectualize perception. Now, the intellect and its activity is commonly held by the unphilosophical to concern higher and more mediate activities than mere perceiving—appreciating, not just hearing, music; criticizing, not just looking at, pictures; theorizing in science and philosophy, not just observing. But there is a philosophical use of the word 'intellect' which is what, I suspect, is in Blanshard's mind. It is the sense in which Kant used 'understanding', as denoting the capacity of the human mind to grasp a coherent object. Without the understanding, Kant taught, objects cannot be thought. Without sensibility they cannot be intuited. But for them to be perceived both are necessary, because intuition without concepts is blind. The use of concepts is judgement, even if it is not expressed in words, and that is why judgement is always implicit in the perception of objects. This activity of the understanding is not 'intellectual' in the popular sense of that word, but it is discursive, and H. H. Joachim has deployed much detailed and careful argument to demonstrate that all so-called 'data' are in fact the product of what he calls 'an analytic-synthetic discursus'[1]—argument which no other writer known to me has ever attempted to refute.

The nodal difference is between the immediate and the mediate, and the 'idealistic' philosophers differ from their opponents only in contending that what the latter claim to be immediate, they can show to be mediate; and what they accept as immediate is only an inarticulate mass of feeling, or primitive sentience, which is preconscious, pre-perceptive and (to use Price's term) pre-judicial. Hirst actually agrees with much of this but he holds perceptual acceptance to be pre-judicial (as Price does), though he

[1] *Logical Studies* (Oxford, 1948), Ch. II.

admits that it is the product of some form of processing and modifying of the *percipienda* to produce the *percepta*. The processing depends on 'cues' and sensed relationships. (Here he follows certain important views of the psychologists.) But it may seriously be questioned whether relations really ever can be merely sensed—for they require distinction and discrimination, the apprehension of 'this-not-that', which is, as Campbell avers, the dawn of judgement. Elsewhere I have argued that 'cues' cannot be effective in perception unless they are also 'clues', the use of which is implicitly interpretation;[1] and this becomes clearly apparent when psychologists like Bruner speak of 'hypotheses' derived from them. Even Hirst's own theory, therefore, requires, by implication, some sort of inchoate judgement and interpretation.

Perhaps the best and most decisive summing up of this discussion may be made by quoting once more the words of Sir Frederick Bartlett:

'It has been shown that a great amount of what is said to be perceived is in fact inferred. . . . It is a matter of very considerable interest that even the most elementary looking perceptual processes can be shown frequently to have the character of inferential construction.'[2]

xii. DEGREES OF ORGANIZATION

The view that perception is implicit judgement and inference, that it is an act of interpretation articulating the more primitive level of sentience, is not as implausible as its critics have sought to make out. Nevertheless, there is some substance in the critics' misgivings. If perceiving is or involves inferring, from what kind of premises does the inference proceed? If they are not data, or constructed out of data, what on earth can they be? What sort of judgements are those of which we are not aware? Is it not nonsensical to allege that we can judge or infer without knowing that we do so? These are persistent questions that all Blanshard's lucidity and persuasiveness have failed to remove from the minds of hard-headed objectors. But answers may be given to them which bring Blanshard's and Hirst's positions nearer together, answers that draw on the notion of organization and may be derived from the evidence already discussed.

[1] See *Foundations*, pp. 407 ff. [2] *Op. cit.*, p. 33.

The percept, we have seen, is moulded both by its context and by the bearing upon it of past experience, and both of these factors are products of a process of organization operating upon the contents of primitive sentience. It operates by the evolution and imposition of schemata, by the combination of contemporaneous structures and by the progressive articulation and augmentation of successive structures. That some of this activity goes on altogether unconsciously, and still more (perhaps) is subconsciously accomplished, is undoubtedly the case—so far we may agree with Hirst. These influences and activities build up as they proceed an awareness of an ordered world of objects, which is not merely a static picture, but a continuous activity of discrimination, distinction, correlation and concatenation of elements; an activity that continuously develops rules of order and connexion, the use of which becomes dispositional. The dispositions are activated in perception which is no more nor less than the act of focusing or bringing to bear, the accumulated effects of this process of organization—the fund of acquired knowledge—upon the present content of sentient awareness.

Thus every perception takes place within a 'horizon', or against a background, of accompanying and acquired knowledge, without which perception is hardly possible. The beginnings of percipient awareness are themselves occasioned by the organizing activity, at lower levels, in physiological processes and the instinctive direction of attention, isolating, distinguishing and correlating the variegated elements within primitive sentience. The simplest act of discrimination, articulating a manifold of this kind, has *the form* of judgement. It is the identification-cum-distinction of 'this-not-that', or the ascription of 'this' (as quality) to 'that' (as bearer of the quality), or the articulation of a relation of 'this' to 'that'. So that the raising of the process to the level of consciousness is the emergence of judgement, which remains implicit only as long as it is not articulated in symbolic notation.

No such simple act, however, is possible in isolation. It is always mediated by background, context, and (if sufficiently developed) former products of the activity of articulation. In as much as this mediation comes to the level of awareness, it is implicit inference. For inference is not merely (if at all) the transformation of sentences (or formulae) in a deductive system—that is at most an artificial symbolization of the results of inference. Actual operative inferring is the reference of a presented object or complex to a structured context (or background) to discover its defining relationships within the system. It is thus the search for, and disclosure of, grounds of connexion within the system. Judgement is but a prior and more elementary phase of the same process. It too refers a distinguished element to the complex within which it is distinguished, and which the judgement articulates (or analyses). Inference merely develops the result by pushing the process further in the direction of articulation, by drawing out the connexions between the discriminated elements and their context in the light of earlier and concurrent discriminations, discovered identities and inter-relations. The key-word is reference. Reference is the primary activity of thinking; and the reference of a sense-content to a structured background (for which another word is 'interpretation') the first beginning of judgement and inference, is the perception of an object.

Let us demonstrate how this is so. We perceive (as has been shown) no isolated or simple data, and every object of perception is already a schematized complex set in a more or less varied background. Strictly we never do perceive single objects (even though complex). We perceive a scene or situation in which objects appear. If we observe a tree or a house, the immediately presented view or aspect is not sufficient to define it for us as a tree, or a house; for, unless we can comprehend the structure of the entire scene and the setting of the object within it, we cannot even interpret (there is no other appropriate word) the perspectival peculiarities of its shape. To do this we must know and take in at a glance the mutual relations, in space, of ourselves and the object, as well as those between the object and other present features of the scene. To see a house as a house, we must at least in anticipation or imagination, become aware of its other possible perspectival views, as well as that presented to us. In Merleau-Ponty's terminology, the house is for us a synthesis or *ensemble*, 'an infinite sum of an indefinite series of perspectival views',[1] an inexhaustible system.

If you object that this is not all at once presented in any one percept, nevertheless it is what, in the object, is implicitly recognized as fact, for what is perceived is a house, not just a façade. What is implicit in the perception, then, is a complex system of spatio-temporal relations between the object in focus and other objects including the percipient's own body. To perceive the

[1] Cf. *The Primacy of Perception*, trans. J. M. Edie (Evanston, 1964), p. 15.

object, therefore, is to interpret certain sentient elements present in experience in terms of this complex system—to see them as a specific factor in that complex.

It is, therefore, at best misleading and at worst sheerly false, to say, with Price and Hirst, that, when we perceive, the datum and the object (the house) 'dawn' on our awareness simultaneously. Attention singles out an element in sentient experience, thus creating or constituting it a 'datum'; but what 'dawns' on us is its character in its setting and its relations to the background, the system of our acquired knowledge. It dawns upon us that it is a house. This may be so rapid that no lapse is detectable, but only because we have become highly skilled, through long practice, at recognizing and interpreting—only because we have ready at our disposal the appropriate perceptual dispositions. Let the object be presented in any way that hinders the rapid operation of our skill and we at once become aware of the lapse and of the process. If the object is presented only momentarily, we find that it has not had time enough to 'dawn'—as experiments in the psychology laboratory show; for, the experimenters tell us, perception is never instantaneous. Or let there be a trick of lighting, or some obscurity in the view, and we peer, question ourselves, entertain 'hypotheses', and bring ourselves, by stages which we can introspect, to the point at which the object literally does dawn upon us as so-and-so.

Merleau-Ponty objects that to regard perception in this way is to reduce us all, in our normal perceiving, to the level of the brain-damaged Schneider suffering from psychological blindness—for it was by such stages of explicit questionings and hypothesizing that he brought himself to recognize familiar objects.[1] Psychological blindness, however, is only an extreme pathological case, which reveals the microstructure of the normal process. There are normal instances in abundance that testify to a similar 'processing of data', some of which have already been found in Bartlett's and Vernon's records. Nevertheless, what Merleau-Ponty says of the directness of our awareness of our own bodies, their intentional mobility, and the spatial world around us, is not to be denied. His very description of it gives evidence of the implicit structures and acts of reference that it entails.[2]

[1] Cf. Merleau-Ponty, *The Phenomenology of Perception*, pp. 131 ff.

[2] Cf. *op. cit.*, p. 137: '. . . to see as a man sees and to be a Mind are synonymous. In so far as consciousness is consciousness of something only by allowing

Perception, therefore, must be understood as the activity of referring a present sense-content to the systematically structured background knowledge of the world; and the successful outcome of this activity is the achievement of recognition. In the process, inchoate hypotheses are subliminally entertained, cues are followed and confidence is finally reached as a result of their mutual corroboration.[1] Conflict of cues and contradiction between hypotheses, what Gombrich uninhibitedly calls 'illogicalities', lead to bewilderment and uncertainty, which stimulate closer attention and further exploration through other perceptual media (other points of view or sense modalities); and, when this occurs and is successful, we arrive at perceptual assurance, and report what we perceive as an observation of fact.

But 'the fact' has not been *given* to us gratis. We have achieved it through a complex activity of schematizing, organizing, reference and interpretation of the scrutinized contents of primitive sentience. 'The fact' is saturated with prior knowledge elaborated in the course of our experience by the same organizing activity. It is shaped and conditioned by its setting, and its character is determined by its context—both of which are cognized, as much by the mind's ceaseless and tireless activity of structuring, as is the fact itself. The fact is utterly dependent for its observable nature, and for its being observed, upon the structure of the known world in which it is perceived, and knowledge of which the percipient has acquired up to that point of his experience.

Observed fact, then, is the focus and crystallization of an ordered knowledge of the world built on a framework of schemata —or, to use only slightly different terminology, concepts. Thus observation is always, in its commonest and most everyday forms, as in its more controlled and systematic scientific use, parasitic upon a conceptual scheme. Fact and theory are correlatively interdependent from the very beginning. They are two facets (the formal and the material) of a single more or less articulated whole of experience, that is constantly being developed and more completely specified by the activity of thinking. It is, then, no

its furrow to trail behind it, and in so far as, in order to conceive an object one must rely on a previously constructed "world of thought", there is always some degree of depersonalization at the heart of consciousness . . . consciousness projects itself into a physical world and has a body, . . . because it cannot be conscious without playing upon significances given either in the absolute past of nature or in its own personal past. . . .'

[1] Cf. pp. 269–71 above.

wonder that we found, when we examined the procedure of scientists and the course of scientific revolutions, that observation and theory were inseparable and interdependent. As the very possibility of perceiving recognizable objects depends upon the structured background of knowledge brought to bear in the perception, the reliance of scientific observation upon the specialized theoretical setting must be even more marked. If ordinary, everyday perception is a skill utilizing acquired dispositions, how much more so is the expert perception of the scientist?

But how is all this related to 'the external world'? The fact, in the account we have given of it, seems to be largely, if not wholly, a product of subjective processes, of the mental set of the subject made up of innate schemata, and of suggestion from prior subjective experiences. What then is the source of objectivity in perception, how is it a knowledge of an independent world?

Here we must realize, and our exposition has already led us to a point where it becomes apparent, that 'the external world' is itself a conceptual construct. It is, as Kant saw, a noumenon constructed in the course of applying the categories in experience. This is not to say, as Kant does, that it is an unknowable thing-in-itself. An 'external world' beyond the reach of perception would be that. The real world is the complex of objects and events built up in the course of our experience by the very means that we have been describing, and its 'objectivity' is the stable coherence of its systematic structure.[1]

Perception is veridical when it fits into this structure without hitch or discrepancy, when the processed evidence 'conspires' to seat it firmly in an appropriate slot, and when apparent success in doing so does not conflict with demands which are being forced upon us by other processed evidence elsewhere. It is, as we have earlier maintained, the consilience of hypotheses and the corroboration of cues that generates perceptual assurance, and it is conflict and 'illogicality' that reveals illusion. The 'subjective' is what, in the course of experience is singular and detached from the system of the coherent world, what is least organized by thought, most haphazard, random and disconnected. The 'objective' is what belongs to the solid web of relationships in the world of established facts. Hegel makes the distinction with admirable

[1] Cf. my 'Objectivity and Reason', *Philosophy*, XXXI, 1956, pp. 69 ff., and Ch. XII below.

accuracy and insight in his explanation of the distinction between sleep and waking:

'But in the waking state man behaves essentially as a concrete ego, an intelligence: and because of this intelligence his sense-perception stands before him as a concrete totality of features in which each member, each point, takes up its place as at the same time determined through and with all the rest. Thus the facts embodied in his sensation are authenticated, not by his mere subjective representation and distinction of the facts as something external from the person, but by virtue of the concrete interconnexion in which each part stands with all parts of this complex.'[1]

xiii. PERCEPTION AND SCIENCE

Perceptual knowledge is thus a polyphasic unity. It is the awareness of a world[2] continuously presenting itself through sentient experience as a system ordered both in space and in time. Each perceived object is a specification of the system which the organizing activity of the mind (what we may now call 'perceptual thinking') articulates out of the mass by applying the organizing or conceptual principles that enable it to 'make sense' of its varied experience. Thus each perceptum is a variation within the system, relevant to its structural organization (a relevant variation[3]), which manifests, or realizes, the principle or principles of organization immanent in the whole.

The fundamental structure of perceptual knowledge, therefore, is the same, though less fully explicated, as that we have hitherto found in science. There is a conceptual scheme ordering a manifold of sentient experience into a coherent and intelligibly manageable world of perceived objects. So, in science, theory serves as an organizing principle for a mass of observed facts. In perceptual knowledge the conceptual scheme is not self-consciously and critically set out as such, as it is in science, but remains largely implicit. Nevertheless, the objects and events of the perceived world are what they are, and are perceived as they are, by the mediation of the conceptual scheme—the structure of the experi-

[1] *Enzyclopädie der Philosophischen Wissenschaften*, trans. Wallace, sect. 398.

[2] Cf. the 'life world' of Husserl and the Phenomenologists.

[3] See *Foundations*, pp. 179 f.

enced world. And, in the same fashion, the facts with which science deals are selected, observation is moulded, and the experimental methods by which they are investigated are directed by the dominant conceptual system. When the perceptive endeavour encounters obstacles, and cues conflict or are inadequate, perception is tentative, hazards guesses, and the object is seen vaguely or inaccurately as something discrepant from what 'dawns' by stages and eventually is recognized. The dawning is the result of exploration of cues and new sensory material until the body of varied indications along with already recognizable factors of the situation fit together and a coherent pattern emerges. So likewise when scientists encounter contradictions in the implications of a theory, or their evidence conflicts, they formulate hypotheses and develop them in the light of past and newly investigated facts until they can evolve a consistent system.

In perception the hypotheses are not explicitly formulated. They are, as it were, the deposits of past experience, the suggestions of acquired dispositions. They are prompted by the shape and tendency of what is already perceptually known. So in science hypotheses grow out of earlier theories, suggested at times by neglected features of them or by the development of implications hitherto overlooked.[1] In both cases confirmation is sought in corroborative evidence and is found when the evidence constitutes a system so interlocked in its parts and elements that it presents a coherent picture, with a single pattern, all of one piece.

This being the case, it is obviously a mistake to separate observation from theory as independent sources of knowledge. It is a fundamental error to regard perception as an independent means of acquiring information against which hypotheses may be checked, and it is no less misguided to think of hypotheses as subjectively inspired visions that must be compared with the separately acquired, theoretically neutral data revealed by sense. Knowledge, whether at the common-sense level of ordinary perceptual experience, or at the level of sophisticated scientific analysis, is not a combination of two separable activities providing two distinct types of information, one of which has to be made to agree with the other by external adjustment. Knowledge is a

[1] Cf. Kepler's attitude to the sun as centre of the planetary system and his discovery of the inclination of the earth's orbit; and Harvey's development of the neglected implications of the valves in the veins.

single developing polyphasic system, that has at all levels a sensuous as well as a structural aspect, and what we call perception—or better common sense—is only knowledge at a more elementary stage of articulation than what we call science.

Scientific observation is itself already more developed than common-sense perception, because it is more deeply impregnated with theoretical interpretation. It is not mere perception but trained, discriminating, theoretically informed perception, of which the untrained layman is innocent, and of which the scientist is capable only after long and arduous practice.

CHAPTER IX

QUESTION AND ANSWER

'. . . there is no science that does not spring from pre-existing knowledge, and no certain and definite idea which does not derive its origin from the senses.' (William Harvey)[1]

i. SCIENCE AND COMMON SENSE

What we call 'the common sense world' is, for the most part, the perceptual world. But it is not a world derived from, or built up by accretion out of, percepts, for no perception ever occurs, or can occur, in isolation, or in its own right; so percepts do not exist independently to be accumulated and related to one another externally. The common-sense world does not so much arise out of perception as perception takes place in the common-sense world. Perception is the momentary focal point of the projection of a world; the focus of the continuing activity of definition through which that world comes to consciousness and is progressively generated out of primitive sentience in the process of articulation from the vague and indefinite to the definite and distinct. Perception is the focus in the sense that the already developed awareness of a world of interrelated objects mediates, and serves as a background of reference for, the achievement of recognition. Knowledge at every level is what Kant said it was: a whole of related and connected elements,[2] though its objects may be more or less vague and the relations between them more or less obscure. Its development in experience is the continuous clarification and definition of the elements and the relations between them, and perception is the point at which we bring this process to bear upon the immediately present situation in which we find ourselves.

As we have seen, perception works by means of cues and often subconscious, or peripheral, indications of relationships, which are spontaneously combined and interpreted, as a rule with great

[1] *First Disquisition on the Circulation of the Blood* (addressed to Jo. Rider).

[2] *Kritik der Reinen Vernunft*, A.97: '*Wenn eine jede einzelne Vorstellung der anderen ganz fremd, gleichsam isolirt, und von diese getrennt wäre, so würde niemals so etwas, als Erkenntnisist, entspringen, welche ein Ganzes verglichener und verknüpfter Vorstellungen ist.*'

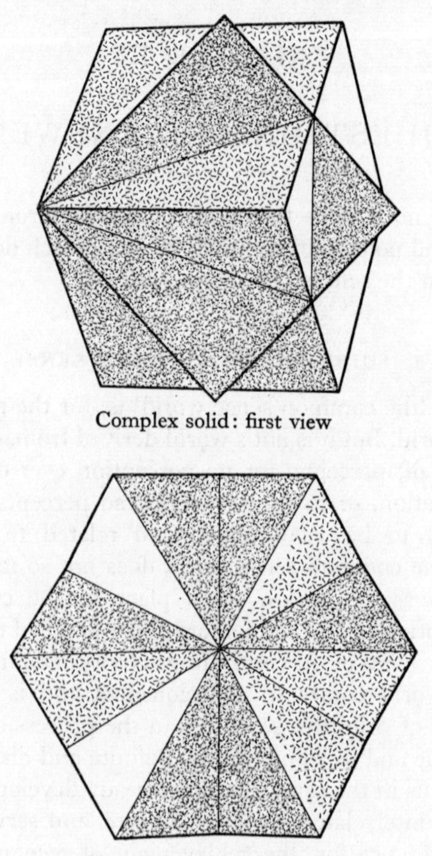

Complex solid: first view

Complex solid: second view

rapidity. But sometimes these cues and indications conflict, as when a child grasps at a bright object beyond his reach; or when two intersecting cubes, seen by an adult from one corner, are made, by the play of light and shade, to look like a kind of Maltese cross or geometrical rosette; or when the sixteenth century anamorphic portrait of Edward VI, seen from the edge of the tablet appears to stand out at right angles to the surface.[1] In such cases we explore from new angles and with other senses, until the

[1] Cf. Gombrich, *op. cit.*, p. 213: '. . . in the original peep show, the head will look surprisingly plastic, as if protruding from the oblique panel. . . . Having difficulty even in imagining the shape of the distorted profile that is equivalent to the normal view, we interpret what we see as a configuration parallel to our eyes'.

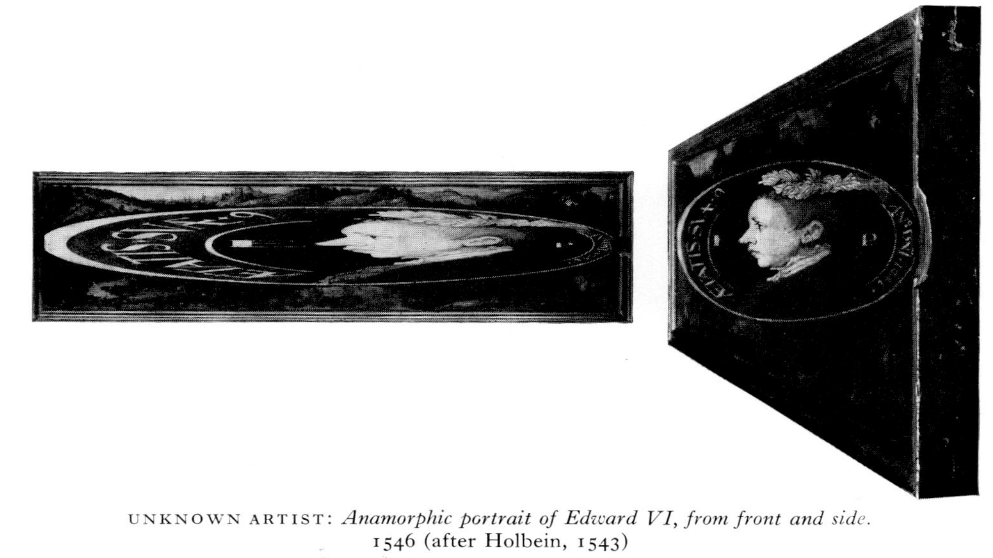

UNKNOWN ARTIST: *Anamorphic portrait of Edward VI, from front and side.* 1546 (after Holbein, 1543)

PLATE 3

cues and indications agree together and we see the objects as consistent wholes. So the 'life-world' of everyday perception is generated and becomes the prevalent backdrop of our waking consciousness.

But these elementary conflicts are not the only ones. The sun rises each morning and sets every evening and in the interval of darkness we perceive no continuance of its existence. So the Greeks believed, at one stage, that it was extinguished every night and rekindled every morning, on the analogy of the domestic fire. The majority of the stars do the same at night, they rise and they set, yet some near the pole move round in a circle. This suggests a revolving spherical firmament carrying them all around the earth, without need to postulate extinction and rekindling. Once the heaven is so envisaged, the sun takes its place with the other heavenly lights and revolves likewise.

Men first explained the anomalies of the perceptual world by means of plausible myths on specious analogies with more understandable experiences. But these stories themselves contained incongruencies. The chariot of Phaeton was never seen. The fiery arrows of Jupiter in the lightning were never found after the storm. These inconsistencies stimulated men to seek more systematic explanations and the first halting essays in scientific theory were made. The heavens were first seen as a revolving globe; but the precession of the equinoxes and the aberrations of the planets revealed contrary motions in this rotating sphere, and new hypotheses were needed. As the implicit conflict of cues in misperception leads to subconscious hypothesizing and perceptual exploration, so conceptual conflicts lead to explicit hypotheses and scientific speculation. And with the rise of scientific hypotheses, their acceptance and popularization, the way in which the perceptual world appears to man changes. Where once he saw the sun extinguished by its descent into the sea, he next sees it as descending behind the edge of the earth. Where once he saw the dome of the heaven, today he looks out into boundless space. Current belief is moulded by accepted doctrine and the common-sense world becomes an amalgam of perceptual experience with popularly accepted (usually half-digested, and somewhat outdated) scientific theory. But it remains a relatively coherent whole of interrelated facts, its persistent discrepancies tantalizing the knower and constantly stimulating him to further investigation. The non-scientist's doubts and inquiries are, for practical purposes, quieted by his faithful acceptance—on authority—of the teachings, so far as he can grasp them in popular form, of the practising scientist.

Common sense is, therefore, a world view made up in part by direct perception, and in part by popularized doctrines evolved from the attempts to remove the contradictions among appearances —the first not remaining uninfluenced by the second, as that is integrated into the interpretative conceptual scheme. So the structure of the common-sense world is moulded and its perceived contents are ordered by evolving scientific theories in a way which is clearly and simply exemplified in early astronomical notions. Let us consider them a little more closely.

Man's life, like that of all the animals, is bound up with, and dependent on, the alternation of day and night and the succession of the seasons. The changes of the heavens must, therefore, have been of interest to him from time immemorial. The night sky presents to naïve perception the appearance of myriads of luminous points set in a dark background, with the moon as a larger luminous body. The first impression is one of a completely random scattering of lights, with no immediate indication of motion. But longer and more careful attention reveals to the observer, first patterns of arrangement (constellations), and next changes of spatial relationship—a movement of rotation can be detected. Over a protracted period this movement may be traced in detail and is found to vary among the heavenly bodies. Some few move faster than others. Some change direction periodically, some move in circles, others along arcs and segments, others in loops. All this, at first, presents a bewildering variety of appearances the only superficial regularity being the diurnal alternation. But if we relate these movements to the position of the sun at its rising and setting, and if we admit the suggestion that the heaven is a great sphere centred upon the earth as a smaller sphere, and if we think of the former as rotating about the latter on an axis which passes through the pole star, we can reduce the variety to order. Now mark out the path of the sun around the celestial sphere during the course of the year as the ecliptic, and we have a framework in which the great majority of motions of the heavenly bodies becomes intelligible. This framework, or organizing conceptual scheme—not itself directly observable, but thought out as an organizing system into which to fit the perceivable phenomena—was the earliest astronomical theory. Once adopted the heavens are *seen* as a sphere revolving about the earth, the sun is seen as moving, not only diurnally across the sky, but annually around the zodiac. And so the year, the seasons, and the heavenly movements are combined in the common-sense awareness of the world, and the poet sings of the time

> 'When that Aprille with his shoures sote
> The droghte of Marche hath perced to the rote,
> And bathed every veyne in swich licour,
> Of which vertu engendred is the flour;
> Whan Zephirus eek with his swete breeth
> Inspired hath in every holt and heeth
> The tendre croppes, and the yonge sonne
> Hath in the Ram his halfe cours-y-ronne . . .'

But this common-sensical astronomy, which is at the same time genuine science, leaves much to be desired, for it cannot consistently accommodate the movement of the planets. So as scientific thinking progressed the simple 'two-sphere' view of the universe became elaborated into a system with the small sphere of the earth at the centre of a nest of greater spheres each carrying one heavenly body or more. This was complicated further by making the spheres revolve on different, mutually inclined, axes (Eudoxus). Then the number of spheres was increased and some were thought to rotate in opposite directions to others (Callipus and Aristotle); then little spheres (or epicycles) were introduced rolling along the larger spheres (Ptolemy); and so the system grew, always in the interests of explaining the movements of the planets consistently with that of the stars.

Even when this seemed to have been accomplished comparatively satisfactorily, there remained defects, for comets did not fit into the scheme and had to be accommodated in the sublunar sphere as meteorological phenomena. Also, increasing accuracy of computation based on the epicyclic scheme revealed ever more incompatibilities until the whole framework was reoriented and made heliocentric and the process continued, as we have seen in earlier chapters, in the pursuit of unified and consistent systematization.

The appearance of continuity in change offers us another such example. Continuity is a pervasive characteristic of the perceptual world. Visual space is manifestly continuous; colours merge insensibly one into another; the pitch, loudness or volume of sound varies continuously. We fill in the blank in our visual field occasioned by the blind spot in the retina by continuing the form or colour of the surrounding appearances, and continuous movement is seen where rapid discrete succession or alternation of lights occurs before us. Yet when we try to give a conceptually intelligible account of continuity we are plunged at once into all the paradoxes of Zeno. Poincaré has shown that contradictions arise whichever approach one makes to the perceived continuum, whether from the side of intensity of sensation, or from the side of sharpness of discrimination.[1] Thus a pressure, A, of 10 gms. cannot be distinguished from B, of 11 gms., nor a pressure of 11 gms. from C, of 12 gms.; but A is distinguishable from C; and we have the contradiction $A = B = C$, yet $A < C$. Or, if we cannot distinguish separate but neighbouring points with the naked eye and we can see that they are distinct through a microscope, yet through the microscope we can also see intervening points continuous with both. Consequently, we find the part homogeneous with the whole, not only qualitatively but in number—which is impossible. Accordingly, science has to be called in to redress the conflicts of the common-sense idea of continuous extension, and mathematicians construct the notion of a mathematical continuum, while physicists conceive a discrete atomic structure in matter to resolve the conflicts between permanence and change.

Earlier we observed how the classical conceptions of atomism and mechanism contained ineradicable contradictions, but these were already scientific concepts, and as science had stepped in to overcome (or explain) the discrepancies of sense-experience, so more sophisticated theories were required to overcome the contradictions entailed by relatively more straightforward scientific models. This interplay between common sense and science, as well as the stepwise progression within science itself, is illustrated throughout the history of intellectual advance.

So science arises out of the world view of common sense, which is a rough and ready system of related facts sufficiently coherent for ordinary practical purposes, but never quite self-consistent enough. Science is the result of the persistent pursuit of intelligible order in this world picture and it is absorbed back into the common-sense view making it more sophisticated though never fully

[1] See *Science and Hypothesis* (New York, 1952), pp. 21–5.

systematic. Yet there is, throughout, continuity of development from unreflective perception to common sense, and from that to science, bringing more precise methods of observation, measurement and experiment, all factors in the one sustained process of systematization of a total experience.

At every stage and in every phase we have a whole of related and connected elements. Perception occurs in an organized perceptual field spatio-temporally continuous with the 'life-world', itself a structured totality. The common-sense view of the world is a more theoretical and largely conceptual system providing a comprehensive frame of reference for perceptual interpretation. It is not separable altogether from science; but it is not a precise system, and includes, much at random, elements derived from direct perception, common experience and folk lore, as well as those from science, in loose conjunction, with no more than half-hearted and at best unreflective efforts at coherent combination. Still, it is relatively and generally integrated and is the product of the same urge to coherent understanding which gives rise to science. At each stage in this series, the structural and the sensuous aspects of knowledge are interdependent. There is no perception without both, and the common-sense world view is just the (somewhat loosely) ordered combination of perceptual knowledge, modified by successive attempts to order it more satisfactorily. Science emerges as the continuation of these persistent attempts, stimulated by questions arising out of common-sense concepts.

Thus we have a continuous series of systems each a polyphasic unity at a different level of development. We must next investigate more in detail their character and interrelations. But we shall find throughout the succeeding exposition that this general serial pattern of successive polyphasic systems is dominant, and provides the key to the solution of problems.

ii. QUESTION AND PRESUPPOSITION

How do questions arise? Collingwood, in his *Autobiography*, rejects what he calls 'propositional logic' and calls for a logic of question and answer. In *An Essay on Metaphysics* he makes a somewhat abortive attempt to develop such a logic. Propositions, he maintains, are answers to questions and they cannot be properly understood, nor can their logical relationships be discerned, unless one knows the questions which they answer. But questions, again, arise out of presuppositions which may be stated in propositional form. Collingwood allows us to call a statement a proposition (and so either true or false—though he prefers 'right' or 'wrong') if it is the answer to a question; but if it is not, if it is ultimate and *a priori*, he denies that it is a proposition and calls it an absolute presupposition. Without such absolute basis, he holds, no science is possible, and the questions that give rise to science arise originally out of absolute presuppositions. Thus a science is a sort of linear deductive system, beginning with absolute presuppositions, leading to questions arising from these, thence to answers, from which new questions arise—and so on.[1]

In the *Autobiography* and his account of historical method in *The Idea of History*, Collingwood gives a rather different description of what he takes to be scientific procedure. But let us for the moment consider only how questions arise. Collingwood never really tells us. He gives one example, which is trivial and not very enlightening, of seeing a line stretched horizontally, taking it to be a clothes-line, and thus presupposing that it was put there for some purpose. If, and only if, we presuppose this, the question arises, 'For what purpose?'[2] But this does not tell us how the question arises any more than it tells us how the thinker arrived at the hypothesis that the line was a clothes-line. (Collingwood remarks that it 'came plump into my mind . . . all at once and unheralded'). In fact the statement: 'That line was put there for some purpose' is meaningless unless we can specify the situation to which it refers. What line is being indicated, in what surroundings and in what circumstances is it being observed, and what is the context in general? Perhaps, for this reason, Collingwood refers to it rather as a 'thought' than as a 'statement'; for a thought is part of an active process of thinking in which the present object of thought would inevitably be mediated and interpreted by reference to a complex context. And this is the important feature of the whole matter, for if you detach the presupposition from its context it loses all meaning and no question arises from it, unless perhaps the general question concerning what it could be referring to. Absolute presuppositions, as Collingwood defines and describes them in the *Essay on Metaphysics*, look much too much as if they were isolated and prior to all substantive context, certainly to all questions. But if they were, they could have no significant implications and no questions could arise. They would themselves be

[1] Cf. *Essay on Metaphysics*, Ch. IV. [2] *Ibid.*

meaningless—mere empty abstractions like the meaningless notion of pure being, which Collingwood so sharply criticizes in an earlier chapter, as the putative subject-matter of metaphysics.[1]

In the *Autobiography*, however, Collingwood gives a better and more suggestive account of question and answer logic. There he objects to propositional logic because it assumes that 'truth and falsehood, which are what logic is chiefly concerned with, belong to propositions as such'.[2] Against this position he asserts that truth, in the sense in which a philosophical theory or an historical narrative is called true—and we may add, also a scientific theory—belongs 'not to any single proposition nor to a complex of propositions taken together, but to a complex consisting of questions and answers'. This is but another way of saying that the kind of truth that is being sought belongs to a body of knowledge that is itself an active process of discovery and development. Questions arise in this process only when strictly relevant to problems.

'Each question,' he says, 'and each answer in a given complex had to be relevant or appropriate, had to "belong" both to the whole and to the place it occupied in the whole . . . There must be that about it whose absence we condemn when we refuse to answer a question on the ground that it "does not arise".'[3]

Here there is no suggestion of isolable absolute presuppositions and what he writes of historical method in *The Idea of History* is more in keeping with what is said here than with what we find in the *Essay on Metaphysics*.

Can we now develop the implications of Collingwood's statements in the *Autobiography* to find a more viable notion at least of how questions arise?

We have suggested above that questions arise whenever conflicts occur in the attempts we make to organize our experience by means of some system of schemata, or concepts. If so, they do not arise, like the phoenix, from the ashes of meaningless absolute presuppositions. What is presupposed, to have significance, must be a body of already acquired knowledge developed from primitive sentience by way of common sense. To this extent Empiricism is right in its assertion that sense is the seed-bed and origin of factual knowledge. So it is, but what sense offers is not a succes-

[1] Cf. *Essay on Metaphysics*, Ch. II.

[2] *An Autobiography* (Oxford, 1939), p. 34.

[3] *Ibid.*, p. 37.

sion of data to be used as building bricks, but a vague incoherent blurred sentient content that the activity of the mind (what issues eventually as thinking) progressively organizes. The presupposition of science, therefore, at least in one meaning of the phrase, is the relatively unscientific world picture of common sense; or, if we think of the particular area of some special science, the relevant realm of discourse within the common-sense world that is the special concern of the science. This presupposition may be explicated and may be expressed in statements, and they may legitimately be called propositions, because none of them will be 'absolute'. They will all be relative to the context of experience which gives them meaning as statements and warrant as assertions. So far as the evidence warrants they may be justified—and just how they are justified we shall presently consider. So far as it does not, they will not be. They may always, in the proper circumstances, be regarded as answers to questions. For knowledge is never fixed or settled, it is a persistent search, a constant development and Collingwood is certainly right to maintain that question is its cutting-edge,[1] arising out of difficulties, prompting hypotheses, and stimulating new efforts towards discovery.

The presuppositions of science may also be seen in another way, not altogether divorced from that which identifies them as common sense. A science is a conceptual system ordering a specific area of experience, and every statement of law, or of fact coming under law, which belongs to the system, derives its significance from its relations to other laws and facts, to the general principle of organization and, in short, to the over-all structural form of the system. These, then, will be presupposed in any such statement, and they are not common-sensical ideas but strictly scientific. If there is in any good sense to be an *absolute* presupposition of a science, it will be the ultimate principle of order on which its entire structure is based. But here we must not anticipate the results of future discussion.[2] Presuppositions of this kind are not unconnected with common sense because they result from the clarification and refinement of the common-sense world-system which science effects, but they are, more obviously, not so much temporally as logically, prior to scientific statements. Common sense is temporally prior as well as (in one respect at least) logically prior. It is logically prior in the sense that questions logically arise

[1] Cf. *Speculum Mentis* (Oxford, 1924), p. 78.
[2] Ch. XI below.

from its obscurities—but these are not the logical *grounds* of the answers which science discovers.

Questions arise from presuppositions—whether single or in groups (what Collingwood calls 'constellations')—when the implications of the statements expressing them, consequent upon the ramifying relations that connect them with the context of articulate awareness to which they belong, give rise to conflicts, or reveal discrepancies, in the texture of that awareness. One example, which may be fictitious but may also have some foundation in historical fact, is the case of an untutored observer, who lives near the mouth of a river and notes the way in which its banks and estuary change from time to time and its channel varies. These observations would not, of course, be unprocessed 'data' but would be the natural interpretation of sensory experience in terms of the common-sense world of objects and events. Reflecting on the observed changes and being ignorant of their causes, he may, quite naturally, imagine that the silt is precipitated out of the water and redissolved into it by turns. Taking this along with the common experience of dissolving solids (like salt and sugar) he would be faced with an apparent anomaly on the level of naïve common sense—the liquefaction of solids and the solidification of the apparently liquid. There are, of course, numerous other familiar examples—melting wax, freezing water, and the like; and, as Descartes notes in the second Meditation, we take the liquefied or solidified matter to be the same throughout the change. If the same things can appear in such opposite forms, what is the underlying substratum that can assume these different appearances? What are things ultimately made of? What is the ultimate substance of things? Thus arises the original question of science, raised by Thales in the sixth century B.C.

Or, again, take the thirteenth-century example of the flying arrow. Aristotle's theory of motion, so fruitful in interpreting falling bodies, rising flames, and the rotating heavens, required that a body be propelled if it were to move. But what propels the arrow once the bow-string loses contact with it? The air rushing in to prevent a vacuum, is the suggested answer; but, then, how is the arrow brought to rest? Again, by the resistance of the air. We have a series of contradictions. First, the demand that the moving body be violently impelled if the movement is not 'natural', conflicts with the observed fact that the missile flies out of contact with the original source of impulse. Next, the force of the air allegedly sucked in behind conflicts with the resistance of the air opposed to the arrow's motion—yet the two do not immediately cancel each other and prevent the motion altogether. So the question must still be faced: What maintains a projectile in flight after the initial impulse? This is an especially instructive example because it includes an hypothesis immediately suggested to provide an answer. So we shall return to it shortly when we consider, as our next step, how hypotheses are discovered.

Thirdly, consider the case of Anderson and the positive electron. Hitherto only negative electrons and positive protons had been recognized. Tracks in the cloud chamber now reveal a particle with electron mass and positive curvature—at once a double contradiction. Positive curvature should imply mass 1000 times as great as that of an electron and, for the degree of curvature of the track, a particle of that mass would have a range of 5 mm., whereas the track in the photograph is 5 cm. long. The question arises: Is a positive electron possible? Note that the experience of vapour trails in a cloud chamber was not new, nor was it unprecedented to find electron trails with curvature in either sense (depending on their direction of passage across the magnetic field). Not even an electron trail with definite positive curvature was new, for Millikan had seen one. The conflict was in the possible interpretation of these phenomena in terms of accepted concepts—and it was from this that the question arose.

The answer, in all cases, is a suggested hypothesis, offered for examination, for testing and consequent rejection or acceptance. If rejected a new hypothesis is sought until one is finally established. So our 'question and answer logic' proceeds in four stages: (i) a conflict appears in the application of our conceptual scheme; (ii) a question arises; (iii) an hypothesis is mooted; (iv) a theory is verified. Under (iv) there is a sub-stage in which unsatisfactory hypotheses are rejected, and the relation of successive approximations to the right explanation is not without importance. In this section we have looked at stages (i) and (ii). In the following we shall consider (iii) and (iv). But let us first take stock of the position we have reached.

A logic of question and answer cannot be concerned with propositions in isolation—for out of them no question arises. The same is true of absolute presuppositions, whether admitted as propositions or not; for questions arise only out of dynamic systems of advancing knowledge; they arise when and because application and development of the system discloses imperfections, conflicts and hiatus. A question is a demand for the resolution of a conflict, or the bridging of a gap, and it issues in a new (initially hypothetical) theory. The logic is thus a logic of discovery and of progress. While the first answer to the question is a new hypothesis, the question is not finally answered until the hypothesis is established. Therefore, the logic of discovery must include the logic of justification. For the truth that science pursues is not the truth of propositions, either in isolation or in sets. It is the truth of an experienced whole in dynamic evolution. So what we seek is a logic of progress, not a logic of fixed relations between arbitrary formulae. Moreover, as the conceptual systems within which questions arise are systems of experience, in which the elements are mutually dependent and the totality is prior to the parts, they are systems in which relations are internal. Hence the logic of such systems cannot be one of purely external relations. Because question and answer are the forms of living and advancing thinking, the logic of question and answer cannot be purely formal, though it will recognize the distinction of form from content and will concern itself primarily with the first. But it cannot wholly ignore the features of the subject-matter which determine the ways in which questions arise. Further, it can also not afford to ignore the manner in which our minds work in seeking explanation and enlightenment. Just as no satisfactory epistemology of perception can afford to ignore the evidence of psychology, so no satisfactory epistemology of science can safely ignore the psychology of thinking. Question and answer is dialogue and its logic is therefore dialectic. It is the method of discovery and development in knowledge, the logic of which is dialectical logic.

iii. THE ORIGINS OF HYPOTHESES

'Questioning is the cutting edge of knowledge; assertion is the dead weight behind the edge that gives it driving force.'[1]

Supposition and questioning, Collingwood tells us, are at bottom the same thing. Thus question and answer coalesce. The answer, in the first instance, is a questioning answer that specifies the general question. It is not sufficient, when faced with a

[1] Collingwood, *Speculum Mentis, loc. cit.*

problematic situation, say, a broken down car,[1] to repeat fruitlessly the general question, 'Why won't the car go?' It must be specified: 'Is it No. 1 plug?' And here we have an hypothesis.

So Kepler, finding that the orbit cannot be a circle (due to conflict of evidence), asks 'What shape is it?' but does not persist in fruitless repetition of the general question. He proceeds at once to the specific question: 'Is it an ovoid?' which is both question and suggested answer. So Harvey, deciding that the heart cannot at once draw air from arteries and lungs while it expels blood and 'fuliginous vapours', asks 'Does the heart function like a bellows or like a force pump?' and specifies this again to 'Is its main action in systole rather than diastole?'—an hypothesis is emerging.

But what is the source of the hypothesis? Does it come 'plump into the mind'? Hardly; for it must be suggested by the body of fact and theory already accepted. Assertion is the dead weight behind the cutting edge that gives it driving force. This body of accepted fact, however, is complex. It is both fact and theory, or ordered fact, interpreted observation—an already structured arrangement of the relevant experience. It is from a misfit in this structure that the question has arisen and it will be by a readjustment that it must be answered. So reference must be made back to the accepted system in order to find an hypothesis.

Questions arise out of the already established body of knowledge when incoherencies occur, and it is the same source—the only available source—that provides the suggestions for tentative answers. Just as in perception elements in the presented complex serve as cues from which subliminal hypotheses are formed, so in explicit thinking, aspects of the known facts—the dead weight of affirmation—give indications of the direction of new advance. We shall not, of course, be misled by our own (or Collingwood's) metaphor into thinking of 'the known facts' as something inert, merely lying at hand to be picked up uninterpreted. They are never theoretically neutral. For what the scientist is doing in proposing a new hypothesis is reconstructing a system already articulate in some degree. We should expect, therefore, that his hypothesis will be both suggested by, and itself a modification of, features within the hitherto accepted theories.

In the examples we have examined this is apparent.[2] The Peripatetic explanation of the flying arrow derived its series of

[1] Cf. Collingwood, *An Autobiography*, p. 32.
[2] See Chs VI and VII above.

hypotheses all from the accepted system. Nature's abhorrence of the vacuum supplied the first, but when that failed, because it could not account for the eventual cessation of the arrow's motion, a new, and more subtle, appeal to the system was made. Motion must be either natural or violent. Could the two not temporarily combine, so that a violent impulse might impart to a missile an evanescent, quasi-natural impetus, which would gradually die out but not be extinguished immediately? That impetus was thought of as a kind of natural motion is apparent from Buridan's identification of the 'natural' revolution of the heavens as impetus; and is seen in Galileo's references to it, for he treats the earth's rotation upon its axis as natural motion and the impetus which it imparts to falling objects (so that they fall vertically and not aslant) as equally natural.[1]

The accepted theory of planetary orbits with which Kepler worked was that they were circular, complicated, if at all, by eccentricity and epicycles. So he begins with the assumption of a circular, eccentric orbit. But that will not work. The orbit, he finds, cannot be a circle. He therefore modifies the circular form by adding a bulge on one side and accounts for it by means of an epicycle—this gives him the ovoid. Again, he falls into contradictions, so once more he modifies his assumed figure. It differs very little from an ellipse. Could the orbit not be a perfect ellipse? Each new hypothesis arises out of the previous one through reconsideration of the total system and readjustment of the relations between its elements in an effort to remove the discovered discrepancy.

The semilunar valves of the heart, Harvey argues, cannot both impede the flow of vital spirits from the aorta back into the left ventricle of the heart and also allow fuliginous vapours to pass. What then is their function? Can it be to prevent the regress of blood during diastole of the heart while the right auricle is refilled with blood from the *vena cava*? Likewise, the mitral valves cannot allow fuliginous vapours to pass in one direction—into the lungs—while at the same time they prevent air from doing the same thing. Do they not, therefore, serve to regulate the flow of the blood in one direction only, from the right ventricle through the lungs to the left auricle? Colombo had already discovered this so-called lesser circulation. Further, Fabricius had shown that

[1] See *Dialogues Concerning the Two Chief World Systems*, pp. 134, 142 and 154.

the valves in the veins obstruct the blood flow in the direction away from the heart—allegedly to prevent its accumulating in the extremities. But if they do so, how could the movement of blood be back and forth as Galen had taught? If it flows only one way from the heart through the lungs, only one way from the heart into the arteries, only one way from the veins into the right auricle of the heart, does it not circulate perpetually? So the main hypothesis develops from unfolding the implications of past hypotheses, both negatively and positively—negatively revealing contradictions which prompt questions, positively in the discovery of possible answers.

Lavoisier was troubled by the inconsistencies of the Phlogiston Theory which could not account for, but rather contradicted, the augmentation in weight observed regularly to occur after combustion and the calcination of metals.[1] Could this augmentation be due to absorption of something from the atmosphere? The question arises out of the puzzling facts and the refractory theory. It is at once converted into an hypothesis which is further supported and specified by reference to the work of the British pneumatic chemists, and is then investigated in detail by Lavoisier experimentally. The weight behind the cutting edge has already been provided by Hales, Black, Priestley, Cavendish and de Morveau.

The laws of reflection and refraction as known to Newton and investigated in the first Book of the *Opticks* do not account for the appearance of Newton's rings as described in Part I of the second Book. Here is a puzzling phenomenon that had been observed before Newton undertook a closer study of it. The question arises: Why are the rays of light transmitted through an interval between two surfaces at certain distances and reflected at others? Newton investigated and measured the intervals and the corresponding diameters of the rings and found them to be periodic. He found the squares of the diameters to be in arithmetic progression: the luminous rings in the order of the odd numbers, the dark rings in the order of the even. So from the known and discovered facts Newton propounds the hypothesis that its passage through the refracting surfaces puts the ray into a state 'which returns at equal intervals' and disposes the ray to be easily transmitted through the next refracting surface, and in the interval between the returns, to be easily reflected. This raises the question

[1] In particular by Guyton de Morveau whose paper Maquer reported to the Academy in February 1772.

how the ray is put into this 'state' by passing through the refracting surface, and Newton answers (though tentatively), that possibly, by impinging on the refracting surface, it sets up a vibration or wave which travels through the medium and is superposed upon the vibrations of the ray itself, so as periodically to reinforce them, and periodically to counteract them.[1]

The hypotheses in both cases are gathered from the facts demonstrated in Book I and described in the first part of Book II. The colours of the rings conform to the laws discovered in Book I, the periodicity is described in Book II. The first hypothesis is scarcely more than a summary restatement of the facts; but the second is a construction arranging and interpreting them so as to make them intelligible. We may note further how Newton in enunciating the second hypothesis argues by analogy from the circular ripples resulting from the splash of a stone dropped into water, and from the vibrations of sound in the air. It is the known facts, the assertions, that drive the cutting edge of questioning, and the question Why? or How? is at once converted into a supposal, or an assertion voiced in a questioning (or problematic) tone.

That new hypotheses emerge from old systems we have found to be commonly the case and we have seen two good reasons why it should be so. First, perception, or observation, itself depends for its import on interpretation in terms of an established system of schemata, any revision of which, to resolve conflicts, can only draw upon the already known in the search for new patterns. Secondly, science proceeds under the aegis of a conceptual scheme which can be modified where incompatibilities appear only by appeal to observations already deriving significance from that scheme itself. What is sought is a new pattern of organization; but neither can that be found *in vacuo* nor can it be extracted from chaos. It can only be discovered by adjustments of the existing order and must be suggested by features already present in the hitherto accepted arrangement of facts. So we find continuity of development even when that is most revolutionary in appearance.

Can we detect the form of argument that leads scientists from the questions arising out of incongruous presuppositions to the hypothesis of a new structural arrangement?

[1] *Opticks* II, Pt III, Prop. XII.

iv. ANALOGY AND ENUMERATION

In many instances, the first step is an analogical argument, taking a clue from an already known structure that accommodates elements similar to those raising the question at issue. Newton, we have seen, argued by analogy from ripples spreading in water and of sound vibrations, to a possible wave-like activity in the ray of light. Harvey and his predecessors argued by analogy from the action of a bellows to the action of the heart, first assuming that in diastole the heart might draw in air; then, when the facts revealed the ineptitude of the parallel, the analogy still held from the expulsion of air by the bellows on contraction to the expulsion of blood by the heart in systole. Though Harvey does not state this analogy explicitly it remains in the background. Darwin argues from the analogy of artificial selection among domestic breeds to natural selection among wild varieties. Similarly, by analogy with distribution of floating negative poles in a positive magnetic field, J. J. Thomson argued to the distribution of electrons in the atom.

The form of analogical argument needs no comment. Two cases which resemble each other in one respect are taken to resemble each other in another related aspect. But the reasoning does not claim finality. At best it points to grounds for conjecture. In other words, it suggests an hypothesis. The known provides the analogue and the analogy the hypothesis.

At this stage induction by enumeration may serve some useful purpose, but the circumstances in which it does so are limited and the outcome of the process is, as is always conceded, uncertain. In the history of the major sciences cases where it has been used are relatively infrequent, but they are by no means completely lacking. One finds little trace of it in astronomy, physics, or chemistry, and much less in most branches of biology than might be expected. Scientists do not as a rule go about wondering whether or not all ravens are black. They make their comparisons between specimens in order to discover systematic connexions and when they have found them they base their definitions on these—not *vice versa*. The questions they seek to answer are not raised by the discovery of frequent conjunctions so much as by exceptions which reveal flaws in the already constructed system. But once the question has been raised, in some types of cases, the discovery of frequent conjunctions may suggest an hypothesis (as it did to Lavoisier[1]). For instance, when a pathologist is looking for the cause of a disease he is apt to note the repetition of concomitant circumstances and to try to discern frequent conjunctions. When he does so he suspects a connexion and forms an hypothesis the test of which is not simply to seek new and similar cases, but to develop the implications of the hypothesis in other ways. To that procedure more attention will be given later. What is here being maintained is simply that some forms of investigation include the search for conjunctions as a stage in the process of discovering a new hypothesis.

As a rule the concomitances are already known and form part of the 'dead weight of assertion'. Then they are analysed and sifted by something akin to Mill's methods in order to exclude irrelevancies until the crucial conjunction is isolated. At this point statistical considerations become important and if the frequency of the conjunction is such as to preclude mere chance the hypothesis that A may be B is adopted for further investigation.

Suppose what is sought is the cause of malaria. Case histories are examined for similar antecedents and accompanying conditions. For instance, it is known to be the case that the disease is most common in warm damp climates and especially in the vicinity of swamps. The methods of agreement and difference are then applied to narrow down the concomitances until it is established that the disease occurs only where mosquitoes breed and that the presence of anopheline mosquitoes are a necessary condition. In all this it is obviously important to notice whether the association of the disease with swamps, mosquitoes, anopheles, and so forth, is only occasional or occurs frequently, and percentages may prove revealing. Finally, when it is clear that there is a constant (or very frequent) conjunction between the presence of anopheline mosquitoes and the occurrence of malaria, the hypothesis is mooted that malaria may be caused by the bite of this particular variety of mosquito. It is, however, as yet no more than conjecture, and we have much farther to go to establish the fact.

Whether or not the causal connexion is suspected, the actual concomitance is often known before the investigation begins, and so is part of the accepted structure of knowledge. At times, however, it is not known, or is not seen as relevant until something draws special attention to it; and at times an investigator will

[1] See p. 175 above.

deliberately set himself to find some constant attendant circumstances where none has yet been identified. But in every case the problem must have been set by some difficulty already encountered in the existing theory. Once a question has been raised concomitances may be sought among relevant observables and when they are noticed they suggest a possible answer. But this, if it is discovery at all, is only its beginning. It is but the emergence of hypothesis.

The function of enumerative induction is largely to determine which facts, among a multitude of conditions surrounding the matter in question, are the most relevant and require closest investigation. It is an instrument of classification and, like classification generally, a method of organization. But both enumerative induction and classification are at best only preliminary stages of theorizing, and the serious business of science begins after they have been done.

Even so, the search for concomitances is no mere matter of counting. The instances counted are not all exactly alike. The relevant similarity is always set in a different context, and the differences are as instructive as the similarities. For instance, malaria occurs in mosquito-infested regions, but these differ in aspect: some are swampy, others relatively dry, and there are several varieties of mosquito. The frequency of occurrence of the disease varies with some but not with others of these factors, but essentially with the prevalence of one special variety of mosquito. Apart from these differences between the areas compared the important connexion would go undetected. So even in induction the structure of the facts is implicit in our selection of cases.[1] The various observations reveal this structure piecemeal, and as more of them are made, the general plan of organization comes gradually to light. This precisely is the effect of Mill's methods. They set the limits of concomitance and map out its variations until a point is reached where more systematic investigations can begin which will discover the universal principle of connexion involved. But even before this, in the preliminary researches, scientific thinking is already proceeding by deduction from the phenomena.

When inductive enumeration has yielded its result the serious business of science begins. The pathologist investigating malaria will examine the complex conditions of the organism during and after the disease. He will draw upon his knowledge of physiology

[1] See p. 347 f. below.

to decide what produces the symptoms. He will trace these to a depletion of red corpuscles in the blood and will examine the blood microscopically. When the microscope reveals the parasite, the scientist will press his investigation in the direction of its life cycle. The same sort of investigation will be required into the habits of the mosquito and the entry into its digestive tract of the malarial parasite, until finally a body of facts embracing the ecology and ethology of the mosquito, the life cycle of the parasite and the physiology of the human organism will have been put together forming a systematic set of relationships which constitutes an explanatory theory of the same kind as we saw evolving in the researches of Kepler, of Harvey and the rest. But this procedure goes beyond the mere formulation of hypotheses and includes their confirmation and establishment—a matter to which we have yet to turn our attention more fully.

The use of enumerative induction, even in a preliminary stage of investigation, is, however, exceptional and is for the most part limited to certain special sciences, mainly therapeutic in aim. Physicians looking for cures to diseases, technicians seeking remedies for mechanical faults in machines, experimenters attempting to improve the efficiency of apparatus, sometimes find occasion to resort to it, but it is not typical even of their methods and still less of the methods of scientists working out explanatory hypotheses for anomalous experiences. They tend to examine directly the structure of the facts as already presumed, to adjust the existing theory in what seems to them to be a fruitful manner and then to make deductions which will guide them to the necessary modification or reformulation of theory.

The system as already accepted is reviewed in the light of the discrepancy which has raised new questions and from that system an hypothesis is drawn, by analogy and by the suggestion of systematic correlations. Something of this kind seems to have been what C. S. Peirce called 'abduction'—what can be drawn out of the already known to suggest a new direction of advance.

V. ABDUCTION

On the subject of abduction, though he constantly reverts to it in his discussion of scientific thinking, Peirce is almost as constantly vague. He insists that it is a form of inference, but he tells us little of what that form actually is. Yet what he does say seems essentially sound, as far as it goes, and is conformable to the kind of view here being advocated. Abduction and induction, Peirce holds, are the opposite poles of reason—and, we may note, as the opposite poles of a magnet are mutally complementary moments in a single magnetic field, so these two kinds of reasoning are two aspects of a single process of discovery. Peirce, however, asserts that 'the method of either is the very reverse of the other's'; whereas I wish to maintain that this opposition in the direction of movement is only apparent, due mainly to differences of emphasis, and that the nerve of the inference in both cases is the same underlying implicit structure of the interpreted evidence. Moreover, what Peirce means by induction here is obviously not a method of simple enumeration, as the following passage reveals:

'Abduction makes its start from the facts, without, at the outset, having a theory in view, though it is motivated by the feeling that a theory is needed to explain the surprising facts. Induction makes its start from a hypothesis which seems to recommend itself, without at the outset having any particular facts in view, though it feels the need of the facts to support the theory. Abduction seeks a theory. Induction seeks for facts. In abduction consideration of the facts suggests the hypothesis. In induction the study of the hypothesis suggests experiments which bring to light the very facts to which the hypothesis had pointed.'[1]

Here it is apparent that 'surprising facts' have raised a question and that the known structure of the facts (presumably those bearing upon the anomaly) suggests an hypothesis—so much is abduction. From the suggested hypothesis consequences follow by deduction which suggest experiments and these bring to light the confirming evidence. This part of the process Peirce brings under the head of induction, but obviously it is not simply enumeration of concomitances, but something much nearer hypothetico-deductive procedure. How, then do the facts suggest the hypothesis? How does the hypothesis yield the experiments? And how do the experimental results confirm? These are the questions that require answers, and the answers to which, when found, will reveal the process of discovery and confirmation as a single process founded on a single logical basis.

But to these questions Peirce does not address himself. He does

[1] *Collected Papers*, Vol. VII (Harvard, 1966), p. 137.

say that the assumption fundamental to science is that the anomalous fact is somehow rationally explicable;[1] and he does say that the hypothesis must be suggested by the facts.[2] Beyond that, all he requires is that the hypothesis should be capable of experimental testing and that it must be such as to explain the initially puzzling facts. He also asks that the hypothesis should be such as would naturally occur to the intelligent mind and that it should be 'simple'. His other requirements are not logical so much as practical, in ensuring economy and expedition in experimentation. Throughout, Peirce insists that the hypothesis is a guess, and the clear implication is that until confirmed it is not yet a discovery; but he thinks that human guesses are as it were inspired because of the inherent aptness, reasonableness and good sense that prompts them—our 'natural instinct for truth'.[3] The guess, in short, is no shot in the dark, but is founded on some already formulated conception of the world and on the intelligent (i.e. scientific) reading of the facts at our disposal. 'When . . . the hypothesis seems to us likely . . . this likelihood is an indication that the hypothesis accords with our preconceived ideas.'[4]

All this, somewhat vaguely and indirectly, points to the doctrine that a question arises out of a discrepancy in an already accepted system, such as would normally lead us to expect certain eventualities that do not in fact occur.[5] The actually experienced facts, when interpreted in terms of the prevalent conceptual scheme conflict with it in certain ways. So we guess at an answer to our question. But the guess is guided by insight, by an intelligent appreciation of the facts, which can only mean, if mere empty verbiage is to be avoided, the appreciation of them as ordered in the light of conceptual schemata hitherto accepted by common sense, or by common sense as modified by scientific discoveries, but chiefly by the scientific theories current at the time of the investigation.

In fine, the structure of the known, implicit in the facts as observed is the source of the hypothesis—the suggested answer to the question, which itself has arisen from the structure of the known. What in that structure is incomplete or discrepant leads to the question; what in it suggests a new pattern points to the answer. The known, however, is not a loose collection of discon-

[1] *Loc. cit.* [2] *Ibid.*, and also p. 122. [3] *Ibid.*, p. 140.

[4] *Ibid.* Peirce proceeds to caution us, however, against the pitfalls of preconceived ideas and urges circumspection and 'an eye to other considerations'.

[5] Cf. *op. cit.*, pp. 113 ff.

nected facts, but a system of knowledge ordered by conceptual schemata. The facts are facts, and have significance as facts only in their context and by interpretation within the conceptual system. It is the immanence of the structural principles in every factual situation that makes it potentially fruitful of new hypotheses; and when they arise they immediately modify the system in some degree and constitute a potential new step in development. 'Deduction from phenomena', so far from being a dubious description of scientific method, is operative virtually at every stage.

Newton's rings, indicating non-uniform reflection and transmission at a surface, raise the question of the relation between the rings and the thickness of the transparent film. Their periodicity suggests the hypothesis of a corresponding periodicity in the energy of the light. The hypothesis is deduced from the phenomena, but not from phenomena in mutual isolation or in separation from the background system which gives them observational significance. The hypothesis is no more nor less than the first step in the further articulation of that system.

vi. CONFIRMATION

Deduction from phenomena does not stop at this point, for the hypothesis is so far only a conjecture and before it becomes a discovery it must be substantiated. In order to confirm it the hypothetico-deductive process is brought into play. The hypothesis is already at hand and consequences are now deduced from it which may be tested by experiment and observation. But again the deduction is from the phenomena, even though, in its course, mathematical calculi may be employed. In deducing the projection of a picture by a lens, Newton uses pure geometry; in deducing periodicity from his measurement of the rings, arithmetic; but it is, as we saw earlier, always the structure of the facts and their relation to their theoretical background from which the conclusions follow. It is impossible to deduce anything from any hypothesis unless what it states is interpreted in the light of systematic relations with its context of knowledge, both observed and theoretical.

The essential point, however, is that the results of duduction from the hypothesis are always different and various consequences, and it is not the simple observation of some one of these that confirms the hypothesis. Let that one consequence be observed as many times as you wish, it still counts only for one piece of evidence—and may even be misleading. Harvey, it will be remembered, protested against the alleged evidence of dissection because in every dissection of a lifeless cadaver the essential facts may inevitably be missed.

Accordingly, the deductions from the hypothesis must be multiple and must yield a variety of possible facts to be tested. They guide the devisal of experiments, and these, to be conclusive, must be different, yet must be systematically related. Thus Lavoisier deduces from the hypothesis that in calcination something is taken from the air, that if calcination takes place in a confined space the volume of air in that space will be reduced during the process. He deduces further that if the calcined body can then be reduced in a confined space the volume of air should be augmented to the same extent as it was previously reduced. He also draws the inference that if something from the air is fixed in the metal the weight of the calx should be greater than that of the metal, and that the reverse should be observable in reduction. It remains only to devise suitable apparatus to test these conclusions experimentally; and this devisal is also a development of what the facts in their mutual relationships imply. When that is done the confirmation of any single conclusion would be (in this example) insufficient by itself. But the confirmation of them all, in their mutual interlocking implications, is conclusive. We can then assert (like Harvey) 'it cannot be otherwise'. The structure of this procedure is obvious enough and there is no need to continue to illustrate how the systematic corroboration of the observed facts is enhanced and finally secured by testing the properties of the air before and after calcination, as well as those of the gas collected during reduction.

The theories of Braithwaite, Popper and Wisdom are therefore sound enough so far as they allege a process of deduction from higher to lower level hypothesis and the testing of the latter by observation. But the deduction is never simply analytic, nor is the testing inductive in the old sense. Empirical it undoubtedly is; but it does not consist merely in the discovery of numerous similar conjunctions the frequency of which as positive instances lends probability to a general statement. The empirical tests are all different. Each confirms a different consequence of the examined hypothesis, which is established when, and only when, they all 'conspire' to form a single coherent and consistent body of evidence. Confirmation, therefore, is constructive rather than inductive, and the constructive process is as much deductive as it is empirical. It consists in the development of that same structure of fact which suggested the hypothesis in the first place, by processes of analogy, experimental articulation and final synthesis of observations. Only when these processes have all been completed may discovery be claimed; hence the context of confirmation and that of discovery are one and the same.

So the logic of discovery and the logic of justification coalesce. The first is the drawing out of suggestions implicit in the established system; the second continues the process by deducing possible experiments and observations from the consequent hypotheses and concludes by the construction of new bodies of evidence. There is no clear-cut separation between these processes for, as the hypotheses are deduced from the phenomena, so are the observable consequences deduced from the hypotheses. Why do valves in the veins obstruct the blood flow in one direction? Is it because the blood circulates? Such intimations arising from known facts (like Harvey's suspicions aroused by the valves in the veins) are developed and reinforced by further investigation of related implications—e.g. experiments with a tourniquet on a man's arm, observation of the working of the animal heart, investigation of relationships between the valves and the vessels in the heart itself, and between them and the passages connecting the heart with the lungs. So the initial supposal grows from a tentative to a confident hypothesis and from that to a confirmed theory.

We must investigate the logic of this process further, but we may note at this point that a scientific theory is a conceptual scheme ordering a realm of observed facts in each of which the principles of order systematizing the scheme are immanent or implicit. They make the fact recognizable as what it is and give it significance as an exemplification of the theory. Accordingly the theory is a polyphasic unity—a single whole the principle of whose unity manifests itself in a varied multitude of instances. Further the theory is not a fixed or static scheme. It cannot properly be represented by an abstract deductive system fixed once and for all by postulates and stipulated rules. Such a system may be devised to symbolize certain relationships within it and may serve limited purposes, but the theory itself is only a phase in a growing system of knowledge, the product and manifestation of a dynamic activity of thinking, which does not cease when laws are formulated, but which goes on to apply them, correlate the results of such application, draw out further implications, raise new questions, form new hypotheses, devise consequent experiments, and effect new developments. Every science is such a polyphasic unity, containing, as theories within it, subordinate unities of the same general character, each relatively self-contained, yet none wholly independent of the rest. It is, like knowledge as a whole, a system of systems, and, like knowledge as a whole, it is constantly on the move, constantly being transformed and developed.

The logic of this development displays itself in the activities of discovery and confirmation of hypotheses, both of which, as we have seen, rely upon the structural character of the whole body of knowledge as determining lines of inference—a body of knowledge within which the elements are observed (or perceived) facts, each and all intelligible only in reference to the totality with its ordered structure. It is therefore the logic of structure that we need to understand, the logic that enables us to infer from the relation of A to B and C to D, to that of E to F, and this we can now investigate on the basis of the evidence we have collected.

NOTE: WHAT IS MEANT BY 'DISCOVERY'?

To what do we refer when we speak of discovery in science? What is it that is discovered? From the historical examples already examined it is plain that what is discovered is always an hypothesis or theory, never merely an object. N. R. Hanson, in a paper of singular interest and scintillating erudition,[1] entitled 'An Anatomy of Discovery', misses this point. He classifies discoveries into numerous kinds, four of which he explains in detail as (*a*) discovery of an *X* (a local object, or process, or singular event), (*b*) discovery of *X* (a class of objects or events or a universal process), (*c*) discovery *that X* (a fact, or state of affairs) and (*d*) discovery of *X* as *Y* (discovery of an object or event which is misunderstood—'a misunderstood observational encounter'). Hanson goes on to delineate types of discovery, by what he conceives to be differences in the method of discovery, as (*a*) 'tripover' or accidental discoveries, (*b*) 'back-into' discoveries, such as the dis-

[1] Published posthumously in *The Journal of Philosophy*, LXIV, 11, June 8. 1967.

coverer at first resists, (*c*) 'puzzle-out' discoveries, where expected results are sought, and (*d*) 'concluding that' discoveries, in which distinct theories are unified. The first classification is not meant to correlate with the second, though some degree of correlation seems to pertain in some cases. These ingenious classifications, called by Hanson, respectively, 'Informal Logic of Discovery Talk' and 'Taxonomy of Discovery Activity' both seem to overlook or misunderstand the essential character of scientific thinking. Let us consider selected examples.

Discovery of an *X*, where *X* stands for a local object, process or event, say, a planet, a coelacanth, or an atmospheric component, is never just the discovery of a specimen, as if it were something isolable, set in a glass case in a museum, and forthwith labelled. It is always the discovery of a new implication of a theory, or the assimilation into a theory of a new feature of experience—the interpretation of a new observation in terms of a conceptual scheme. All Hanson's examples are of this kind. Herschel did not just discover Uranus. He did not point his telescope at the heaven and see through it a new object labelled 'the planet Uranus'. He set out to map the heavens afresh and in minute detail. In doing so he observed a discoid object where no object had previously been recorded. He was not the first to observe it, it had been seen earlier by Flamsteed and Lemonnier (who made eight observations of it in 1768–9), and, after its orbit had been computed, it transpired that there had been twenty previous sightings of the planet. But few of the earlier observers had noticed it as anything unusual, and Flamsteed thought it was a fixed star. Herschel's observation was different because he was thinking and working systematically. Even so, it was not enough to see an object previously uncharted; it was necessary to know that its appearance as a disc distinguished it in character from a fixed star; and it was necessary, further, to observe it over a period to see whether it changed position. Only after this had been ascertained could it be identified as a body belonging to the solar system; and not even then was it identified as a planet, for Herschel at first reported it to the Royal Society as a comet.

At what point then can we say that Uranus had been discovered—when sighted (but not really observed) by earlier astronomers, when taken to be a fixed star by Flamsteed, when observed as an unexpected object by Herschel, when identified as a comet, or when confirmed as a planet? Surely only when confirmed. But it could not be confirmed except by being assimilated into the general Newtonian astronomical system. It had to have been observed on several occasions, its position in the heavens had to be calculated for each observation, and the entire orbit had to be computed. Until all this had been done the discovery was not complete. What then had been discovered? Not just a planet, but a whole complex structure of facts; and not just a particular structure of facts, but the extension of a whole system of astronomical theory. The current 'paradigm' had been further articulated.

Nor was this, as Hanson alleges, a mere 'trip-over' discovery. There was little about it that was accidental. True it is that Herschel was not looking for a new planet. But he was deliberately mapping the heavens in detail, and it would be remarkable if he expected to find nothing new by so doing. If that had been his expectation he would surely never have undertaken the project. He was looking for whatever might be there. Whatever he found, which had not previously been charted, would naturally have been unexpected; but once the planet was seen, it caused a strain in the accepted system. All known planets were already accounted for. Here was a planet-like object which the known system did not predict. The question arose: What was it? An immediate answer could be given in terms of existing theory. As it appeared discoid, it must be either a comet or a planet. Which was it? There were two alternative hypotheses between which a decision could be made only by ascertaining the path followed by the newly observed body. When that was done the comet hypothesis was excluded and the planet hypothesis concurrently confirmed. The procedure is no more accidental than is Kepler's, who as little expected the Martian orbit to be an ellipse as Herschel expected to find a new comet.

The fishermen who netted the strange fish in 1938 off the East African coast were not looking for, nor expecting to find, any living member of a species thought to be extinct, nor did they 'discover' a coelacanth. That was done by Professor Smith of Rhodes University, who identified the specimen in relation to criteria provided by elaborate theories of biology, taxonomy and paleontology. The catch was a matter of chance, but the discovery was the assimilation of a new datum into a theoretical scheme. What, then, was discovered—a fish, a specimen, or the *fact that* the genus *Latimeria* is not extinct? Did Herschel discover Uranus, a new planet (a type of heavenly body), or the *fact that* there were more than six planets? Were these discoveries of an *X*, or of *X*, or *that X*? The same analysis obviously applies to Becquerel and radioactivity, Anderson and the positron, Nicholson and the ninth satellite of Jupiter. The historical circumstances of these discoveries differ, but the scientific method involved is in all cases similar. It is the interpretation of observed data in terms of a conceptual system.[1]

When something *X* is discovered as a *Y*, it is always in the transition period of conceptual change between two successive systems (major or minor). When Galileo discovered Saturn's rings as handle-like appendages the Aristotelian–Ptolemaic system was in process of developing into the Newtonian. After Newton, no astronomer would have assumed that a planet could have handles, he would at once have thought in terms of orbits (or rings). The misinterpretation was due to the lack of an adequate explanatory system (aided, perhaps, by a faulty telescope). The same was true of Millikan's misinterpretation of the positron as a proton, of Blackett's identification of mesons as queer electrons, and of Anderson and Nedemeyer's mistake of the mu meson for a pi meson. Here Hanson has forgotten the effect of theory on observation, though elsewhere he himself is most insistent that observation is theory-laden. So his 'informal logic of discovery talk' reveals only that discovery-talk is for the most part loose, imprecise and devoid of insight—a complaint voiced by Hanson himself.

His 'taxonomy of discovery-activities' is as little reliable. Already we have disposed of the myth of 'trip-over' discoveries. What is accidental about any discovery is only the historical circumstance, not the systematic thinking that converts an accident into a discovery—that extends theory to cover new observations. And we shall be in no perplexity about 'back-into' discoveries once we realize that the scientists in every case, as Hanson himself admits, are 'attempting to conserve the larger conceptual framework within which they are working'.[2] Whenever a scientist does this he

[1] By now, it must be clear to the reader that 'observed data' are no theoretically neutral empirical content, as is assumed by G. L. Farré in his paper, 'On the Linguistic Foundations of the Problem of Scientific Discovery', *Journal of Philosophy*, LXV, 24, December, 1968. Farré's distinctions between 'what is the case' (wic), 'matter of fact' (mof) and 'statement of fact' (sof), though he constructs from them an intriguing new variety of gobbledegook, will not survive our analysis of perception in Ch. VIII. His further remarks on discovery, with some adjustment and correction, fall easily into place in the scheme that has been developed in this chapter.

[2] *Op. cit.*, p. 338.

resists new ideas. So Priestley 'backed into' the discovery of oxygen (or, if you like, discovered it as dephlogisticated air). So Dirac struggled to explain away '—W', and Millikan the appearance of a positive electronic vapour trail. So Fitzgerald laboured to overcome the null result of the Michelson–Morley experiment and Lord Kelvin refused to believe in the existence of X-rays. The lack of adequate explanatory systems coupled with allegiance to the one which has hitherto proved most successful, not only causes data to be misobserved (or even overlooked), but also causes theorists to reject anomalous observations—until the structure of mounting evidence offers a new schema into which both old and new can fit.

'Puzzle-out' discoveries, as Hanson calls them, are simply cases in which scientists are articulating a paradigm—working out in fuller detail the implications of theory. What they discover has usually in some form already been theoretically anticipated. So we may ask whether it really was Hertz who discovered radio waves or Clerk Maxwell—whether Pluto was discovered by Lowell and Pickering, or by Tombaugh—whether Segré or Dirac discovered anti-matter. Once the dialectical logic of scientific advance is understood discoveries of every complexion fall naturally into place.

Finally, those who see farther because they stand on the shoulders of giants are the synoptics who weld the partial adjustments of old theories, that proliferate in times of revolutionary change, into a new and embracing pattern. We have described this process in Chapter VII and need not repeat our account of it. These are Hanson's 'conclude to' discoveries which 'subsume and reticulate'. They are the discoveries (if that is even an appropriate term) which consolidate revolutions and establish new conceptual schemata.

The varieties that Hanson lists vary, when they do, only as they are typical of different phases of the dialectical process: the articulation of a formulated schema, the discovery of aporiai, the initiation of a new schema (great or small) to accommodate data hitherto anomalous, or the integration of several minor schemata, freshly elaborated, into a single new epoch-making system.

What is discovered? Always a theory, the extension of a theory, or the modification of a theory. The discovery of anomaly is hardly a discovery—it is only the beginning of one. It raises the question to which discovery is the answer. But the discovery is not made until the answer has been established as the right answer—until the new theory has been justified. The context of discovery, therefore, is no different from the context of justification. At most they are different phases of the same process and the logic which is appropriate to both is the logic of question and answer. We must pursue further the character of this logic to seek more light on its constructive character and the dynamic of its progress.

CHAPTER X

THE LOGIC OF CONSTRUCTION

i. THE CONCEPT OF STRUCTURE

Structure, organization and order are closely related, if not identical, concepts. Structure suggests most strongly spatial arrangement, order temporal or serial arrangement, while organization is the most general of the three terms. But they are by no means mutually exclusive. The tracing out of a spatial arrangement is serial in order, as is evident in the laying of a mosaic pattern or the scanning process which gives rise to a television picture. In contemporary ideas of physical space-time spatial and temporal order have become inseparable, and we found, when discussing the metaphysical implications of scientific concepts, that serial order in a continuum was fundamental to both structure and organization.[1] Order may be characterized and defined in contradistinction from chaos, which is its opposite and the defining feature of which is random change. Our earlier discussion[2] showed that random change was change of direction unrelated by any regular principle to prior changes. Order, on the other hand, is change directed by some such principle, and it becomes random to the extent that the direction is the more frequently reversed. Reversal rather than something less drastic is necessary, otherwise the principle of order will not be thoroughly disrupted, and anything less than complete disruption would be less than total chaos. The more frequent the reversals the more completely random is the change, and as the frequency increases to infinity so the continuum approaches homogeneity.[3] Completely homogeneous continuity was, however, found to be impossible and random change to be a relative concept dependent upon that of order for intelligibility. Order, then, will be marked by its degree of departure from complete homogeneity.

But here we find two opposite tendencies asserting themselves; for, on the one hand, homogeneity is sameness and departure from it involves difference; on the other hand, excessive and wholly irregular difference is just what we think of when we speak of random change. It follows, therefore, that order must involve

[1] *Foundations*, pp. 462 and 466. [2] *Ibid.*, p. 463. [3] Cf. *ibid.*

change, or difference, but not sheer difference without some element of sameness, similarity or recurrence. Sameness manifesting itself through differences is precisely the character of regulated change, or variation governed by a principle. The changes are all related to an identity of some kind—which is what is meant by saying that they are governed by a single (or identical) principle. Thus arises that relevant variation which is characteristic of polyphasic unity—a sameness that displays itself in a variety of phases or manifestations.

Accordingly, the discovery of structure, pattern or order in any manifold, will involve the discovery of identities related to differences in precise and determinate ways. Similarities, recurrences, periodicities in the manifold will be pointers to these relationships and will serve as clues to the principle of order fundamental to the structure. The discovery of structure, moreover, has proved to be the main requirement and the central endeavour of scientific procedure. To grasp its logical character will therefore be of paramount importance for the understanding of scientific method.

ii. FORMALISM, LOGIC AND PSYCHOLOGY

According to prevailing usage, a logic is an algorithm of one kind or another and to produce a logic appropriate to some set of relationships in the real world is rather to invent some special kind of calculus than to discover by analysis the form of the relation of which the logic is required. Logic today is seen as an intellectual tool and not as a theory of the principles of inference. Any attempt to discover the latter is thought to belong to the philosophy of mind or to epistemology, if it is not frowned upon as psychologism with no legitimate place at all as a philosophical discipline. In this chapter I shall not attempt to construct an algorithm (for reasons to be outlined presently), but I shall seek to develop a theory of the formal character of structure, which has been found throughout our investigation to be fundamental to scientific discovery, empirical confirmation and experimental proof. If this theory is classed as epistemological no great harm will be done and as such it will naturally be relevant to philosophy of mind. But if objections are raised on the score of psychologism, some demur may legitimately be made.

Psychologism is strictly the attempt to explain away logical connexions in terms of psychological causes, which is a pernicious and self-refuting practice. It is quite another matter to take cognizance of the form and procedure of live thinking actually occupied in precise analysis and cogent inference. Cogency, indeed, is a crucial issue. The inference compels; the conclusion from the premises is inescapable—is this due to psychological causes or is it compulsion of some other sort? Psychological compulsion is usually what we attribute to emotional stress, to which rationality either is altogether foreign or contributes only in some confused and distorted way. Logical compulsion on the other hand is rational in essence—but what does that mean? It means at least that denial of the conclusions will involve contradiction of the premises or of something established earlier. But to discover such contradiction would require the same sort of inferential thinking as did the original conclusion, and its cogency is what is in question. Rules of inference and 'laws of thought' must be invoked. The rules of inference, however, cannot be arbitrary, for if they were, the choice of rules would be merely subjective and no compulsion (unless of the psychological kind dependent on emotion) would result; and 'laws of thought' is a phrase which is once again suspect of psychologism.

If an argument is such that its conclusion must be asserted once the logical nexus is seen, it can only be because in *some* sense our minds work in a way that leaves us no option. It is true that in fact one may acknowledge a connexion between premises and conclusion and yet assert the contradictory of the conclusion; but if anyone did so it would be thought either that he had not really seen the connexion or that (if he had) he was being merely perverse and irrational. If he claimed that his mind worked that way (joking apart) he would be regarded as insane. So logical cogency must, in some sense and to some extent, depend on the way in which our minds normally work. Psychology, therefore, cannot be wholly irrelevant, for rational procedures are an important part of our psychological make-up. A creature psychologically incapable of reasoning could produce no logical theory, not only because logical theory is itself reasoning, but also because it is the fruit of reflexion upon the way we normally think. Logical compulsion, therefore, is also a kind of psychological compulsion. It is the way we are compelled to think when we think normally—i.e. free from fortuitous influences of a purely emotional and impulsive kind.

Logic cannot avoid giving some attention to the psychology of thinking if it is properly to accomplish its task, for it seeks the general characters of objective thought which justify movement from premises to conclusion. These characters can be discovered only by thinking, and justification can be defined only in terms of subjective intellectual satisfaction with the results of such thinking. Validity in logic is what excludes contradiction, and this satisfies because our minds are so constituted as to seek in all spheres, practical as well as theoretical, the elimination of conflict. But this is only a minimum requirement which might be satisfied by mere repetition that would fall far short of ultimate satisfaction, because another and more important demand is for progress from the known to the unknown—and that it should be reliably achieved. Again, the criterion of reliability can, in the last resort, only be subjective, though it is the same for all who think normally. We shall find presently that the psychological demand, like the logical, is for necessity, and the psychological satisfaction is achieved through the discovery of structure. Without this the logical demand would be both unintelligible and futile. What logic seeks are the principles that make our thinking valid—that is, intellectually satisfying. Examination of the psychology of thinking should then give some guidance in the search for these principles.

Nevertheless, logical cogency cannot be wholly attributed to the character of our mental functioning, otherwise it would still be a subjective matter, whereas objectivity is its distinguishing mark. In that case it must derive, at least in some measure, from the nature of the subject-matter thought about. It is this which somehow forces us to the conclusions that we reach. If it should ultimately transpire that there is some connexion between the nature of the subject-matter and the way in which we normally think, the cogency of logic may turn out to be much the same thing as 'stubborn fact'—but discussion of that possibility will, for the present, have to be deferred.

But prevailing logical doctrines look askance also at the suggestion that logical procedures are in any way dependent upon the material to which they may conceivably be applied. Whether a logic will prove useful in application to some specific subject-matter is (according to current theory) entirely fortuitous, depending upon an accidental isomorphism between the structure of the logic and that of the factual relations to which it is applied. It has been claimed that logics can be invented which have no apparent or immediate application to anything at all, just as algebras and geometries have been evolved for which no special physical or practical application was (at the time) known or expected. These systems are held to be purely formal calculi, completely uninterpreted in terms of factual data, and applicable to any material (if at all) only by fortunate coincidence.

What is formal, however, presents or represents a form, which cannot be a form of nothing; so the question arises of what it is that a formal calculus presents the form. Three answers are possible: (i) it is the form of a language; (ii) it is the form of some subject-matter of knowledge; or (iii) it is the form of thought or reasoning. The second and third are nowadays usually rejected out of hand, yet if the first is accepted the other two seem both to be immediately involved. Language is the medium of communication and to be intelligible it must speak about something. Accordingly it must be so structured as to be able to refer to and represent the subject-matter. One would expect it to be adapted at least in some measure to the nature and configuration of that about which it discoursed. Furthermore, language is a means of expression, and what it communicates is what the speaker believes, or how the subject-matter appears to him, or what he thinks about it. His language must, therefore, be in some way and in some degree assimilated to the way in which he thinks—the way his mind works. So far as he thinks sanely, reasonably, intelligibly, the form of his language must express the characteristics of his thinking that justify and answer to these epithets. Consequently, if logic is formal, the form with which it is concerned can be linguistic only to the extent that it is related both to the subject-matter of thinking and to the character of the thinking itself. It can, I think, be demonstrated that current logical theories are all developments from a philosophical position—explicitly stated by Wittgenstein in his *Tractatus*—which held that logic expressed (or 'showed forth') the form of facts, and which presumed the facts to be atomic in form with an internal structure centred in bare particulars.[1]

The facts as revealed by contemporary scientific theory are very definitely not of this form,[2] nor is the structure of perception, nor of scientific thinking as discovered in the foregoing chapters. The kind of logic implicit in them will therefore, in all probability turn

[1] Cf. my paper 'The End of a Phase', *Dialectica*, Vol. 17, 1963.
[2] See *Foundations*, pp. 24, 44 f., 51.

out to be very different from that with which contemporary logicians busy themselves. We have found that the method of science is constructive rather than inductive, and now as we turn to consider what the logical character of such construction may be, we meet *in limine* an obstacle arising from the very notion of logic as it is entertained and exercised at the present day.

If we were to attempt to discover—or, perhaps it would be more appropriate to say, construct—a logic, appropriate to the scientific procedure so far described, by following the prevailing methods, we should seek to state axioms, set down undefined terms, define constants and operators enabling us to write down formulae, and devise rules for the transformation of sentences. But clearly we could not produce in this way the sort of thing we require. At best we might hope to construct, by such means, some kind of abstract model of the type of system prevalent among scientific theories, and even that would be expecting too much. For what we seek is the principle of inference which validates process from a partial configuration of elements to its completion; and it is obvious that we cannot find this by exploring among indefinable particulars, or by manipulating the undefined terms which are supposed to represent them. Unless we are in possession at the outset of the principle of order on which the configuration is constructed we cannot properly begin. Such a principle is no indefinable particular but functions as a universal. It would have to be presupposed in any axioms or rules of procedure formulated for the proposed logical calculus. In consequence these may not be arbitrary and could not be laid down prior to any knowledge of the system but can only arise out of it, by abstraction.

A structural principle, a rule of order, or a mathematical function—the type of principle we should need—is actual only in the totality which it organizes. It may be potential (or implicit) in any particular that exemplifies the rule, but on account of the partial and incomplete character of the fragment the structural principle can never be fully actualized in anything short of the whole. We may, therefore, define a structure as a manifold or complex, the elements of which are mutually determined by interrelations fulfilling a principle of relatedness the complete expression of which fully actualizes and determines the configuration of the whole complex. The principle determining the form of a circle is equidistance of all points on the circumference from one central point. This principle immediately determines the whole figure and guarantees its completion, which no arc or segment by itself can do. Yet the principle is immanent in every arc and segment. A curve, which is part of the circumference of a circle, may be continued in any way to complete some figure of a different shape, but adherence to the formula $x^2 + y^2 = r^2$ (where x and y are co-ordinates and the origin of the axes is centre) inevitably gives us a circle. The principle determining the whole is thus logically prior to every part and it is in the whole that the explanation (intelligibility) of the part must be sought. One cannot, therefore, begin from parts that are structureless and hope to find the principle of construction by accumulating large numbers of them. Nor can one hope to find that principle by breaking up any incomplete portion of the structure with which one may be presented into undifferentiated units.[1]

It follows that explanation is never merely synonymous with analysis, for intelligibility is impossible without synopsis of the whole within which the explicandum falls. But again because the whole is polyphasic, because it is differentiated and can be a structure only so far as it has differentiable parts, explanation cannot be simply synonymous with synthesis either (if that means merely collecting together simple units into groups or 'classes'). The elements must vary relevantly to some organizing principle, which must, therefore, be immanent in them. Explanation, accordingly, requires both synthesis and analysis; for an organizing principle, stated as a simple formula, is a mere abstraction, and until it is specified in its proper instances it cannot be properly understood. A whole, or totality, what might suitably be called a manifold, becomes intelligible, therefore, so far as the organizing principle which permeates it is made manifest in its constituent parts, and the parts are intelligible only so far as they are seen in their mutual relations as actualizing the principle of structure which determines those relations. Because this is so, one can begin neither from primitive undefined elements nor from pure principles of order, but only from a relatively unarticulated (or partially articulate) manifold, to be made intelligible by a

[1] For this reason Bertrand Russell was wrong to demand simple parts as the foundations of all manifolds. He misunderstood Leibniz' position in consequence, for the Leibnizian monad (despite Leibniz' own protestations) is far from simple. Leibniz calls it so to emphasize its wholeness and self-completeness and to make it independent (in certain ways) of all others. Each monad is a world in itself, reflecting in its own internal structure the entire universe.

process of concurrent articulation and organization,[1] bringing to clearer light at once both the specific detail and the generic (or universal) principle of organization.

This is exactly what we found occurring in the development of perceptual dispositions, and in the consolidation of perceptual assurance whenever the initial phenomenon was vague, ambiguous or indeterminate. Knowledge, both perceptual and scientific proceeds, not from the particular to the general, nor from the self-evident simple to the complex and specific, but from the vague and confused to the articulate and precise. It remains throughout and at all stages 'a whole of related and connected elements', at different levels of development. At every stage in the advance of science it is such a whole, in which the area of investigation has already been organized to some extent and up to a point; in which some principle of order has already been adopted, though not necessarily worked out in complete detail; or it may be a relatively unsatisfactory principle which leaves obscure some portions of the field or, as the articulations are developed, leads to incompatibilities. The details are then re-examined by the scientist in the attempt to discover either how the accepted rules of interpretation may be more consistently applied, or, when continual frustration occurs, whether he can detect in the relationships of the phenomena a new rule or principle of organization which will reveal a new, more complete and more comprehensive pattern.

iii. SYSTEMATIC THINKING

A process of this sort seems to be typical of human thinking, for it is very like what Sir Frederick Bartlett's experimental results reveal.[2] He set problems to his subjects for solving, requiring the discovery of a structure from inspection of a given fragment. The first problem requires the subject to select words from a group and arrange them in a column according to a pattern partially indicated by a fragmentary example.

[1] What H. H. Joachim called 'an analytic-synthetic discursus'. See *Logical Studies*.

[2] Sir F. Bartlett, *Thinking* (London, 1958), Chs 2 and 3.

A, GATE, NO, I, DUTY, IN, CAT, BO, EAR,
O, TRAVEL, ERASE, BOTH, GET, HO, FATE.

ERASE
FATE

Solution:

A
BO
CAT
DUTY
ERASE
FATE
GET
HO
I

Another problem was to continue the series: 1234, 2134, 2143. We need not consider the variations of order and method used by the subjects in their attempts at solution, but we can note significant common characters in the problems and the requirements for successfully solving them.

In every case one or more rules of procedure had to be inferred from the given fragment of the required order—e.g. in the first example above, the rule is: 'Write the words in alphabetical order of initial letters, so that the number of letters in each word differs by one from its immediate successor.' In the second example, the rule is: 'Reverse the order of each pair of digits in turn until the original order is reconstructed' (1234, 2134, 2143, 2413, 4213, 4231, 4321, 3421, 3412, 3142, 1342, 1324, 1234). Once the rule is extracted the order of steps in the procedure to the solution is either strictly determined or becomes progressively more determinate as it continues. If latitude of choice exists at the beginning, it becomes progressively less as the subject proceeds. Bartlett remarks that though the majority of subjects overlooked the evidence that could be used to determine the order of steps, this was important because a wrong step might lead to unfruitful digressions. What was needed to determine order was an appreciation of 'directional properties' of the system. But, for our purpose, one very important feature of the procedure is that the steps become progressively more determinate until, on reaching the conclusion, no further manipulation is either possible or necessary.

We know we have reached the conclusion and that it cannot be otherwise without disrupting the whole system.

In a later experiment, subjects were required to substitute numbers for letters in an addition sum in disguise, thus:

DONALD
GERALD
ROBERT

given only that $D = 5$ and that each number from 1 to 10 has its corresponding letter.[1] The special interest of this example is that successful solution depends on a key step. Having seen that T must be nought—$D(5) + D(5) = 10$—the key step is to note that $O + E = O$, hence E must be 9, for it cannot be zero, and $N + R > 10$. Once this step has been made the rest follow, each dictated by its predecessor. If $E = 9$, A must be 4 and $L + L + 1 > 10$. R must be an odd number because 1 is carried from $D + D = 10$. It cannot be 1 (or L would be 5), nor 3 (for $D + G > 5$), nor 5 nor 9; hence it must be 7, and L must be 8. $5 + G + 1 = 7$, therefore G must be 1. N must be greater than 3 and cannot be 4, 5, 7, 8 or 9, so it must be 6. This makes $B = 3$ and we are left with $O = 2$.

$$\begin{array}{r} 526485 \\ 197485 \\ \hline 723970 \\ \hline \end{array}$$

The rules for procedure are provided mostly by arithmetic, but the key step determines the order in which they must be applied.

Bartlett remarks that 'the steps in sequence which the thinker achieves are closely comparable from start to finish, or proceed by some lawful function; but the items which are linked by the steps differ from stage to stage'. The rules which express these regularly varying steps are extracted from the limited range of data presented. These again serve as a clue to the structure of the system, and (he asserts) it can be shown that, whenever a rule is formulated as one to be used, it is derivative from this structure (at least, as the thinker takes it to be).[2] He records that subjects instituted a definite search for points of similarity between given items in the available information. The search is satisfied only

[1] *Op. cit.*, pp. 51–9. [2] *Ibid.*, p. 93.

when the points of agreement, consistent with observed differences, have been so identified that they issue in a rule (or rules) giving a number and order of steps which are universal—'the same for all'. Their importance is more significant for the subject than the frequency of their occurrence. This implies, says Bartlett, that the agreements identified are seen as belonging to the system.

This description is in close accord with the account, given above, of the conception of structure, to discover which, we said, it would be necessary to search for identities expressing themselves in and through differences. The identities alone are not enough. They must be seen as 'belonging to the system'; and the differences must be related to them according to the ordering principle of the whole. Strictly, this ordering principle is the object of the scientist's search, so the discovery of resemblances in the mass of available evidence is only the first indication of that principle, and recognition of the ways in which the differences are relevant to the similarities will be equally essential to its complete discovery. To see the agreements as belonging to the system, therefore, implies that some inkling of the structure of the system is already present. And this is the case if, as we have maintained, the matters under investigation are already organized in accordance with an already existing and hitherto accepted theory. In Bartlett's examples of problem-solving, the presented material is not wholly chaotic but has certain recognizable features relevant to some known order (for instance, the order of the alphabet, or of the natural numbers, or the rules of arithmetic).

The discovery of similarities, agreements, or recurrences is the starting-point of inductive reasoning, but the goal of induction is not mere generalization. It seeks to develop from this discovery a hypothesis as to the principle of structure or order to which the similarities provide a clue. The clue is given more effective expression in a rule of procedure, and the process of following the rule is deductive. If it succeeds it leads with increasing inevitability to a conclusion that is necessary within the system. The process of following the rule works out, by successive steps of construction, the articulation in the factual material of the principle of order and so displays the system in explicit form—a demonstration which establishes the hypothesis.

In short, any sort of 'inductive' process resting on the discovery of like cases is significant just so far as the similarities are evidence of systematic relationships which afford rules of procedure (or 'inference tickets'). This is the basis of all valid generalization, and Bartlett especially notes that once the rule has been extracted it is treated as extensible to other instances.[1] Of course, in actual practice, scientists seldom succeed first time in their endeavours either to discover the right hypothesis or to establish those that they entertain. But the logic of their procedure is evidenced most clearly in their successes and is the measure both of success and of failure.

It is immediately obvious that this pattern of thinking corresponds in large measure with that of Kepler's search for the martian orbit. He has to translate a number of observed positions at stated times into points on a geometrical figure described about the sun. For this he needs a rule. He tries to find one which would describe a circle, and this requires him to regard the sun as eccentric and to discover the extent of the eccentricity. No such rule is found to cover all the data accurately and he seeks a new rule, now for an ovoid, but without success. Finally his discovery, that the secant of the angle subtended at the midpoint of the *lunula* by the sun's distance from the centre of the orbit corresponds with the greatest width of the *lunula*, gives him the formula (or rule) for an ellipse, though at first he does not recognize it ('*O me ridiculum*'). When he does, the steps follow consecutively determining the orbit as an ellipse with the sun at one focus.

But not all thinking, even in science, is within so well marked a closed system as the examples to which reference has so far been made. This is, perhaps, the typical form of mathematical thinking or of any to which mathematics contributes the core. There is much besides that we are more apt to view as 'inductive', and which Bartlett describes as 'adventurous' thinking. Here the evidence is more diffuse and precise rules of procedure more difficult to extract from it, as alternatives are frequently possible.[2] Yet in all cases the thinker strives to reach a definite terminus on the basis of the presented evidence, which lies beyond the limits of that evidence, and which he takes to be universal for all reasonable persons—that is, any who 'is not mentally defective, or mentally ill, or abnormally prejudiced'.[3] In all cases the effort

[1] *Op. cit.*, p. 94.

[2] If, however, alternatives open up in too many directions or in exponential progression, the thinker abandons his method and changes his tactics.

[3] Bartlett, *op. cit.*, p. 97.

is to reveal structural characteristics, and, we are assured, no situation is ever totally devoid of structure unless it is one in which changes are continuously occurring wholly at random.[1] It is to reveal the structural characteristics of the area under investigation (Bartlett avers) that the scientist plans experiments. 'His aim . . . is not simply to extend and define knowledge within a system that is to be treated as already completed, but to help to develop and understand systems which are so far literally incomplete'.[2] The system, in short, is not altogether 'closed', and the process of discovery is not restricted to a determinate number of steps in a determinate order, though one sequence may prove better than others.

The work of Harvey and Lavoisier is more illustrative of this more 'adventurous'[3] kind of thinking. It begins from the available information, seeks to detect in it evidence of structure from which some kind of rule of procedure for further investigation may be extracted, and then, by following that rule amasses more evidence developing the structure more completely until the system is sufficiently coherent (or approximates nearly enough to closure) for the conclusion to be beyond reasonable doubt. What supports it, however, is not just a multiplicity of similar cases, but the variety of interlocking evidence that constitutes the system.

Finding the current physiological theory at variance with itself in many points, Harvey sought for a new system. He was already in possession of many of the facts—the imperviousness of the septum, the structure of the mitral and semilunar valves, the valves in the veins, the lesser circulation, and so forth. What was needed was a rule for construction that would relate all these in a consistent way. The critical insight was to see the action of the heart as in systole. This fact is first investigated and confirmed by experiment. The rule then adopted is to investigate the direction of flow in as many parts of the body as possible, particularly the relation between the flow in the arteries and that in the veins and to interpret the findings in the light of the propulsive action of the heart. It transpires first 'that since the blood is incessantly sent from the right ventricle into the lungs by the pulmonary artery, and in like manner is incessantly drawn from the lungs into the left ventricle . . . it cannot do otherwise than pass through (from

[1] Bartlett, *op. cit.*, p. 110. [2] *op. cit.*, p. 98.

[3] Bartlett uses the term more loosely to include everyday practical thinking and much which is far from rigorous.

the right ventricle to the left) continually'. It remains to determine the direction of flow in other parts of the body, and that is shown to be from the arteries to the veins in such quantities that it could not be supplied simply from the *ingesta*. The conclusion follows that the blood circulates; for, in the light of all this evidence and its systematic interrelations, no other possibility is open and 'it is necessary to conclude that . . . the blood is impelled in a circle'.

Lavoisier's research also followed the course more suitably described by Bartlett as adventurous thinking. The same characteristics are detectable in it as have already been noticed, but the system within which he was working is looser and less determinate. Consequently the order of steps dictated by the rule which emerges from the evidence is not so rigid. The evidence from which he begins is that provided by the British pneumatic chemists and the traditional processes of calcination and reduction; and the key insight is the realization that combustion involves some portion of the atmosphere in integral chemical reaction. The procedure then is to measure and correlate quantities of 'air' absorbed and emitted in various reactions, in experiments so designed as to converge upon the precise relation between combustible substance and the 'highly respirable' portion of the atmosphere. So the gas emitted during metallic reduction with charcoal is examined and compared first with that produced by 'effervescence' and then with that produced by reduction without charcoal. The reverse process of calcination is then investigated and compared with the burning of phosphorus and sulphur. All these processes give evidence of reducing the quantity of air in which they occur and of continuing only as long as the quantity of available air permits. Finally quantitative correlations clinch the argument and the hypothesis that a significant portion of the atmosphere combines with the combustible material is confirmed. Each set of experiments establishes something which leads on to the next stage in experimentation; and, though the procedure is not absolutely rigid, it is sufficiently consequential to justify our bringing it under Bartlett's description.

That description applies no less to Darwin's marshalling of the evidence for evolution. The available evidence was copious and varied, and it suggested the hypothesis that specific variation might be due to progressive selection (as witnessed by domestic varieties). The key insight was the inevitable struggle for survival imposed by the geometrical increase of offspring in any species and the limited means of subsistence. This suggests lines of investigation to look for evidence of change in the direction of selective advantage and of common ancestry among divergent species. The final result is a system of convergent evidence from widely differing contexts all supporting the central hypothesis in various ways.

The cogency of the arguments is rooted in the structure of the evidence. In closed systems this structure is revealed by application of rules of procedure suggested by the elements of structure detected in the premises. In more open systems the procedure is more exploratory and aims at discovering the elements of structure and developing a pattern of evidence seriatim on the basis of which a more rigorous quantitative closed system can subsequently be erected.

The pattern of thinking is thus to examine the available evidence in order to detect in it elements of structure and to develop this to a point at which a rule (or rules) of procedure can be extracted from it enabling the thinker to complete the structure. If the system so discovered is a closed system and the rule discerned is the right one, the steps of the procedure are progressively determined to a conclusion which closes the system and is therefore final. This is clearly the case with Kepler, in a great measure with Harvey, and, to some extent, it is always the case, though the finality of the conclusion is proportional to the openness of the system. As we suspected, therefore, logical cogency, the outcome of system, is the same thing as factual necessity, the determination of facts by their mutually interlocking relationships. For the facts as observed are already ordered in some degree, and the rule of procedure is inferred from the evidence by detecting in it elements of structure. The rule once extracted determines the course of the thinking that develops the system in theory, and it is due to the systematic character of the thinking that its outcome is necessary. The conclusion is necessitated at once both by the nature of the fact and by the nature of the thinking.

iv. NECESSITY AND CAUSALITY

Necessity in scientific reasoning derives, not from tautology or the mere definition of terms, but from ordered structure in a closed system. The system is closed if the principle of order determines it completely, and is open if this principle determines it only partially, demanding further research for more complete systematization. Closure, therefore, is a matter of degree and the consequences of closure vary in degree accordingly. Necessity results from the inevitability of the steps of the argument and their order, which the rule of procedure prescribes. Descartes had sound insight when he insisted that reasoning must follow a precise order (*debitus ordo*) if it is to reach an assured conclusion. But this prescribed order is determined by the structure of the system as a whole and not simply by the relations between adjacent members. These relations are seen as necessary only in the light of the entire structure. So far as they express and exemplify the determining principle of order, there is, within the system, no alternative to them—unless the system is one which provides for alternative expressions at certain junctures (as is the case in some biological systems). When alternatives are permissible they are nevertheless restricted according to rule, for each of them is still a manifestation of the organizing principle, and an element of necessity persists.

It follows that a causal hypothesis, alleging a law of connexion between specific events, is established only when a system of relationships has been demonstrated that requires the kind of connexion alleged. In such a system there will, in consequence, be a constant conjunction between events of the relevant sort, but the necessity of their conjunction will not lie in its mere repeated occurrence. It will be inherent in the system, and the constancy of repetition will be derivative from the necessity. A causal law, therefore, is not a statement of observed constant conjunction merely. It is the enunciation of a principle of structure or organization among objects and processes within which the relation between the events in question is necessitated, because if it did not occur the entire system would be disrupted.

On the side of theory, the reason for asserting necessary connexion is the systematic interrelation of the evidence and not simply its frequent occurrence. In contemporary science a causal law usually takes the form of a mathematical equation or formula expressing a principle of order and actualizable in just such a polyphasic system as would determine connexions between its constituents. But necessary connexion being relative to system is *pro tanto* relative to the degree of completeness of the system. What may be regarded as necessary in a limited sphere may prove to be contingent in the light of wider evidence. The law of planetary motion that is necessary in the Newtonian system is only an approximation in the Einsteinian; and what may be true of a chemical process in an inorganic context may not be true of it in the internal environment of an organism. The cogency of the causal law is proportional to the comprehensiveness both of the theory to which it belongs and to the system of facts to which it applies.

V. PROBABILITY

If we accept this view of necessity and contingency, a corollary follows with respect to the meaning of probability. We drew attention in Chapter II to two main types of meaning that have been given to probability, and though there are varieties among each of these, we sought to reduce them to the dichotomy of subjective (or what Carnap and others have called 'psychological') and objective. This reduction seems to me to be justified, for Carnap's distinction between 'psychologism' and 'qualified psychologism'[1] really amounts to the same thing. What Carnap regards as qualified psychologism is the interpretation of probability as degree of *rational* belief, and this again is understood by most contemporary writers to mean belief justified by logical relations between evidence and conclusion. Such relations are usually held to be either of two kinds, mathematical or inductive, the first of which is wholly objective, in the sense that it is not dependent upon the subjective inclinations of the thinker. The second depends on the nature and extent of the experienced evidence and is 'subjective' only in the sense that this will vary with the circumstances of the judging subject. Probability, with reference to empirical evidence, is therefore relative to the extent of our knowledge, and is 'subjective' in no other sense. But to understand probability in this way is by no means psychologistic, for the logical relations between the known evidence and the conclusion are taken to depend upon the objective nature of the facts observed, and not upon the psychological condition of the observer. The epistemologist is strictly concerned only with these logical (or 'objective') relations and with degrees of belief only so far as, being based upon the recognition of these relations, they are rational.

The relevant logical relations are either mathematical or

[1] See *The Logical Foundations of Probability* (Chicago University Press, 1950), pp. 37–51.

inductive, and in either case the mathematical calculus of chances presumes indeterminacy among the events the probability of occurrence of which is to be calculated.[1] This is implicit in the so-called Principle of Indifference—the assumption that the events concerned are all equi-possible. If we adopt a Humean view of experience, or a view that assumes evidence always to be theoretically neutral, a mathematical calculation of the probability of occurrence of events can be based upon the frequency of past observations. But this, as we saw, would be no justification for projection of the judgement into the future. That would involve the acceptance of some principle of inductive argument which would require independent justification.

The whole trend of the argument in this Part, however, has been to establish the fact that empirical evidence is never theoretically neutral, and that empirical judgements derive their meaning as well as their credibility from a conceptual scheme ordering a systematic body of knowledge which is the context in which they are made. Scientific statements of natural laws, or of the application of natural laws to particular facts, are now to be seen as having reference to, and deriving cogency from, a precisely elaborated system of facts of which observed regularities are minimal evidence. The degree of their probability will therefore depend upon the extent (or comprehensiveness) and the coherence (or degree of integrity) of the system of evidence. This brings us back to the original etymological meaning of the word—'provability' where to be proved is to be tried (*probari*) and not to be found wanting—for in natural science a proposition, or hypothesis, is 'proved' to the extent to which it is seen to be required by a systematic body of evidence.

Probability is truth in some degree—though to call it such is to anticipate the discussion of truth that I am reserving for a later chapter.

'All propositions', wrote Keynes, 'are true or false, but the knowledge we have of them depends on our circumstances; and while it is often convenient to speak of propositions as certain or probable, this expresses strictly a relationship in which they stand to a *corpus* of knowledge, actual or hypothetical, and not a characteristic of the propositions in themselves. A proposition is capable at the same time of varying degrees of this relationship, depending

[1] Cf. p. 33 above.

upon the knowledge to which it is related, so that it is without significance to call a proposition probable unless we specify the knowledge to which we are relating it.'[1]

Keynes is entirely right in this statement of the matter. But he assumes that some of our knowledge, to which the probability of a proposition would be relative, is acquired 'direct', implying that we have it independently of hypotheses and conceptual schematizing. It is the sort of knowledge, he says, with which the theory of probability is not concerned, and its truth would presumably be independent of its relation to the *corpus* of knowledge to which the passage quoted refers. This assumption and its consequences in Keynes' theory, or in any other theory that makes it, is what I am seeking to deny. But so far as the probability of a proposition rests upon the body of evidence by which it is supported, the degree of completeness of that body is the measure of its approach to truth. It should follow that if the body of evidence could be made absolutely complete both in comprehensiveness and in logical coherence, the proposition following from it would be wholly true. The implication of this consequence we shall consider later. Here we need only remark that in empirical science the body of evidence is never complete in either of these ways, and scientific theories are, therefore, never more than probable, even if usually in high degree.

On the other hand, within the system of hitherto established knowledge, a proposition may be so firmly supported that it can (for convenience, as Keynes asserts) be spoken of as certain. The degree of its necessity, we have already noticed, is proportional to the comprehensiveness and integrity of the system to which it belongs. But within that system it may be incontestable. This is true, however, of relatively few propositions, even within a restricted system of empirical facts; and where information is lacking, determining rules may not be applicable. Consequently some degree of indeterminacy may enter into our reasoning so that the truth of related propositions can be assessed only statistically, by the use of a calculus of chances. This happens when we are dealing with matters like the vital statistics of a population or the incidence of conditions the precise causes of which are largely or entirely unknown.

The reasons for this ignorance vary in different cases, and in

[1] John Maynard Keynes, *A Treatise on Probability* (London, 1963), pp. 3 f.

statistical theories of contemporary physics it seems to result more from the difficulty of articulating the system, the general character of which is known, than from an inability to discover general laws. When the energy system is treated as a whole the laws of its configuration emerge, but particulate details are indeterminate within limits and for reasons that can be precisely defined, and they can be approached only by means of elaborate statistical calculations.

Contemporary theories of probability, so far as they treat of the mathematical calculus of chances among events regarded as equipossible and indeterminate, are of great value and significance in cases such as these. And theories that deal with inductive probability are at fault only so far as they consider the observed data to be theoretically neutral. As constant conjunction is an indication of systematic interconnexion, a calculation of probabilities based upon the frequency of its occurrence will not be misplaced. Taken as a probability estimate of the existence of systematic connexion, it may quite legitimately be projected into the future, for if there is such a connexion the conjunction of characters observed in the past may rationally be expected to persist in future instances. Such calculation, however, does not amount to proof and the actual existence of the connexion has still to be discovered. The assessed probability may be a good ground for presuming the connexion and for proposing the hypothesis. But it would still have to be established by further investigation, and perhaps by devising special experiments, in order to amass a body of structurally interrelated fact, of the kind exemplified by the work of Lavoisier or of Darwin, before confidence in it could be dignified by the adjective 'scientific'.

The conception of probability as a logical relation (put forward by Keynes and Carnap) is, however, subject to a serious difficulty to which A. J. Ayer has drawn attention,[1] and from liability to which the view here presented might not be considered free. If the degree of probability of a proposition is relative to a given body of evidence, it follows that with reference to different amounts, or differently selected, evidence the degree of probability of a given proposition, would vary. But each assessment would depend on precise logical relations and would therefore be necessarily true. If this is so, there would seem to be no rational

[1] See 'The Conception of Probability as a Logical Relation' in *Observation and Interpretation*, ed. S. Körner (London, 1957), pp. 12–17.

ground for preferring one assessment to any other, either in theory or in practice. To recommend the assessment based upon the larger body of evidence would be without warrant, as that assessment would be no more reliable (and no less) than any other based on more meagre information.

One might be tempted to argue that the fuller the evidence the higher would be the probability of the proposition, and that the more probable is always preferable to the less. But this argument is fallacious, for more evidence might well reduce the probability of a proposition which, on the strength of certain selected facts, seemed more likely. For instance, if a politician directing the policies of a government is known to be of aggressive character, it is probable that he will try to settle international disputes by threats rather than by negotiation; but if in a given situation threats would align against him more powerful forces than he could muster, he would more probably resort to negotiation in spite of his aggressive disposition. The evidence of the former taken alone makes more probable what the larger body of evidence makes less. It thus appears false that probability is a measure of the extent of the evidence, or of its self-consistency (for neither body of evidence is markedly incoherent).

If items of evidence are taken as purely fortuitous and unconnected with one another and if they are related to the conclusion simply by generalizations from commonly experienced regularities, the calculation of probability will be no more than a summation of pro and con. So far as it is mathematically accurate in each case it will be indisputable and the reason for preferring one estimate, on the basis of one set of evidence, to any other based on some other set, would remain obscure.

M. H. Foster and M. L. Martin contend that Ayer has overlooked Carnap's distinction between the logic of induction and its methodology. The former lays down the principles for assessing probabilities but does not (and cannot) dictate preferences. But a necessary, if not a sufficient, condition of rational belief is laid down in the methodological rule that the probability estimate based on the total amount of available evidence is to be preferred.[1] This riposte does not seem to dispose of Ayer's point that no such methodological rule appears to have any rational backing. A probability estimate calculated on a partial selection of the avail-

[1] See *Probability, Confirmation and Simplicity*, eds Foster and Martin (New York, 1966), p. 21.

able evidence would be mathematically, and therefore logically, just as true as one from all the available evidence; and if one demanded (even in principle) all possible evidence as the only sufficient condition of an adequate probability estimate, the result inevitably would be either 0 or 1.

This puzzle arises however only if the proposition to be tested and the evidence relative to it are assumed to be isolable items of information related to one another only externally (by frequency of conjunction, or the like) and subject, therefore, to a mathematical computation of chances on the presumption of equal possibility. The situation is very different as soon as one realizes that no proposition of any significance has a separable, self-contained meaning, but is tied by a network of implicative threads to a complex background of knowledge. Discovering the evidence for its truth consists in tracing these connexions with other already established facts, and the degree of its probability depends upon the closeness of the interweaving and density of the mesh that comes to light. Different bodies of evidence on the strength of which different probability estimates may be made are not, therefore, mutually independent. Relatively isolated facts (such as the politician's aggressive temperament) have themselves only relative weight or importance with reference to the larger complex in which they must function and probability estimates based upon them will be similarly relative. We must therefore prefer that assessment which relies on the largest possible and the most fully integrated body of evidence obtainable.

Further, increase in the volume and relevance of the evidence validates a new proposition in each case, because the actual meaning of the proposition changes as further implications are brought to light and what with more evidence is more probable is a richer content of knowledge. That the premier, being irascible, will use threats, may on the face of it be probable; but that the premier, being prudent as well as irascible, will use threats only if it is to his advantage, is more probable and is a new and expanded proposition. In the light of this fuller insight, the first proposition is less probable, but it is not wholly false, for it has been preserved and developed in the second, which is therefore preferable because more complete. The second hypothesis is not is just an independent alternative to the earlier one, but is a better version of the same body of fact.

In as much as frequently observed conjunctions of attributes or events are indicative of structure in a body of fact that is being investigated, numerical calculations of probability are helpful. But because it is the structure that we seek to discover and not just the relative frequency of conjunctions, the amount of evidence on which the calculations are made is as important for the probability estimate as is its content. The larger the body of evidence the better indication will it give of the underlying structure, and the more fully will it reveal the principle of order involved. Mathematical estimates should, therefore, not be stated baldly (as Carnap recognized) but always in conjunction with the evidence adduced, $c(h, e)$ being the degree of confirmation of the hypothesis on the given evidence. Thus there is a second order value depending upon the extent of e, as well as the first order value indicated by c. At the limit, when e becomes all-inclusive, c becomes 0 or 1, and we no longer have probability but certainty.

vi. INDUCTION AND DEDUCTION

Bartlett records that for his subjects the 'importance' of agreements detected in the evidence weighed more than their frequency. The recognition of importance, however, is already a discernment of elements of structure. In the examples we have described of scientific research this recognition of agreements having importance for structure is noticeable. For Kepler it was the correspondence between the width of the *lunula* at mid-point and the secant of the subtended angle (5° 18′); for Harvey, the construction of the valves both in the heart and in the veins. These were not all similar except so far as the direction of flow that they permitted with respect to the heart was uniformly in a circle; for Lavoisier it was the augmentation of weight in all cases of combustion coupled with the constant association of calcination and reduction with absorption and emission (respectively) of some form of 'air'. This detection of similarities indicating structure unites the inductive with the deductive form of reasoning, for the structure gives ground for the rule of procedure that validates the deductive process and is itself inductively inferred from the similarities. Deduction, therefore, issuing as it does by application of a rule extracted from the structure perceived in the evidence, is deduction from the phenomena as Newton maintained.

Accuracy in detecting the principle of order determining the structure of the facts, and the consequent adoption of the right rule of procedure in constructing the supporting evidence for a theory, is what validates deduction from phenomena. It is the principle of order that provides 'the nerve of the inference', and this is implicit, or immanent, in every feature of the pattern that inheres in the facts under investigation. It follows that inspection of a number of examples will facilitate first the recognition of similarities of relationships and secondly that of the ordering principle to which they serve as a pointer. When this has been seen and a rule of construction has been derived from it, the detailed pattern of structure can be developed, and as this development proceeds it reveals more fully at each step both the nature of the organizing principle (which if at first inaccurately determined may in consequence be more closely adjusted to the increasing evidence) and the configuration, the structure, of the facts. Accordingly, the next step in the development can be more definitely prognosticated. The schematic order, implicit from the start, becomes more clearly discernible as examples multiply. Hence the plausibility *prima facie* of traditional theories of induction and the soundness of Aristotle's claim that by close inspection of particulars the universal becomes apparent. But accumulation of instances is no mere repetition, and so far as it is does not serve to enlighten. To be scientifically valuable, each observation must reveal some difference, some nuance, which supplies new information and offers a fresh clue to the form of the total structure.

This is what we found in the work of the scientists we have examined. Their observations are not of repeated instances of the same fact or event, but of different, mutually related facts and events. Kepler is concerned with different positions of the planet at different times, as well as with different and concomitant positions of earth and sun. Harvey describes the action and structure of different parts of the vascular system, indicating their mutual relations. Lavoisier experiments with different substances in different combinations and situations and with different arrangements of apparatus. He draws his conclusions from the interplay of the various results he obtains in these experiments, and also relates them to those of his predecessors and contemporaries. The same is eminently true of Darwin and of Anderson, both of whom draw their evidence from the most diverse sources, and the details of their procedure need not be tediously repeated.

Induction, accordingly, is never mere generalization from

repeated instances—never argument simply of the form 'AB frequently, therefore AB always'. It is always construction depending upon growing insight into the principle or order implicit in the evidence and determining the nature of the facts. It is therefore at the same time always deduction—from phenomena.

On the other hand, deduction is never merely tautological transformation of sentences or formulae. Even 'A is not not-A' is no sheer tautology. It is the abstract, formal statement of the minimal condition of system, and means that not-A is distinct from, opposed to, and complementary of, A. Deduction proceeds on two indispensable conditions: (i) insight into the principle of order governing a structured totality, and (ii) knowledge of the interrelations among the particulars. The first is to be had only by a kind of intellectual intuition (already described) to which the particulars serve as cues (the source of plausibility both of Aristotle's and of Descartes' theories); and the second cannot be obtained apart from observation of the phenomena. Deduction is therefore always at the same time, to some extent, induction. This is true even in mathematics, as Henri Poincaré has clearly shown. The purely analytic reasoning in mathematics, he maintains, is no more than verification of rules of procedure derivative from the principle of order of the series of natural numbers. These rules can be established, however, only by mathematical induction and recursive definition. In the last resort, the result of a recursive proof is grasped only by insight into the truth of a synthetic proposition which Poincaré maintains is synthetic *a priori*.[1]

Induction and deduction, in consequence, turn out to be merely two sides of the same logical process, which, perhaps, it might be better to call construction. If this is true of so basic a science as mathematics, it is likely to be true of all scientific reasoning, and here again we have Poincaré's testimony:

'Mathematicians therefore proceed "by construction", they "construct" more complicated combinations. When they analyse these combinations, these aggregates, so to speak, into their primitive elements, they see the relations of the elements and deduce the relations of the aggregates themselves. . . . For a construction to be useful and not a mere waste of mental effort, for it to serve as a stepping-stone to higher things, it must first of

[1] See *Science and Hypothesis*, Ch. I.

all possess a kind of unity enabling us to see something more than the juxtaposition of its elements.'[1]

This description would apply with equal aptness to the procedure we have found empirical scientists following. They too proceed by construction, which they then analyse into primitive elements deducing concurrently the relations between the elements and the over-all pattern of the construction from the governing principle of order. We shall see in the next chapter that these constructions, to be useful, 'serve as stepping-stones to higher things', as we have already seen that they reveal more than the mere juxtaposition of their elements.

vii. SCIENCE AS SYSTEM

We can now explain and justify more explicitly the point on which we insisted at the end of the last Part. Current doctrine (Braithwaite's, for instance) defines a scientific theory as a deductive system. In doing so it is not wrong; but the deductive system is not merely a series of sentences related by transformation rules such as might be represented by a calculus. It is a system of facts organized by a principle or law, or a set of such principles themselves interrelated according to some higher principle. Strictly the theory is the organizing principle, and the facts it orders constitute its domain or province. The principle provides the rule of inference by which the scientist can conclude to states of affairs which do, or which under stated conditions would, obtain from a knowledge of other states affairs belonging to the same system. The theory operating as such an ordering principle is thus manifested or exemplified in the facts which comprise the system. It is variously manifested, in mutually complementary ways, and so displays the sphere of experience which it organizes as a polyphasic unity.

A system of such systems all similarly interrelated constitute a science, and ideally all sciences, knowledge as a whole, are related to one another in a single system organized by a single principle. This, at any rate, is the ideal contemplated by scientists and the goal towards which they continually strive. 'The supreme task of the physicist', wrote Einstein, 'is to arrive at those universal elementary laws from which the cosmos can be built up by pure deduction.'[2] The advance of science, both internally to each

[1] *Op. cit.*, p. 15. [2] *The World as I See It*, p. 125.

discipline, and with respect to their mutual relations, has been marked by successive unifications of hitherto apparently separate branches: astronomy, with mechanics, mechanics with optics and electro-dynamics, physics with chemistry, and both of these with biology. Unifications such as these have occurred at different stages, or critical points in the progress of science, and at each such stage, every science presents itself as a polyphasic unity. The progress of science as a whole, moreover, consists of a series of theories, each developing out of its predecessor. It is a succession of theoretical systems (polyphasic unities), the mutual relation of which is next to be examined. The task of the next chapter will be to exhibit the structure of scientific advance—the dialectic of scientific development. It will explicate further the character of polyphasic unity in its dynamic or progressive aspect, and display the advancement of knowledge as itself a system extending over the course of time.

CHAPTER XI

THE DIALECTIC OF PROGRESS

i. COMPREHENSIVENESS AND CONSISTENCY AS MARKS OF ADEQUACY

At the end of the last chapter we defined a scientific theory as a conceptual principle or schema organizing an assignable body of experience or range of phenomena. Its satisfactoriness as a theory will be proportional to the extent to which it does successfully organize the phenomena delineated without conflict or inconsistency. No theory ever does this with complete success, because no theory is ever absolutely complete either in its coverage or in its articulation. Every theory represents a stage in the development of knowledge and none can be finally and totally satisfactory short of omniscience.

Stated in isolation from the facts which its function is to order, a theory or conceptual schema is a mere abstraction and, apart from all reference to the phenomena, would strictly be meaningless. The phenomena, on the other hand, as the objects of observation, derive their character and significance from the theory. The concrete, polyphasic unity, which is scientific knowledge, therefore, is the integral combination of observation and theory as a dynamic whole—an experience which is constantly growing, like the living organism which enjoys it. It is a dynamic system and is constantly growing because it is imperfect and its imperfection gives rise to the aporiai, the contradictions, that impel the theorist to new researches, so that his endeavour is constantly towards greater comprehensiveness, in extent and in detail, with closer unity, integrity and coherence.

ii. OBJECTIONS AND CRITICISMS

A theory of this kind, it may be said, is simply a revival of the coherence theory of knowledge the difficulties of which have long since brought it into disrepute. But no such objection can be admitted as it stands, first because the best known and most weighty criticisms of the coherence theory are none of them conclusive, and secondly because that theory, as generally understood by its critics, is held to define truth as a body of mutually implicative and consistent propositions without reference to empirical data. That in itself is a misrepresentation of the theory as advocated by philosophers like Bradley, Bosanquet, Joachim or Blanshard, and is certainly inapplicable to what is being advocated in this book. In common with these philosophers I am indeed denying that sense-data are the touchstone of truth, but I retain sensuous perception as the foundation and starting-point of science and I am far from denying that observation plays an important part in the establishment of scientific theories. Nor do I believe that any of the above-named philosophers would have made any such denial. That the theory here advocated is in part a coherence theory I would not seek to gainsay; but if it is to be given that label, it is, at all events, a coherence theory that includes in the body of coherent knowledge identified as science all the available empirical evidence. It differs from contemporary empiricist theories primarily in its conception of the nature of empirical evidence and its relation to theory. Where the empiricist regards the results of observation as independent of theory and as (at least potentially and in principle) neutral with respect to alternative interpretations, I have tried to show in detail that this cannot be so.

No mere group of mutally consistent propositions may be considered 'true' simply because they are mutually consistent. But their 'truth' or 'falsity' will not depend upon their correspondence or failure of correspondence with any external body of fact. We already know that there is and can be no external body of fact—external, that is, to all theory. If it is external it must be because it is an interpretation or an ordering of facts by some different principle. That is, it must be a rival theory. No body of mutually consistent propositions can correspond (or fail to correspond) to any single extraneous fact; for no single extraneous fact can exist without implying some theory, either inconsistent with the one under consideration or coherent with it. We have found that no isolated fact or observation, however interpreted, is sufficient by itself to overthrow a well-established theory, and that what does not at first seem to fit can be, and is, usually ignored or explained away. It is as if one tried to judge the correctness of a curve in mathematics by comparing it with a single point. One may find that the curve does or does not include the point, but, if it does not, it is not because either fails to 'correspond' to the other, but because in order to include the point (along with all others plotted) a different curve is needed which conflicts with that previously tried. Otherwise we have a curve the equation for which is not satisfied throughout the whole of its length—one which is not consistent with itself. So with theories, if they organize the relevant observations without internal conflict they are scientifically satisfactory, but if in the attempt to do so they generate internal contradictions, they are not.

Moreover, a set of mutually consistent propositions is something fixed and fossilized—a view of science which I am at pains to reject. For I am anxious to insist on the dynamic character of scientific knowledge and to reject the conception of a theory as a static body of propositions, each and all with a fixed and unalterable significance. I maintain that scientific theories are constantly growing and developing, and that scientific statements derive their significance from the context in which they are made—subject to the conceptual scheme which informs the theory to which they belong. Their significance therefore changes as the scheme develops so that the advance of science alters their import and their truth. Failure to take cognizance of this progressive character of scientific knowledge lies at the root of one of the commonest criticisms of coherence as a test of adequacy. This is the objection that a scientist might light upon a beautifully coherent and elegant theory only to discover that there is another, equally coherent, that explains the facts just as well. In that case he must find some new criterion by which to choose between the two. The same difficulty, however, dogs the heels of the current empiricist theory of inductive method, according to which there are always innumerable hypotheses that can be made to fit any given body of observed facts, and mere appeal to the available evidence will not help us to decide between them. Some other criterion is needed, therefore, than a direct appeal to the facts,[1] some criterion, for instance, like simplicity. But what constitutes simplicity in a scientific theory is a much disputed question that empiricist doctrines have failed to answer, and it proves more easily disposable in terms of coherence. Nevertheless, in the actual history of science rival claims have often been made by theories apparently equally coherent and equally consonant with observation. By what criterion, then, we

[1] Cf. S. F. Barker 'On Simplicity in Empirical Hypotheses', *Philosophy of Science*, Vol. 28, No. 2, 1961, pp. 162 ff.

have to ask, has the one which eventually prevailed gained precedence? The best answer seems to be that that theory survives which has greater and wider explanatory power—in other words, which remains most self-consistent over the more comprehensive range.

There are two kinds of situation in which dilemmas may arise. The first is that in which two theories prove equally satisfactory because they are, in fact, isomorphic. This was the case with Schrödinger's wave mechanics and Heisenberg's matrix theory. Here no difficulty arises, for one is not dealing with two equally coherent rivals but with only one system in alternative forms.

The second type of situation arises when a science (or some branch of it) is in a phase of transition, rival theories tend to proliferate, and until one proves sufficiently comprehensive in its explanatory efficacy, two or more in a limited sphere may seem equally satisfactory. Thus Tycho Brahe's theory that the sun revolved round the earth and the other planets round the sun, at first explained the apparent motion of Mars just as well as the Copernican theory; but it was ruled out by its inconsistencies in a wider context, from which the Copernican theory, as it was developed, proved free. Similarly, the electromagnetic theory of matter put forward by Lorentz and Larmor explained very coherently the FitzGerald contraction, a hypothesis convenient in its turn as an explanation of the null result obtained in the Michelson–Morley experiment. But it gave way to Einstein's relativity theory, because that was more comprehensive in explanatory capacity. Einstein's theory does not disprove or refute FitzGerald's. The 'contraction' still has a place in relativity theory, but instead of being an absolute measure of length, it is a measure merely relative to a moving frame of reference. This idea of length as relative is more universally and more consistently applicable for no absolute scale can be singled out. Thus the Einsteinian theory is preferable. In all such cases the criterion for choice between rivals, given comparable coherence,[1] is comprehensiveness and scope, so that the less adequate view, in the course of time, comes to be seen as something postulated *ad hoc* to explain only a limited range of phenomena. In short, the final criterion of adequacy is comprehensiveness of explanatory power. But what, after all, is it to

[1] Strictly, however, there is no equivalence of coherence. The less comprehensive theory breaks down when it is applied more widely just because in the wider context it gives rise to incoherence.

explain facts other than to show them in their systematic interrelations, set within a system of knowledge which is in the last resort self-supporting. Explanatory power is precisely the capacity of a theory to organize into a system without conflict a wide-ranging body of facts; and the measure of explanatory efficacy is the extent of the body of experience so ordered.

The question of simplicity may be cleared up in much the same fashion. It is not easy to say what constitutes simplicity in a theory. The term has never been, and probably cannot be, satisfactorily defined. The two-sphere theory in astronomy is far simpler than the Copernican and it explains quite nicely the diurnal rotation of the heavens. But when called upon to explain the apparent motion of the planets, it becomes overwhelmingly complex. By the time the two-sphere theory had developed into the Ptolemaic system, through successive efforts to explain more of the facts and remove persistent incoherencies, it was clear that the heliocentric system gave promise of greater simplicity.

The same was true of the Phlogiston Theory, which at first seemed to give such admirable and simple explanation of the facts of combustion and calcination. But as the articulation of the system advanced and contradictions multiplied, the theory became hopelessly involved and Lavoisier's alternative, in essence much more complicated, seemed simpler. It was more complicated because it was destined to substitute a host of elements where four had earlier sufficed and because it revealed those four as more complex than had been suspected. Yet the new conception of chemical combination gave a much simpler, because more coherent, explanation of a far greater body of facts concerning combustion and oxidation.

Prima facie, relativity can hardly be described as a simpler theory than the classical dynamics. But by the time the latter had been elaborated to explain the facts of electromagnetism, the theory of the aether had become intolerably clumsy as well as sadly self-contradictory. The introduction of the principle of relativity and the substitution of the Lorentz transformations for the Galilean removed the incoherencies and simplified explanation of the facts; and further generalization, while it complicated the formulae for the law of gravitation, simplified the concepts of mechanics and dynamics by reducing them to geometry.

Simplicity, in short, is a function of comprehensiveness when the condition of consistency is to be fulfilled. A simple theory that gives a coherent explanation of relatively few salient facts, when its scope is widened and new applications reveal contradictions, becomes highly complicated by efforts to keep it consistent with itself. The superseding theory which can give a consistent explanation over a wider range then appears more simple though it may be much more complicated than the earlier theory was originally.

Attempts to explain simplicity in terms of the number of primitive terms or initial postulates required by a theory all break down because these appurtenances are irrelevant to the issue. What is required of a theory is coherence and what will coherently systematize the area of experience under scrutiny is the simplest theory. As we have seen there may be more than one, but extension of scope will eventually decide between them. 'Simplicity', therefore, is only a misnomer for coherence. Philosophers fail to define it because they seek for it in the wrong place, either in some form of abstract simplification (as Descartes did when he found simplicity chiefly in mathematics), or in some kind of reduction of the number of premises logically presupposed. The true source of 'simplicity' is the coherence with which a large and varied body of facts can be unified within a single system by a single principle of organization. 'Simplicity' is thus a misleading term and 'unity' might be more appropriate.

Some critics have objected that a very systematic and self-consistent theory such as Newtonian physics may, when new facts come to light, prove false, so that the criterion of satisfactoriness cannot be simply its systematic and consistent form. There is, of course, much truth in this objection. Self-consistency is a function in part of abstraction, and the more we leave out of consideration the easier it is to be apparently self-consistent; and when attention is drawn to some of the factors omitted discrepancies in the theory make themselves felt. It is for this reason that alternative geometries can be developed, each *prima facie* internally consistent. They are all abstract and their shortcomings are revealed, with the consequent internal conflicts, only when applied—that is, in a wider context. Euclidean geometry served very well in physics until its application to electromagnetic phenomena called in question the basic presuppositions of measurement, such as the existence of an absolute space-time frame and the transference of congruency. Then, although as pure geometry it remained self-consistent, as physical geometry it revealed internal incoherencies.

Furthermore, even within the limits of pure mathematics, no system is self-sufficient with respect to internal consistency. Gödel has shown that self-consistency cannot be proved within the system, but only by reference to a meta-system going beyond the axiom-set the internal consistency of which has to be established.

But, as stated, the objection implies a superficial view of scientific procedure such as, throughout this essay, I have been trying to combat. First we must note that all scientific theories have some degree of systematic coherence, for this is the condition of their being regarded as scientific in any sense. But none is complete, and the degree of coherence is limited. As the implications of the theory are developed these limitations become apparent and contradictions arise demanding new efforts at systematization. It is only on this account that a theory is ever superseded and no theory has ever been abandoned simply in spite of its self-consistency. Aristotle's broke down when the implications were developed of his theory of motion and led to plain contradiction, as has already been noted. The Newtonian system, magnificent though it was, contained internal conflicts from the very first, some of which eventually proved insuperable.[1] The corpuscular theory of matter contained incoherencies between the ideas of impact and elasticity and both of these were in conflict with the implication in the Newtonian theory of gravitation of action at a distance. Then came the concept of the aether with demands for high rigidity coupled with conflicting demands for high fluidity, and all the complex inconsistencies of the behaviour of light which emerged from the classical theory of electromagnetics.

Classical physics was, and for many purposes still is, a remarkably comprehensive conceptual system, with unprecedented explanatory power, built up by successive feats of spectacular scientific achievement; but it was never wholly free of internal inconsistencies and as these became increasingly obstructive to further advance its inadequacies forced scientists to the construction of new theories. These, when they eventually emerged, were adopted precisely because they were able to render the classical doctrines more complete and harmonious, a fact for which we have the testimony of some of the best philosophical and scientific authorities.

The judgement of Max Planck runs as follows:

[1] Cf. Ch. VII, Sect. iv, pp. 214–15.

'The Theory of Relativity seemed at first to introduce a certain amount of confusion into the traditional ideas of Time and Space; in the long run, however, it has proved to be the completion and culmination of the structure of classical physics. To express the positive result of the Special Theory of Relativity in a single word, it might be described as the fusion of Time and Space. . . . Looked at in this way Einstein's work for Physics closely resembles that of Gauss for Mathematics. We might further continue the comparison by saying that the transition from the Special to the General Theory of Relativity is the counterpart in Physics to the transition from linear functions to the general theory of functions in Mathematics. Few comparisons are entirely exact, and the present is no exception to the rule. At the same time it gives a good idea of the fact that the introduction of the Theory of Relativity into the physical view of the world is one of the most important steps towards conferring unity and completeness. . . . The Principle of Relativity has advanced the classical physical theory to its highest stage of completion and . . . its world-view is rounded off in a very satisfactory manner.'[1]

And here is Ernst Cassirer's testimony:

'The general theory of relativity shifted these "independent and permanent relations" to another place by breaking up both the concept of matter of classical mechanics and the concept of the ether of classical electrodynamics; but it has not contested them as such, but has rather most explicitly affirmed them in its own invariants, which are independent of every change in the system of reference. The criticism made by the theory of relativity of the physical concept of objects springs thus from the same method of scientific thought which led to the establishment of these concepts, and only carries this method a step further by freeing it still more from the presuppositions of the naïvely sensuous and "substantialistic" view of the world.'[2]

As yet, similar statements with respect to quantum theory are more doubtful, unless one concedes the validity of Schrödinger's claim for the possibility of accounting for the behaviour of particles

[1] Max Planck, *The Universe in the Light of Modern Physics* (London, 1957), pp. 17–20.

[2] *Einstein's Theory of Relativity* (New York, 1953), p. 386.

wholly in terms of waves, or of Landé's for the possibility of accounting for apparent wave-like phenomena in terms of particles. Here, perhaps, science is still in a stage of transition between rival conceptual schemes and conclusive decisions have yet to be made. For all that, Cassirer has argued that quantum theory has done no more than clarify and make more precise basic concepts (like 'material point') which classical physics left obscure.

'If today,' he says, 'the sacrifices which have to be made . . . seem greater and more difficult than ever before, it nevertheless appears to me that atomic physics has not destroyed the bases on which physical knowledge rests; rather it has made them known more clearly than ever before, in their characteristic individuality and conditional nature.'[1]

At the same time, the older theories are not proved sheerly false by those that succeed them. Each in its own way is true, or adequate, in some degree. Each has had its contribution to make, as we have already illustrated in Chapter VII. Nor are the older theories wholly abandoned when newer views are adopted. The positive contributions they afforded are preserved and developed in more adequate theories but they are modified and transformed, seen in a new light and interpreted in a new context, which gives them different and wider significance.[2]

iii. SCIENCE AS A SCALE

The development of a science, and of science as a whole, constitutes a continuous series of systems (or theories) differing progressively in their degree of adequacy. Alternative to this term I have used 'explanatory power', or 'satisfactoriness', 'coherence', 'consistency' and 'comprehensiveness', all to indicate those characteristics which signify the success of a theory in fulfilling its function of rendering our experience of the world intelligible. The extent to which any or all of these notions are equivalent to 'truth' we shall consider later. The account of the historical development given in Chapter VII above has made it clear that

[1] *Determinism and Indeterminism in Modern Physics* (Yale, 1956), p. 196.

[2] Cf. the identification in quantum theory of the classical principles of Least Time and Least Action and the new significance that they gain thereby. See de Broglie, *The Revolution in Physics* (London, 1954), pp. 35–41, 162, 168 f., 261–3.

science advances through a continuous series of conceptual schemes differing each from the next in degree. The series is continuous, each theory growing out of its predecessor and each constituting a gradation in a scale of systems progressively increasing in comprehensiveness, articulation of detail and systematic integration.

The process, we have seen, moves from one view or conception of the world, via questions that emerge when its implications are developed in the course of articulation, to new hypotheses. These are developments or modifications of the old in specific ways that have been suggested by the experienced phenomena themselves. Finally there emerges a new system organizing the experienced facts in a new way and established by a structure of interlocking evidence, which displays the pattern of the facts and actualizes the order determined by the new principle of organization. The hypotheses, as we saw, are themselves the specified questions—suppositions which are the growing points of the developing system. And the process of confirmation is simply the working out in detail of the developed structure in accordance with a rule (or rules) derived from the implications of the available evidence. The logic is throughout the logic of developing structure a continuous process now revealed as a continuous series of systems each a specific version or conceptual ordering of the relevant experience, but each improving upon its predecessor by resolving the difficulties implicit in the predecessor's shortcomings.

The systems (or theories) are thus stages in a scale of progression, while each is at the same time a specific example of a theory of a definitive range of phenomena, be it of motion, of the heavens, of the physical world, or (more usually) of all these taken together. But because each new theory arises out of the contradictions inherent in the old and effects a resolution of them, there is always an element of opposition between them. For the superseding theory is the counter to, and nullification of the incoherencies of the earlier one and so establishes itself in opposition to it, while, at the same time, it is a development of it and in definite ways preserves what in the earlier theory was of positive value.

This is all clearly illustrated in the history of science. The Ptolemaic astronomy aimed to serve the same purpose as the Copernican. They were both efforts to reduce to intelligible order the observed phenomena of celestial movement, each is a specific example of astronomical theory, and the latter developed from the former as we have already described. They are thus consecutive gradations in the development of astronomy. Yet they are also in diametrical opposition, for the first is geocentric and the second is heliocentric. The first is based on the Aristotelian antithesis of natural and violent motion, while the second entails principles of inertia and gravitation. Violent motion continues only under the impulse of forces; inertial movement continues in a straight line with no forces acting. Natural motion is purposive (to regain a natural place), whereas inertial and accelerated motion are purely mechanical. Nevertheless, the development from one system to the other is an unbroken continuum—in fact it is only a phase of a longer series of systems, going back to the Greeks, that progressed from Pythagoras to Eudoxus, from Eudoxus to Calippus, from Calippus to Ptolemy, and from Ptolemy to Copernicus by way of Aristarchus of Samos. Copernicus is only the critical point in a continued development with Tycho, Kepler, Gilbert, Galileo and Newton as its major figures.

That the development is continuous is certified by the fact that earlier ideas are preserved in later theories. Circular motion is not abandoned, for the earth rotates on its axis. Natural motion becomes inertia; impetus becomes momentum, violent motion acceleration.

> 'Nothing of them that doth fade,
> But doth suffer a sea-change
> Into something rich and strange.'

The same relationships obtain between Newtonian and Einsteinian dynamics. Both theories subserve the same scientific aim, they are, however, sharply opposed. Newtonian gravitation is action at a distance. Relativity tolerates no such paradox. Newtonian space and time are absolute, Einsteinian relative. Newtonian forces act in straight lines between mass points, Einsteinian space absorbs all forces and substitutes for them curvature of the metrical field. Newtonian inertial movement is in a straight line, Einsteinian movement of every kind is along a geodesic. But, again, old ideas are still preserved. This geodesic motion (along what Eddington calls the 'natural track'[1]) is akin to the old Aristotelian natural motion, and also has affinity to inertial movement. The

[1] See A. Eddington, *Space, Time and Gravitation* (Cambridge, 1920), pp. 69–70.

classical laws have not altogether disappeared from relativity physics; they remain as limiting cases when certain quantities vanish from the relevant equations. And the antithesis has been bridged by the idea of the field, which entered classical physics with electrodynamics and merges in relativity physics with the space-time continuum.

The quantum theory of matter and energy is similarly related to the classical theory. Opposition between them is obvious and the former arises quite clearly from contradictions appearing in the latter. The Rutherford atom, according to classical notions, should have collapsed into the nucleus, and it was only when continuous emission of energy was denied that the Bohr model gave promise of solving the problem. For classical mechanics momentum and position are independently determinable, for quantum mechanics they are conjugate quantities. In classical mechanics wave motion and particle motion are distinct, in quantum mechanics they overlap.

Nevertheless, there is continuity, as has already been observed, between Jacobi's theory and Schrödinger's, and through the part played in the wave function of the Hamiltonian operator.[1] And when particles are conceived as wave-packets and orbiting electrons as standing waves, quantum theory begins to look more like a broadening unification of classical ideas than a contradiction of them.

So we have a series of theoretical systems forming a continuous development, each in its own way complete in itself, each serving the same scientific purpose, yet each opposed to its immediate successor and predecessor, despite continuity of evolution and the preservation in the later theories of key ideas from the earlier. The theories are all exemplifications, or specific realizations, of the same generic type; they each represent a stage or gradation in a scale of development, in which the later is the more adequate, the more comprehensive and the more self-consistent. Each successive theory is a better theory, fulfilling more satisfactorily the purpose at which they all aim. Yet in some sense each is in opposition to its neighbours, while it preserves from its predecessor an element of truth, to which it gives a new significance and which it renders more fruitful.

It is its inadequacy to the universal function of science that causes the breakdown of the more primitive system. Internal

[1] See de Broglie, *op. cit.*, p. 169.

contradictions come to light because experience reveals features with which the theory cannot cope. The reality of which it claims to give an account is more complex and more intricately interrelated than its principles allow, and the more complete whole thus makes demands upon it which it cannot meet—or which by its efforts to meet it is thrown into internal conflict. Thus it is the presage of the more adequate concept which operates as the stimulus and propelling force of development. The more adequate theory gives the better account of the facts, it is more coherent, more comprehensive, more 'true'. It reveals the degree of truth of the less adequate, contradicts those aspects in which that falls short, yet continues and more satisfactorily fulfils the progress towards the ultimate aim of science. Consequently the procedure by which the advance is made is governed by the totality which is its outcome, and, just as we found the whole logically prior to the part, so we find now the end logically prior to the process. For process is the development of the whole and the serial unfolding of its parts. However, as the nature and significance of the part is not self-contained, but is dependent on and determined by its place in and contribution to the whole, its aspect changes as the development unfolds, and what it appeared to be in relative self-sufficiency proves inadequate and false. But it is not destroyed altogether in the process. It is only transformed and preserved with new meaning in a more highly organized and more widely efficacious totality.

When one looks at the process in closer detail, the same pattern asserts itself. Copernicus set out the heliocentric system with numerous Aristotelian and still some Ptolemaic features. Tycho Brahe modified it in an attempt to retain its advantages without sacrificing those (as they seemed) of the older system. Kepler then restated the heliocentric position substituting elliptical for circular orbits and incorporating a new hypothesized celestial dynamics anticipating the law of gravitation. Each theory is again a development of its predecessor, each contradicts its predecessor in some sense, yet each is a new version of the same theory—a new attempt to organize the same phenomena.

In still greater detail we find Kepler, in his search for the martian orbit, assuming first that it is a circle, then an ovoid, then an ellipse. These three figures form a series, each a modification of its predecessor, the ovoid by imposition upon the circle of an epicycle turning backwards, the ellipse by generalization of the circle. If the circle is regarded as the perfect figure—Aristotle's reason for assigning it to the movement of the heaven—then an ovoid is opposed to it as irregular and imperfect. The ellipse, again, is opposed to both ovoid and circle. It is once more a regular geometrical figure, as opposed to the former; and in relation to the latter it is more general as opposed to the more particular. Yet all three are species of one genus. They are all closed curves such as might be the orbit of a planet. But only the last is adequate because it alone incorporates all the facts without conflict and harmonizes with dynamic theory.

Meanwhile Galileo, exploiting ideas of natural motion and impetus, developed the new theory of trajectories and falling bodies. While Kepler developed the celestial aspect of the new system, Galileo developed the terrestrial. Each in his sphere provides an example, within lesser scope, of the sort of developmental structure exhibited by the major development. Newton's encompassing edifice, over-arching even more than the work of these two, is a system of systems relating to Aristotle's in much the same way as the lesser unities within it relate to their lesser predecessors. It is a fresh stage in the scale of development, with all the characteristics already described, and is, besides, hierarchical (as are its predecessors), through its inclusion of the lesser contributors, the preservation (in modified form) of past concepts, and its consummation of past aspirations.

Similar illustrations of this pattern of developmental structure are provided, first, by the Harveyan achievement following on the work of Vesalius, Colombo, Fabricius and Cesalipino; secondly, by the integration and fruition of the work of the British and French eighteenth-century chemists in Lavoisier's synthesis. Thirdly, in Darwin's impressive synthesis, the work of his predecessors of the preceding half-century are combined and rendered scientifically fruitful. It would be possible to elaborate in detail how each of these developments constitutes a scale of forms of the same general pattern as those outlined above. But the details have already been recorded in Chapters VI and VII and need not be repeated.

iv. DIALECTIC

The dialectical character of this development is apparent from the character of its dynamic. An incomplete system, or, what is the same thing in alternative form, a manifold that is imperfectly systematized, involves internal conflicts.[1] The rules extractable from its elements, when followed out in different contexts lead to results which mutually collide and negate one another. This is the stimulus to research which seeks to modify the conceptual scheme in such a way as to reconcile the conflicts and resolve the opposition. For this reason rival theories confront one another as opponents and that which succeeds in resolving an opposition is the negation of the incomplete and self-refuting. But this negation (as we saw) has a positive aspect. It preserves that in the faulty theory which is valuable while it places it in a new setting, a developed conceptual scheme which gives it new significance, and at the same time explains both the short-comings and the explanatory value of the old. A continuous advance by way of opposition and reconciliation is the precise meaning of dialectical process. It is now apparent that this dialectical character belongs specifically to the question and answer logic advocated by Collingwood and outlined above as the appropriate logic of science insofar as it is a growing, exploratory and advancing system of knowledge.

The development of scientific theory is thus a scale of forms, mutually overlapping, mutually contrasting, mutually complementary, in ways that have been instanced. The successive theories differ in degree of adequacy, oppose each other in dialectical controversy, and successively bring to fruition the continuous effort towards unified explanation of phenomena. Each theory is an organized scheme in which a principle of order is being followed, and each exemplifies, or realizes, the principles more fully and intricately than its predecessor.

It is, moreover, this principle of order of which we said in Chapter IX that it was, in the best sense of the phrase, the absolute presupposition of science. But taken in this sense, the absolute presupposition is proleptic for it is not realized except at the end of the development, though it is implicit from the first, presupposed, if only tacitly, and operating as the stimulus to further advance.

[1] It must not be forgotten that there are, strictly speaking, no fragmentary systems in experience, because every object cognized is interpreted in the light of the existing body of knowledge—the funded and accumulated experience as hitherto ordered. A defective theory is, therefore, not so much 'incomplete' as vague and confused. What was asserted earlier must be constantly borne in mind, that the progress of learning is from indefinite to definite, not from fragment to whole, and that at every stage we have a whole of related and connected elements.

Looked at in this way we can see that this absolute presupposition is equally immanent in common sense as the starting phase of the development. This fact accounts for the ambiguity we noticed earlier which enabled us to say—as now appears, appropriately—both that common sense is the presupposition of science from which its original questions arise and that, nevertheless, its absolute presupposition is the ultimate structural principle that gives significance to its details and renders them ultimately intelligible. For this principle also underlies common sense and is the essence of such intelligibility as it possesses. Without it, the relevant questions would never arise and science would never come into being.

The total pattern of dialectical progression in the advance of science answers exactly to the account of system set out in *Foundations* (Chapter XXII). System, there, proved to be a serial order progressively unfolding an organized structure as a continuous scale of forms each expressing more adequately than its predecessor the principle of organization of the whole.[1] This principle is universal to all its manifestations, so that each phase of the process is a particular specification of it. And it is this universal principle that is immanent in every phase from the very beginning of the development, and is progressively being brought to explicit expression throughout its course.

The law of gravitation and the sun's pull upon its circulating satellites is implicit in the very first proposal of the heliocentric system (as Kepler realized). It manifests itself, though only obscurely, in the observed motion of the planets. It becomes further explicated in Kepler's and Galileo's laws and is made explicit and developed in detail in Newton's system. The circulation of the blood is implicit in each partial discovery prior to Harvey, the impermeability of the septum, the lesser circulation, the structure of the valves; and in the procedure of his own research: the action of the heart, the relation of flow in the arteries and in the veins and the quantitative assessments. Likewise, the principle of oxidation is implicit in the discoveries of the pneumatic chemists and emerges progressively in Lavoisier's experiments first with nimium, then with sulphur and phosphorus and finally with mercury.

[1] Cf. *op. cit.*, p. 467.

V. THE UNITY OF SCIENCE

But while the dialectical progression is apparent even in these relatively limited episodes of scientific discovery, the more general progress is from one dominant conceptual system to another, and each of these constitutes an entire world view implying a metaphysical system. Thus the major scale of forms is from one dominant scientific scheme to the next by way of major scientific revolutions. This scale is a progression of world views and its completion would be a single, all-inclusive coherent, intelligible system in the light of which all phenomena, all events and objects in the universe would be explicable—the ultimate rational comprehension of all experience. This is the goal and ideal of science, so far to seek as ultimately to be (in all probability) beyond the limits of human capability, but nevertheless the ideal towards which approach is asymptotic.

The advance of science has, in consequence, a two-dimensional character. It moves from indefinite to definite, and at the same time from a relatively loose and disparate collection of theories towards a single unified system. These are but two aspects of the same movement from naïve common sense to systematic and coherent theorizing. We have seen how an ordered world of distinct perceptual objects arises from primitive sentience and develops into a common-sense world view, how this, when obscurities and contradictions frustrate understanding, develops into a reflective effort at scientific explanation. A conceptual scheme emerges, is articulated, reveals faults and is reconstructed through an advancing series in successive revolutions. This is one aspect of the matter. The other is the unifying effect of this advance. Newton's system united celestial with terrestrial mechanics, Maxwell united electrodynamics and optics, Einstein brought all these into a single system of relativity physics. The quantum theory has unified physics and chemistry and, in a continuing process of interconnexion, physics, chemistry and biology are rapidly becoming a single proliferating science. The already recognized continuity between physiology and psychology augurs a still further-reaching amalgamation, and so science progresses always towards greater unification and more ramifying integrity.

It does not, however, follow that this widespread interconnectedness justifies the reduction of all phenomena to a single level of activity—quite the reverse. The unification of science does not warrant or entail the interpretation of biological or of mental phenomena as merely physical events. Physical events they may well be or involve, but, as I have argued at length elsewhere, self-enfoldment and new dimensions of complexification give rise to new types of whole and correspondingly new laws of action, so that different levels of integrated complexity produce different orders of reality, which may be mutually continuous without being mutually reducible. This is because the principle of organization in every totality determines its character throughout and in detail, and that principle is fully realized only in the completed whole. The ultimate key to explanation, therefore, is that which explains the most complex and highly developed, not that which applies only to the lowest level of complexity, even though the former is and must be, in a sense implicit or potential from the start.

vi. HIERARCHY

The dialectical scale is, moreover, hierarchical. Not only does each successive conceptual system include and interrelate within itself lesser schemata, but each preserves and in a sense consolidates features derived from earlier systems. Each theory, or system of theories, is accordingly a summing up of the entire development prior to its own emergence. The principle of explanation of the more developed theory makes intelligible the errors of the less. Not only does one recognize more clearly why Ptolemy was wrong in postulating epicycles as soon as one realizes that the solar system is heliocentric, but one understands more precisely how the apparent movements of the planets involve the kind of complication that prompted Ptolemy's hypothesis when one knows that they result from the rotation of the earth upon its axis combined with its revolution in its orbit about the sun. Equally the stages by which the later system developed elucidate the character of its explanatory principle. It is virtually impossible to understand the theory of relativity, the successes of which remain a mere mystery, until one appreciates the problems and difficulties arising out of the Newtonian system which the relativity theory finally resolved. As Collingwood maintained, one understands the meaning of a statement (or a theory) only when one knows the question to which it is intended as an answer.

Consequently, the ultimate system is the entire scale of conceptual schemata in dialectical relation, and it is the dialectical principle itself that finally elucidates. For this reason more than any other reductionism is ruled out. What only the whole can render intelligible can never properly be explained in terms of the merely rudimentary, and the source of the genetic fallacy becomes obvious.

vii. AGREEMENT OF RESULTS

We have found that scientific theory is a multiplex structure of polyphasic systems proliferating as a dialectical scale of forms. The tendency of its development is towards ever greater unification over an ever-extending field of constantly increasing diversity; and while its nisus, on the one side, is towards unification, on the other its application and articulation penetrates ever more deeply into the minutest of details. Its primary function is the completest possible rational organization of the perceptual experience from which it begins, and its success is measured by the degree of its integrity, its extent and its coherence. These conclusions have emerged from an examination of particular cases in the history of science and from the general pattern of progression throughout that history, and they converge with the conclusions reached by examining the contemporary results of scientific analysis and speculation in its various branches.[1] Such convergence of evidence from different sources was found earlier to be the essential feature of the confirmation of hypotheses, and should strengthen the credibility of the theory of scientific validity here developed and of the world view of contemporary science itself—the metaphysics of which scientific discoveries form the foundation.

It remains to consider how far this argument is valid or whether it is viciously circular. It does not necessarily follow from the fact that scientists commonly use a certain method, that that method is a valid means of attaining truth—I have myself used this argument in criticizing attempts to justify induction. If by the use of such a method they arrive at conclusions from which it does follow that the method is the appropriate one for science, have we anything better on our hands than a circular argument? And even if this objection is not pressed, we may still ask whether it enables us to claim for science reliability as the truth about the actual world.

[1] Cf. *Foundations*, pp. 489–93.

For we have maintained that observation is itself the product of interpretation, and that interpretation, the organization of sentient experience, develops into scientific explanation, becoming more comprehensive, more articulate and more coherent. But if science is no more than the development of what is admittedly subjective experience, if it is no more than the imposition upon confused sentient awareness of principles of order, what guarantee have we that it has any significant bearing whatever on the nature of the actual world? For those principles of order may be no more than inherent characteristics of our thinking and no less subjective than sentience itself.

CHAPTER XII

SCIENCE AND TRUTH

i. OBJECTIVITY

Is the aim of science the comprehension of things as they really are, or is it no more than the construction of a symbolic notation convenient for action in a merely phenomenal world? And if the first is the aim that scientists profess, can they ever achieve more than the second? These questions cannot be answered directly for they presuppose tacitly metaphysical and epistemological doctrines which have to be examined in their own right. To speak of 'what things really are' presumes potential knowledge of the hallmarks of reality. How is what they really are to be distinguished from what they seem to be? And if we cannot answer that question, how can we allege, in any circumstances, that the world of our experience is purely phenomenal? By reference to what *Ding an sich* could we give it that status? What seems is subjective, what is is objectively real; so the demand is for clarification of the distinction between subjectivity and objectivity and some definition of these terms.

This issue we have already faced in our account of perception. For common sense, what is objective is what belongs to an external world independent of our perceiving, and what is subjective is what is dependent upon our minds, or consequent upon the organic conditions of experiencing. To make this distinction, however, we must already be aware of those conditions as objective, and we cannot know what belongs to the external world as opposed to what is mind-dependent until we have already made the distinction between the subjective and the objective in experience. In primitive sentience there is no such discrimination, and the consciousness of a world of objects arises from it as the result of an activity of organizing and the imposition upon its content of schemata. Thus the common-sense conception of an external world is itself dependent upon our thinking and on the conditions of experience, so that the distinction of objective from subjective cannot be made to rest on the common-sense criteria.

The only reliable criterion we have of the 'objectivity' of things is their stability and coherence in our experience and the persistent interconnexions which they display. Every other criterion proves unsatisfactory. Sensible immediacy will hardly do, for that is the acme of subjectivity. 'Hard' sense-data are non-existent, so objectivity cannot be built out of them. Observation is reliable only to the extent that perception is veridical and the admission of veridicality already presumes the distinction of which we are in search. Objectivity, therefore, can only be understood as what repeated and consistent experience shows to be coherent.

For the scientist objective reality is the most satisfactory theory he can give of the phenomena—and that we have defined[1] as a body of experience organized by a conceptual schema. Eddington, rejecting the use of 'reality' as a merely honorific epithet, maintained that, for the physicist 'the real world' was a symposium of different points of view, carried out according to strict rules.[2] Physical reality is what the body of trained physicists say it is, as they pursue their science according to the rules that have, in the course of long and skilled experience, been found to produce the most consistent results.

In short, our sentient experience in its most primitive form is subjective, because confused, obscure and incoherent; and we reach objectivity by a persistent process of ordering, constructing and organizing—the process properly identified as 'thinking', which leads, through perception and common sense to science, the most objective knowledge of the world we can get. To say this is the same as to say that the aim of science is to comprehend the nature of things as they really are. But, if this is so, how can we know if, or to what extent, it is achieved? If the best account we can get of things 'as they really are' is scientific theory, and if that is no more than the imposition of schemata (no matter how complex and in how many hierarchical steps) upon the contents of admittedly subjective sentient experience, is our 'real' world not just a construction of our own making? In what proper sense can we regard science as a *discovery* of reality?

ii. SCIENCE AND REALITY

The answer can be given only in terms, once again, of the scale of forms. From primitive sentience emerges (by the organizing activity of the living mind) a perceptual world ordered first by

[1] P. 352 above. [2] *The Nature of the Physical World*, pp. 282–6.

'common sense' and then by science in a series of developing conceptual schemes, which, as the process continues, become more highly systematic and intricate. But sentience itself is a product of the subconscious activity of organizing impulse and reaction in response to environmental impingements, an activity that is characteristic of the auturgic organism. In *Foundations*[1] I tried to show that the organism in its self-maintaining adaptation to its surroundings responded to physical influences, which, as physical theory had demonstrated, were internally related to the whole of the physical world. In doing so the living being constitutes itself into a polyphasic system at a special, dynamic, self-maintaining level. It thus becomes a sort of focus of the world by its own adjustment to it.

True it is that we can make this claim only on the strength of scientific theory itself (biology, physiology and psychology), and I shall consider presently to what extent it commits us to a *hysteron proteron*. If, however, we may be allowed to accept it for the moment as true, we see the whole gamut of experience from sentience to the most developed knowledge as a continuation of the scale that constitutes nature itself. This too was found to be a scale of forms,[2] from space-time to energy systems, from elementary particles to atoms, from atoms to molecules and thence to macromolecules and crystals; from these the progression continues to viruses and living organisms, and they by evolution reach a stage at which their physiological activity is so highly organized and so intensely integrated that it comes to be felt. The whole series is one of dynamic process constantly increasing in complexity and elaboration of structure, in which each successive phase is a self-enfoldment of earlier phases, incorporating them and displaying new properties and capacities. Sentience is that critical juncture at which mind emerges from living activity, and the development from sense to intelligence and explicit knowledge makes no break in the continuity.

What we have done in this Part, accordingly, is no more than to trace in more detail the character of the work of the intellect—in one (though not the only one) of its typical forms—that was adumbrated in Chapter XXI of *Foundations*; and what was said of knowledge in the final chapter of that book, is here being elaborated.

Every developing scale is a continuous series of phases related to one another as degrees of realization of the generic essence, or

[1] Pp. 183 f. [2] See *Foundations*, pp. 483–5.

principle, of which the scale itself is the specification. The phases differ, therefore, at once both in degree and in kind. Further, as specifications of the generic essence, they are mutually distinct, while as less and more adequate realizations, they are mutually opposed. These are the marks of the dialectical character of the progression. What is, throughout the succession of conceptual schemata, coming into its own—what is being explicated—is the universal principle governing the entire series, and in that series it becomes continuously more concretely, more elaborately and more minutely exemplified. But the forms in the scale are themselves its specifications and the earlier are preserved and integrated within the later. At any point in the development, therefore, the end of the scale (up to that stage) is at once both a maturer phase and a summing of all the previous phases.

This was found to be true of the entire *scala naturae* and is now seen to issue in that awareness of the world which we experience through perception and is refined and systematized in science. The advance of science, again, is a similar scale and is continuous, through the living and conscious organism, with the series of natural forms. It is, therefore, the further specification and realization of the same polyphasic unity, bringing it to explicit consciousness and elaborating it as a precise system of knowledge.

If indeed the real world is such a scale, then the activity of theoretical construction in which it issues will be the articulation, as an explicitly known system, of that totality which is the world itself. The end of the scale will be the exposition of the scale in its most highly developed, most coherent and most intelligible form. To give just such an account of the world is the aim, and, within the limits of its resources, the practice of science. And if the account which it gives is reliable, then it does achieve its aim (at least in some measure), for it shows the process of nature as issuing in just such a product as science itself is.

How far, now, by accepting from science its own account of nature, have we argued in a circle? Can we legitimately use the theories of science itself to validate its own pronouncements? To state the question in this way is to misrepresent the position, for we have all along maintained that what validates the pronouncements of science is the systematic and coherent character of the evidence. The problem we are facing is whether the whole body of knowledge, describing the world as a scale of forms is no more than an elaborate human invention, or whether it is an account true of an actual universe which exists independently of human thinking. There must, of course, be some sense in which a world of which the human mind was a part could not exist wholly independently. But there is also a valid sense in which any part, or collection of parts, of a complex whole might exist whether or not other parts of that whole existed—a sense in which a lesser whole might precede in existence a greater with which it only partly overlapped, as the sapling is pre-existent to the full-grown tree and the new-laid foundation precedent to the completed house. We shall argue that the relation between the human mind and the rest of the world is such as to justify both of these descriptions.

If the world were a scale of forms such as has been described, it would issue in awareness of itself, through the medium of some such organism as the human individual, which would give a relatively true account of its own nature. The existence of such an account, therefore, may be taken as important evidence of the fact that it is the product of such development. Moreover, as the known pattern of development within consciousness is that of a dialectical scale, it should be legitimate to extrapolate downwards and to postulate a physical world, with a biological superstructure, to account for the occurrence of consciousness. If we conceive of experience as no more than a train of successive conscious states, as Hume tried to do, we have no pretext for any such extrapolation, and we are constrained to confess that our 'impressions' arise from unknown causes. But if reflection upon experience reveals that it is always and at every stage a whole of related and connected ideas, that the gamut of its development is a series of dialectically related gradations each of which is such a whole, and that every such whole on analysis proves to be a scale of forms, it is no wild speculation to suggest that the 'cause' of our ideas is a prior phase of the scale related to other phases, both precedent and subsequent, as are the phases of the conscious series. This notion of 'cause' has already proved to be the most satisfactory,[1] and this way of accounting for consciousness the most intelligible.[2] Add to this the fact that the sciences, developing independently of philosophical reflection, present conceptions of their various fields which, when synoptically viewed by the metaphysician, constitute a dialectical series of the sort demanded, and we have ample grounds for making ontological claims for scientifically

[1] Cf. p. 340 above, and *Foundations*, pp. 472–6.
[2] Cf. *Foundations*, pp. 292–309 and Ch. XVII.

established theories, as well as epistemological claims to truth for scientific discovery.

Our epistemological theory, moreover, will be self-validating, instead of self-refuting (like Locke's). It will be such as to make the knowledge of its truth possible for any thinker whose knowledge comes to be in accordance with the theory. It will, therefore, be free of the epistemologist's fallacy—a rare virtue among epistemological theories and one in itself sufficient to recommend it.

Finally, the conclusion has been reached from an examination of the actual practice and the forms of reasoning adopted by scientists irrespective of their assertions about the nature of the subject-matter, and it has turned out to be in close agreement with the result obtained from independent reflection upon the reasoned assertions of scientists concerning the nature of things at all levels, physical, biological and psychological. It is a conclusion in support of which all the evidence conspires. Can we ask for more by way of confirmation?

iii. CRITICISM AND DEFENCE

Still, the persistent critic may object (i) that we have shown only that knowledge is the progressive clarification and reduction to order of a confused mass of sentient experience. What need, he may ask, to postulate a prior cause? Might sentience not be just self-existent and could not experience, as Bradley held, comprise the whole of reality? Or (ii) our critic may ask what evidence we have to show that the world might not be sheer chaos—the random movement of elementary particles, or a confusion of radiant impulses. This might explain the blurred indistinctness of primitive sentience; and science, following the inherent tendency of our minds to systematize, might do no more than impose upon this chaos schemata which create the semblance of an ordered world. Quantum physics, he may point out, does seem to reveal an underlying indeterminate activity which can be made to fit our determinate schemata only by statistical methods, and to conform to regular laws only by averaging tendencies among vast numbers of events, the precise determinations of which are in principle uncertain.

(i) Beginning with the Bradleian objection, we should first remind our critic that, within its own limits, conscious experience gives evidence of having degrees. We are aware both of declining towards unconsciousness and of the emergence of consciousness from subconscious and preconscious states.[1] We experience the submergence of consciousness under the influence of a slow anaesthetic and re-emergence on recovery from its effects. We experience different levels of attention declining into the dimmest awareness and even to oblivion. Many sensations on the periphery of our attention (the pressure of our clothes, or the continuous roar of the sea, for instance) are wholly overlooked and we would not claim consciousness of them while failing to notice them; yet if attention is drawn to them we have to admit that in some way we had been experiencing them all the time. Moreover, when we recall a name or a poem we learnt long ago, our consciousness of it emerges somehow from the unconscious where it has mysteriously lain dormant, though in the interim we cannot say that we had wholly lost possession of the knowledge. These are but a few of the pointers within consciousness itself that it arises out of and depends for its existence upon preconscious states.

Secondly, we may urge that perception and thinking are activities and are phases of a continuous activity of organizing, the nature of which is to generate forms in a progressive scale of less and more complete unification of differences. The nisus of this activity is not satisfied in the highest developments of our knowledge, nor is it wholly unproductive even at the lowest. There is no cognition without some degree of structure, if it be only figure and ground, and every possible object of awareness is already a complex of some kind. Consequently even the least object of perception bespeaks the activity at a lower level of the same organizing principle as is operative in it. A dialectical scale of forms does not (admittedly) begin from zero, but consciousness, even at the sensuous level, is already too highly organized to constitute a beginning which does not imply forms still more primitive. Sentience itself is a union of multifarious differences in a single potential awareness. Forms in which such differences were mutually external and less integrated are implied by the very possibility of sentience. Hegel constantly emphasizes the fact that feeling internalizes and unites what at lower levels were self-external differents.[2] In the face of these facts we cannot justifiably deny the need for postulating a prior phase in causal relation to consciousness and accounting for its emergence.

[1] Cf. *Foundations*, pp. 294–6.

[2] See *Philosophie des Geistes, Enzyclopädie*, Sects 389, 401–2.

Moreover, if one adopts a theory such as Bradley's, what reason can one suggest for the organizing nisus of thinking? What is it in the first place that befuddles consciousness and prevents it from enjoying completely clear awareness from the start? Bradley attributes its initial and persistent short-comings to the finiteness of human mentality. But what is the source of its limitations, if we postulate no physical and physiological preconditions of its existence? Apart from these, finiteness of mind seems to be a purely arbitrary postulation. For the essential character of consciousness seems to be to transcend imposed limits, and unless we can find some extraneous ground for limitation consciousness itself would not provide any.

A doctrine like Bradley's therefore, leaves 'experience', as it were, hanging in a vacuum, and forces us to raise the questions, whence sentience in the first place? and why is it primitive? On what conditions could consciousness arise? Or, if it is self-existent, why not absolute and complete from the start? Descartes argued of the *res cogitans* that if it were cause of itself it could remedy its own defects and would so be perfect, and for Spinoza virtually the definition of God—the absolute whole of Nature—is *causa sui*. Our awareness cannot be its own cause and still be imperfect; thus its very origins in sentience force us to look for yet more elementary sources.

(ii) Let us turn now to the second objection, that if knowledge is no more than the imposition of schemata upon a disordered experience the world might be something entirely different from what science suggests. Science of necessity presents us with an ordered world, for science is the endeavour to make our experience of the world intelligible and total disorder, in the extremest meaning of the word, is total unintelligibility. If the scientific picture is not true of the world, it must be because the world is either completely chaotic or is ordered on some different principle from that recognized by science, a principle different from that inherent in the human intellect. But in both cases we must regard the human mind either as part of the world and a natural product, or as somehow radically divorced from the world of nature, knowing it by inspection from without. The second alternative takes us back to the seventeenth century and is far from appealing; for it makes the relations of the mind to its object, as well as to the human body virtually inexplicable. Adopting this alternative would commit us to conceiving the human mind as the apanage of a disembodied soul attached in some inscrutable manner to a physical body, a doctrine the credentials which have long since become irreparably tarnished, not only in all the relevant sciences but also in metaphysics. It is the doctrine of the ghost in the machine that can do justice to the facts neither of physiology nor of psychology. Enough has been said in adverse criticism of this position in so many quarters that we may reject this alternative without further ado. I have discussed the problem elsewhere[1] and shall present no further arguments here; nor is the position one which a scientifically minded commentator is likely to advocate, so we may presume that it is not intended by our hypothetical critic.

We must, then, consider the implication of acknowledging the membership in the world of nature of the human mind, first if the order of that world is different in reality from what it is in science, and secondly if the world in reality is not ordered at all. In either case our knowledge would be false, and any action based upon it would be frustrated by its inappropriateness to the real state of affairs. This should produce an incoherence in our experience that would apprise us of our error and bring our knowledge into line with reality. But it is sometimes argued that our scientific knowledge diverges from the nature of the real *because* it is merely an instrument of adaptation enabling us to act more successfully in an alien environment. (This, for instance, is Bergson's argument, at least, with respect to physics.) Such reasoning is obviously incoherent, because successful action in an alien environment would demand an accurate assessment of the actual situation, and any systematic misrepresentation of it would be a misadaptation.

There are, however, more serious difficulties. Once we admit the fact that the mind is a natural product, we must accept the implication that the nature of the world is such as to be capable of producing the mind. A radical divergence between the principle of order governing nature in general and that operative in the mind would militate against any such capability. At best, it would make it very difficult to explain. But there is little need to concern ourselves with any such difficulty, for our analysis of the concept of order has shown that every order resolves itself into a scale of forms of which each of its products will be one. In that case what-

[1] See *Nature, Mind and Modern Science* (London, 1954), *Foundations*, Pt III, 'Mind and Mechanical Models' in *Theories of the Mind*, ed. J. Scher (New York, 1962), and 'The Neural Identity Theory and the Person', *International Philosophical Quarterly*, Vol. VI, 4, December, 1966.

ever the order of nature might be, the mind as one of its products would be a phase in the natural scale and would have to exemplify the universal principle of which the scale was the specification. How the mind could then impose upon its experience of nature a radically divergent system of organization would be unintelligible.

The assumption that the world might lack all semblance of order is in still worse case. For then there would be nothing in reality which could give rise to a mind at all, or even to an organism —the physical prerequisite of mind. The critic who makes the assumption must allow that the mind's activity itself is one of systematizing, for he professes to explain scientific knowledge, in a quasi-Kantian manner, as the imposition upon the world's chaos, of schemata provided by the mind. If the mind is part of the natural world there must then be some systematizing activity in nature—that of the mind—and the question must be faced how a merely chaotic world could give rise to an activity of ordering. This, we must hold, is impossible.

(*a*) If the world is envisaged as a confusion of purely random physical movements (whether of energy or particles) it will be subject to the second law of thermodynamics, that no physical action within a limited system can produce order from disorder. If the range of physical activity is very great, however, a limited amount of order might arise by chance, given sufficient time. The probability of this occurrence will be inversely proportional to the extent of the order in space and time (themselves, be it noted, structural concepts). And the greater the improbability of the events the longer must be the time allowed for their accidental occurrence. The world of nature as described by science is so immense and so complex a series of ordered events that the chance of its occurring in the course of purely random activity would be virtually infinitesimal and would presuppose a lapse of virtually infinite time. Infinitesimal probability is not easily distinguishable in practice from impossibility; and, though it may be in theory, infinite time, apart from order, is inconceivable. For time is an ordered series of events and apart from any such series there can be no time, for no one instant would be distinguishable from any other.[1] Time, therefore, can be concurrent only with the ordered

[1] Cf. A. Eddington, *Space, Time and Gravitation*, p. 157: 'A region outside the field of action of matter could have no geodesics and consequently no intervals. . . . Now if all intervals vanished space-time would shrink to a point. Then there would be no space, no time, no inertia, no anything.' Cf. also my own paper 'Time and Change', *Mind*, Vol. LXVII, 1957.

series, and the notion of that series arising by chance within a disordered chaos is incoherent, unless the latter is relative to some dominant order. We have already found randomness unthinkable apart from order for other reasons to which we shall presently return.

(*b*) Further, the alleged random activity is conceived as physical, and contemporary physics has found the physical universe to be of finite duration. All the various estimates of its age, each derived from quite different evidence, concur in a figure of the order of 4×10^9 years. This is far too short a period to provide for the chance occurrence of a protracted and complex orderly series in a chaotic universe. And the idea of any sort of physical activity in process before that period would not appear to have any physical meaning.[1]

(*c*) These, however, are not the strongest arguments against our would-be critic. The very notion of randomness as the primary physical conditon is denied to him by the science to which he appeals. For elementary particles, with which quantum physics is concerned, are thought of as wave packets—that is, as structured entities—and the indeterminacy of their behaviour is relative to an energy system. According to Pauli's Principle, the energy system is prior to the identification of particles even as separate entities.[2] The random motion of particles presupposes the mutual independence of the particles, and no sense can be made of the proposition that order might perchance arise out of random motion, if that which moves exists only within a structured system—if order is in fact a prior condition of its existence.

An attempt to go behind particles to something still more primitive would fare even worse; for radiant energy is wave motion, which is nothing if not patterned activity. Periodicity is the very epitome of order. System is primary and fundamental to all physical concepts and no meaning, therefore, attaches to the suggestion that the world might possibly be a chaos of random physical activity, which, in the attempt to understand it, our intellect subjects to artificial schematic categories.

The whole discussion, however, rests upon fallacy. Both objec-

[1] Cf. G. Lemaître, *Hypothèse de L'Atome Primitif* (Neuchâtel, 1946); E. A. Milne, 'Fundamental Concepts of Natural Philosophy', *Proceedings of the Royal Society of Edinburgh*, 1943.

[2] See W. Pauli, *The Exclusion Principle and Quantum Mechanics* (Neuchâtel, 1947); H. Margenau, 'The Exclusion Principle and its Philosophical Importance', *Philosophy of Science*, XI, 1944; and *Foundations*, pp. 131–9.

tion and reply draw their evidence from physical theory in their efforts first to attack and then to defend the truth of scientific knowledge. But if science bears no relation to the actual world, no evidence drawn from science can prove either that it does or that it does not. To appeal to quantum physics to show that the world might be other than science alleges is tacitly to assume that science can tell us truly what the world is like. If the real is in truth other than our knowledge shows, no evidence from quantum physics could support the allegation. On the other side, to argue from quantum theory against the view that despite appearances the world might be sheerly chaotic, is to assume that quantum physics is true of the world (which our opponent is denying). So the defence commits *petitio principii.* All this goes to show how futile it is to argue about the confessedly unknowable and to play in philosophy with *Dinge-an-sich.*

(*d*) Finally we return to the metaphysical objection to the idea of randomness as prior to order. A random series was found to be equivalent to a homogenous continuum, because in a heterogeneous continuum the intervals between the changes are inevitably in some way orderly, and randomness increases as these intervals are reduced. In the limit there are no intervals and so no changes and the continuum becomes homogenous. But that again destroys it altogether, because without distinction of parts there is no continuum.[1] Randomness is thus a relative concept to which order must be prior, and the supposed chaos that our objector takes to be primary could in fact only be derivative from order.

The upshot is clear. Not only is the notion of a totally chaotic world unviable, but, so long as the mind is admitted to be part of nature and to be an agency of organization, we are committed to the presupposition that nature itself must involve a principle of organization capable of giving rise to mind and the knowledge of nature that our minds develop. If this were not so no mind could exist and so no science. And this argument involves no vicious circularity, for we have not maintained merely that because we imagine the world to have a certain structure, which would give rise to knowledge of the sort we claim, therefore the world must have that structure. Rather the argument runs that unless the world had the sort of dialectical structure revealed in knowledge, we could not have the sort of knowledge that we enjoy. But as we do undoubtedly enjoy an awareness of the world of this sort, the

[1] Cf. *Foundations*, pp. 462 f.

presumption that reality has a dialectical structure is well founded. Hence the dialectical character of our knowledge is itself evidence of its truth about the world. The argument on pp. 375–7 might be castigated as illicit affirmation of the consequent, and was therefore stated with due caution and qualification (though many accepted scientific arguments are of this form: if h then p and p is true, therefore probably h). But the form of the above argument is *modus tollens* which the most pernickety logician will accept as valid.

iv. KNOWLEDGE AND ITS OBJECT

The relation between knowledge and the world may now be stated in two ways. In as much as the natural world is a single continuous development including living organisms, among them human beings, their thinking minds and the knowledge they acquire of the world they live in, knowledge and nature are not mutually independent. For nature must be such that the human mind and its knowledge can develop within it, and knowledge must be the fuller realization, in reflective consciousness, of the principle of order and its multifarious exemplifications that constitute nature. The scale of forms, we have maintained, is a succession of more and more adequate expressions of the generic essence or principle immanent in the whole system. So late a product as the scientific awareness of the system is therefore the most adequate expression to date of that generic principle. Consequently science and nature must be intimately interdependent. And the best theory will present the world as it really is.

On the other hand, though each later form in the scale presupposes, is generated from, and subsumes within itself the earlier phases, the earlier forms are independent of the later so far as they may come into being and can exist prior to the appearance of subsequent phases. In this way nature, below the level of humanity exists independently of human knowing, and human knowledge relates to it, not as a replica to an original, but as a more highly developed system that has evolved out of a more primitive one.

This theory of knowledge reveals the degree of truth in earlier theories. It supports a causal theory of perception so far as it holds that perception is the structuring, or interpretation, of primitive sentience, which, in turn, is the felt integration of physiological activity in the organism; while it interprets the causal connexion in terms of that between the complex phases of a dialectical series. It justifies the correspondence theory, for knowledge corresponds to its objects to the extent that it develops out of them. It is not, however, a correspondence between 'things' and 'ideas' as single items, so much as a correspondence between entire relational wholes or situations in which the organism finds itself and the conscious elaboration of the emergent sentient awareness. It upholds the coherence theory so far as it asserts that knowledge is the coherent ordering of experience and is adequate and successful to the degree in which it is systematic. In all these various aspects of the theory the epistemologist's fallacy is avoided, for we do not claim to know 'external reality' by any means other than perception and the different modes of thinking, but only through analysis of experience and the implications within it of dialectical process.

The theory agrees with Empiricism to the extent of making sentient experience the primary stage in the development of knowledge; but it sides with idealism in denying the existence of hard data. Latter-day Empiricism tends to equivocate on this issue, and some linguistic analysts have in more recent writing approached partial and disconnected truths by devious paths.

Finally, the theory itself is an example of what it maintains of theories in general and in its own character it conforms to its own teaching. How does it answer the questions with which our investigation first began? What is it that makes a theory scientific and distinguishes science from superstition? And why is it that science can be seen to progress, while other spheres of human endeavour appear only to flounder or regress?

V. VALIDITY AND PROGRESS

The validity of science can be measured only by considering the degree of systematic coherence with which it can account for the widest possible range of experience. What is systematic thinking is scientific and what distinguishes science from superstition or prejudice is its systematic character. Very systematic creeds have, of course, been held which are not, by modern standards, scientific. But all such creeds are scientific, in so far as they are genuine attemps at explanation. Even mythology in its original purpose was such an attempt. Its subsequent failure to measure up to scientific standards was its failure to account consistently for all the relevant facts. But the history of early science, so far as it is discoverable, proves it to be a direct development from explanatory myths.[1]

In the course of our investigation we have come to see progress in science as dialectical and not as a mere accumulation of factual material variously ordered under different rubrics. The capacity of science to progress, therefore, is inherent in its very nature as systematic thinking. The scientist strives to systematize more and more coherently, constructing more and more unified wholes, covering more and more comprehensive fields. A system which does this, we have found, issues as a dialectical progression. Hence science, successfully to be science, must progress. That, however, is largely a tautology, for knowledge fails to progress as it fails to be scientific and what we must seek to understand is the reason for its continued success. This question might be approached from several angles, educational, sociological, or economic, as well as epistemological. But our concern is only with the last. Epistemologically the success of science is nothing other than its systematic character and that is self-generating, so that the more systematic it becomes the more it increases its capacity to develop. In science more than in any other sphere of activity the aphorism is true that nothing succeeds like success.

In other forms of human activity, even though they are properly classed as intellectual, this characteristic is less obvious, because these other forms are more complicated by emotional factors. This is the reason for the frequent and widespread doubt about the reality of their progress. Art and religion are not purely rationalistic but are largely involved with feeling and symbolism. They are not, for that reason, devoid of logic; but the systematic character of their products is more heavily disguised. Morality is still more complex for it involves a degree of conflict in practice between the demands of reason and the impulses of appetite. While moral principles therefore may progress even markedly, moral practice does not necessarily follow suit. In actual fact, there is progress in all these spheres but like all progress it is dialectical and, therefore, is not always recognized, because what is expected is a sort of linear accumulation of achievement, whereas what actually takes place is a continuous complexification of modes of experience none of them in the first instance very simple. It is not my purpose to go into these matters in detail, for they are complex and difficult,

[1] Cf. F. Cornford, *Principium Sapientiae* (Cambridge, 1952); and Bruno Snell, *The Discovery of the Mind* (Oxford, 1953).

and to do them justice would require, at least, an elaborate treatise on the philosophy of mind. I shall content myself with the remark that experience, though we may attend in turn to each of its several aspects, is a single very complex whole, and none of its numerous facets is independent of the others. Intellectual progress cannot therefore be partitioned among the different activities of the mind, and the progress of science, as it occurs, is both the consequence (in some respects) and the cause (in others) of progress in other forms.

vi. SCIENCE AND METAPHYSICS

A word, however, is needed about the relation of science to philosophy—in particular to metaphysics. It must be apparent that though the subject of discussion in this book and its predecessor has been science, the manner has been philosophical. It is the business of the philosopher to reflect upon experience, and reflection upon the practice and results of scientific thinking gives a synoptic view both of the world and of knowledge. The resulting theories are ontological, cosmological and epistemological, and they form the body of metaphysics.

Is metaphysics a science?—the question frequently asked and variously answered since Kant gave it prominence. The answer, in the light of what has been said above, is that metaphysics is scientific to the extent that it is systematic thinking. Does it progress? Again, the answer is that it progresses so far as it keeps pace with the sciences upon which it reflects and takes cognizance of their progress. As science is part of the culminating phase of that dialectical progression in which the ordering principle of actuality in the world continuously realizes its concrete systematic nature, issuing (as it does for us) in the achievements of the human intellect, and as metaphysics is the final reflection upon all this, bringing it to the highest level of systematic thinking, metaphysics may be seen as the summation of the scientific dialectic, as the systematization of systematic thinking, and therefore as itself scientific in the fullest sense of the word.[1]

But metaphysics is not confined to reflection upon the natural sciences, it must include in its scope the social and philosophical

[1] Cf. Hegel, *Phänomenologie des Geistes*, Preface: 'Among the many consequences that follow from what has been said, it is of importance to emphasize this, that knowledge is only real and can only be set forth fully in the form of science, in the form of system . . .', trans. J. B. Baillie (London, 1966), p. 85.

sciences also. To include them in the present study would make it too unwieldy, however necessary and interesting that addition would be. A third volume would be required to accommodate them. This wider scope of metaphysics, however, does not alter its character as systematic thinking nor absolve it from investigating those very general and ultimate presuppositions of our reflective experience which have sometimes earned for it the approbrium of mystery-mongering. Metaphysicians may at times fall to that temptation, but when they do they produce bad metaphysics. So long as the thinking is systematic, attends closely to the form of experience upon which it reflects and does not misrepresent it, the study may claim a place among the sciences. But, of course, metaphysics is a philosophical, not a natural, science. It is reflection upon reflection and therefore a science of the second—or in some cases of even higher—degree.

vii. SCIENCE AND RELIGION

The scale of forms constituted by the progress of science points beyond itself. It is a gradual approach to a unified system of knowledge, which, if fully achieved would be 'the whole truth' about the universe; but it is not fully achieved by science, which is limited to the capacity of the human intellect and the means of discovery at its disposal. The final consummation and completion of the scale would be omniscience, and the limits of science leave unexplained much that is not simply beyond its ken but also implicated in what is already known. The implicit acknowledgement of a necessary yet unattained totality is left, therefore, for elaboration in a different form of human experience.

For instance, modern cosmologies all postulate a beginning of existence for matter which none of them can explain. Evolutionary theories of the universe[1] postulate a beginning of the physical world in a primordial explosion of matter. But what the source of its existence might be it has no way of telling. The primordial concentration of matter could not have existed for more than a split second before the explosion, because no such concentration of matter as the theory envisages could be stable. Steady-state theories which seek to conceive the universe as infinite and uniform in space and time have still to admit the continuous recession of the galaxies and must compensate for this

[1] See *Foundations*, Ch. V.

by the creation of matter in empty space.[1] But, once again, they can in no way account for this creation. For science, therefore, the ultimate source of existence, its ultimate ground and origin remains unknown. Nevertheless, there must be some source and some ground, for without it what is known would not be.

The limitation of established knowledge not only leaves room, therefore, for belief but demands it and science passes over into religion. Religion, however, is not merely belief about the unknown, it is, like science, a totality which embraces the whole of experience. So far as it is belief, it is belief, not just about what lies beyond science, but about the whole of reality. It is belief in the reality of that whole to which science points but cannot fully attain. As belief (a creed) religion cannot, therefore, ignore or neglect science and if it conflicts with the science of the day it is usually because it still tacitly harbours outdated scientific theories. To this extent religion must progress as science progresses; otherwise it becomes obsolete even as religion.

Religion, moreover, is not only belief, it is the personal relation of man to the totality of the universe and the ultimate nature of the real. His beliefs about these will inevitably be reflected in his attitude towards them and that will in some way or other invade every aspect of his life. For religion is not only a creed, it is also, and for that reason, an ethic.

But because religion transcends science and embraces beliefs about what science must presuppose yet cannot prove, it cannot wholly express these beliefs in literal, explicit and systematic form. Consequently, it resorts to metaphors and symbols, which taken literally (as they ought not to be) appear to be at variance with scientific knowledge, or even to be absurd. Further, as religion affects the whole of human life and man's attitude towards reality, it takes a hold upon his emotions and directs his conduct. The symbols in which it expresses ultimate beliefs, therefore, have emotional appeal and are valuable for more than simply the cognitive interpretation to which they may be liable. This fact is an additional reason for apparent conflict between religion and science for which emotional appeal is altogether irrelevant.

In truth, however, there is no real conflict between science and religion if each is taken at its best. Each, in its way, is a claim to apprehend the truth about the world and there are not two truths but only one. As creed, therefore, religion overlaps with science

[1] Cf. *ibid.*

and it is partially identical with it. Its use of myth and allegory is symbolic, and in sound religion the interpretation of the symbols will be in accordance with scientific knowledge. Where religion goes beyond the limits of science literal interpretation may not be possible and then, like Plato, we must understand the myth as expressing, not the truth, but only something like it—because the truth lies beyond our comprehension.

Further, as religion is our attitude towards ultimate reality, it determines the values we acknowledge and respect; and because these relate to what is most fundamental the over-riding attitude to the most basic value is that of worship. A man's religion is determined by what he worships—what for him is the ultimate value in life—that is for him his god, be it the God of traditional religion, or human welfare, or material gain, or Mao-tse-tung. And the soundness of his religion depends upon the extent to which his ultimate values are justifiable on scientific grounds. For only if his creed is true can his evaluations be just. Science is commonly held to eschew values (which is not entirely correct); yet even if it does, religion, in determining values may ignore facts only at its peril.

Does this mean that belief in God, the ultimate value is justifiable only if science supports it? In the last resort, and for the reason just mentioned, we must answer, Yes. There may be many experiences that convince individual men of God's existence but none of these experiences *justify* the belief unless they can themselves be authenticated, and in the end the only authentication that will serve must be scientific. But modern science, you may say, is more likely to incline one to atheism than to belief in God, and if that is to be justified finally only on scientific grounds, is religion not more likely to be wholly rejected than to be established? We should have to agree that *if* science showed sound reason for rejecting all belief in God, religion, except as superstition, would disappear. But the best evidence, even such as we have recorded above, is to the contrary. If science is a scale of forms approaching asymptotically a totality that transcends the limits of the human intellect, and if its conclusions in every branch reveal a world that is itself a developing scale implying a transcendent totality beyond human experience, a religious belief in and attitude towards an infinite being is justified on scientific grounds, and the much advertised 'death of God', if taken literally, is as mythical as the Nordic *Götterdämmerung*.

INDEX